Recent Progress in Optical Technology

Volume I

Recent Progress in Optical Technology
Volume I

Edited by **Roderick Swayne**

*C*LANRYE
INTERNATIONAL

New Jersey

Published by Clanrye International,
55 Van Reypen Street,
Jersey City, NJ 07306, USA
www.clanryeinternational.com

Recent Progress in Optical Technology: Volume I
Edited by Roderick Swayne

International Standard Book Number: 978-1-63240-444-2 (Hardback)

Contents

Preface

Optical technology is a field of study that refers to any physical phenomenon that is related to light or vision. It could either be infrared light or visible light that usually performs a specific function. These signals can be interpreted by a computer as data and this data can be transferred over a network. Optical fibres and even a computer mouse are examples of optical technologies. Optical technology is also used in many computers, where calculations are done using photons in infrared or visible beams, instead of electric current. Using this technology allows computers to greatly increase computational speed. It might someday be possible to have optical technology that analyzes data at speeds many magnitudes greater than that of a conventional computer that uses electric current. One thus understands that the advancement of optical technology has huge ramifications in industries that use such research and technology that in turn affect our daily lives. Therefore there is also a need for research into various projects in this rapidly changing field of study. This also points towards the increasing demand for skilled researchers and graduates in this discipline who will be able to advance with the evolution of the technology itself.

This book is an attempt to compile and collate all available research on optical technologies under one umbrella. I am grateful to all the contributing authors and my family for their support.

Editor

Splice Loss of Graded-Index Fibers: Accurate Semianalytical Descriptions Using Nelder-Mead Nonlinear Unconstrained Optimization with Three-Parameter Fundamental Modal Field

Raja Roy Choudhury,[1] Arundhati Roy Choudhury,[2] and Mrinal Kanti Ghose[3]

[1] Applied Electronics and Instrumentation Department, Sikkim Manipal Institute of Technology, Majitar, Sikkim 737136, India
[2] Physics Department, Sikkim Manipal Institute of Technology, Majitar, Sikkim 737136, India
[3] Computer Science Department, Sikkim Manipal Institute of Technology, Majitar, Sikkim 737136, India

Correspondence should be addressed to Raja Roy Choudhury; ra ch2@yahoo.co.in

Academic Editor: Yong Zhao

A faster and accurate semianalytical formulation with a robust optimization solution for estimating the splice loss of graded-index fibers has been proposed. The semianalytical optimization of modal parameters has been carried out by Nelder-Mead method of nonlinear unconstrained minimization suitable for functions which are uncertain, noisy, or even discontinuous. Instead of normally used Gaussian function, as the trial field for the fundamental mode of graded-index optical fiber a novel sinc function with exponentially and $R^{-3/2}$ (R is the normalized radius of the optical fiber) decaying trailing edge has been used. Due to inclusion of three parameters in the optimization of fundamental modal solution and application of an efficient optimization technique with simple analytical expressions for various modal parameters, the results are found to be accurate and computationally easier to find than the standard numerical method solution.

1. Introduction

Single mode fiber is considered as the most important broadband transmission media for optical communication system. Achieving accurate values of modal field distribution in such fiber is very essential, as it can provide basic solutions for wave equation and many useful properties like splice loss, microbending loss, fiber coupling, and the prediction of intramodal dispersion [1]. However, the various expressions for the fundamental modal field that have been reported so far are not able to predict propagation constant and modal parameters exactly in all regions of single mode operation [2]. The Gaussian approximation shows poor accuracy for lower normalized frequency region although this region may involve single mode fiber operation [2]; however, it can perform satisfactorily only for higher normalized frequency region and give good result near the cut-off frequency of next higher mode [3]. Besides, it is also equally important that the approximation should describe the field in the cladding

accurately, as it is useful in the study of evanescent coupling problem. To overcome these inefficiencies, an exponentially and $R^{-3/2}$ decaying trailing edge fundamental modal field solution in core-cladding interface region has been considered.

To achieve higher accuracy compared to Gaussian function, the Gaussian-Hankel [2], the generalized Gaussian [4], the extended Gaussian [5], and the Laguerre-Gauss/Bessel expansion approximation [6, 7] have been proposed so far. An approximate analytical description with no requirement for optimization has also been presented [8]. But such analytical expression may not work for all specifications of an optical fiber. In the proposed formulation, Nelder-Mead method of nonlinear unconstrained minimization and the process of minimization of core parameter (U) for all specific requirements have been used to achieve an accurate and computationally appropriate result.

Unlike the existing reported fundamental modal solution with one or two parameters [2–8], an attempt has been made to propose a three-parameter fundamental modal field solution for graded-index fiber to introduce more flexibility to solve the fundamental modal solution more accurately, especially in core-cladding interface region wherein the solution has different form (exponentially and $R^{-3/2}$ decaying trailing edge). Ghatak et al. [9] had arrived at simple analytical expressions to describe different optical fiber characteristics by implementing variational technique. Again, the optimization process requires expressions for propagation constant β and core parameter U. The analytical expressions for β and U used for the present study involve many fiber parameters, such as core radius (a), refractive indices of core and cladding (n_{co} and n_{cl}), aspect ratio (S_0), and wavelength ($\lambda = 2\pi/k$), where k is the free space wave number. Hence, any desired specification can be incorporated by varying these parameters. Now, the task of optimization can be carried out by using Nelder-Mead method of nonlinear unconstrained minimization, to meet a particular design.

For graded-index optical fiber at the splices, the power transmission coefficients with transverse and angular mismatch have been estimated by using the methods given by Meunier and Hosain [10] and Hosain et al. [11]. For arbitrarily graded-index fiber, the Gaussian approximation does not give accurate result at lower normalized frequency or in cases where the power law profile deviates from its simplest form [12]. Further, the numerical solution requires rigorous computations and specialized numerical techniques [13]. However, using the proposed three-parameter fundamental modal solution coupled with Nelder-Mead method of nonlinear unconstrained minimization, the algorithm becomes comparatively easier to be implemented on an ordinary personal computer, which provides computationally more efficient result [14, 15] than standard numerical method and yields excellent agreement with exact solutions. This is achieved due to the fact that requisite analytical formulae are deduced beforehand and then parameters of those analytical expressions are found by optimization using Nelder-Mead simplex method for nonlinear unconstrained minimization. Furthermore, Nelder-Mead simplex method for nonlinear unconstrained minimization is a direct search method [16, 17] which does not require any derivative information, so it can optimize nonstationary functions, as needed for the problems under study [18–20]. The proposed semianalytical model can also be used in the study of nonlinear fiber [21].

2. Formulation of the Problem

2.1. Theory. Splice loss can be evaluated analytically with the help of the following equations [22]:

$$\int |\psi_1|^2 R\,dR = \frac{\log(\alpha R/R_0) - Ci(2\alpha R/R_0)}{2}, \quad (1)$$

$$\int |\psi_2|^2 R\,dR = \frac{\sin^2\alpha e^{2\mu}}{R}\left(-R_0 e^{-2\mu R/R_0} + 2\mu REi\left(1, \frac{2\mu R}{R_0}\right)\right), \quad (2)$$

$$\int n_{f2}^2 R|\psi_1|^2 dR$$
$$= \frac{1}{(-4\alpha + 4\alpha S_0)}$$
$$\times \left[2n_2^2\alpha S_0 \log\left(\frac{\alpha R}{R_0}\right) - 2n_2^2\alpha S_0 Ci\left(\frac{2\alpha R}{R_0}\right)\right.$$
$$- 2n_1^2\alpha \log\left(\frac{\alpha R}{R_0}\right) + 2n_1^2\alpha Ci\left(\frac{\alpha R}{R_0}\right)$$
$$- n_1^2 R_0 \sin\left(\frac{2\alpha R}{R_0}\right) + 2n_1^2 R\alpha$$
$$\left. + n_2^2 R_0 \sin\left(\frac{2\alpha R}{R_0}\right) - 2n_2^2\alpha R \right], \quad (3)$$

$$\int n_{f2}^2 R|\psi_2|^2 dR$$
$$= \frac{e^{2\mu}\sin^2\alpha}{R(S_0-1)}$$
$$\times \left[-n_2^2 R_0 S_0 e^{-(2\mu R/R_0)} + 2n_2^2\mu R S_0 Ei\left(1, \frac{2\mu R}{R_0}\right)\right.$$
$$+ n_1^2 R_0 e^{-(2\mu R/R_0)} - 2n_1^2\mu REi\left(1, \frac{2\mu R}{R_0}\right)$$
$$\left. - n_1^2 R_0 REi\left(1, \frac{2\mu R}{R_0}\right) + n_2^2 R_0 REi\left(1, \frac{2\mu R}{R_0}\right)\right], \quad (4)$$

$$\int \left|\frac{d\psi_1}{dR}\right|^2 R\,dR = \left[\frac{1}{2R_0^2}\log\left(\frac{\alpha R}{R_0}\right) - \frac{1}{2R_0^2}Ci\left(\frac{2\alpha R}{R_0}\right)\right]\alpha^2$$
$$+ \frac{\alpha}{2RR_0}\sin\left(\frac{2\alpha R}{R_0}\right)$$
$$+ \frac{1}{4R^2}\cos\left(\frac{2\alpha R}{R_0}\right) - \frac{1}{4R^2}, \quad (5)$$

$$\int \left|\frac{d\psi_2}{dR}\right|^2 R\,dR$$
$$= -\frac{\sin^2\alpha e^{2\mu}}{4R_0^2 R^3 e^{2\mu R/R_0}}\left(3R_0^3 + 3R_0^2\mu R - 2R_0\mu^2 R^2\right.$$
$$\left. + 4Ei\left(1, \frac{2\mu R}{R_0}\right)\mu^3 R^3 e^{2\mu R/R_0}\right), \quad (6)$$

$$\int |\psi_1|^2 R^3 dR$$
$$= \frac{1}{8\alpha^2}\left[-2\alpha RR_0 \sin\left(\frac{2\alpha R}{R_0}\right)\right.$$
$$\left. + 2\alpha^2 R^2 + R_0^2 - R_0^2\cos\left(\frac{2\alpha R}{R_0}\right)\right], \quad (7)$$

Splice Loss of Graded-Index Fibers: Accurate Semianalytical Descriptions Using Nelder-Mead Nonlinear Unconstrained
Optimization with Three-Parameter Fundamental Modal Field

3

$$\int |\psi_2|^2 R^3 dR = -\frac{\sin\alpha^2 e^{2\mu} R_0^2}{2\mu e^{2\mu R/R_0}}, \qquad (8)$$

$$\int |\psi_1|^2 R^q dR$$

$$= \frac{\alpha^2}{R_0^2 (q+1)} R^{q+1}$$

$$\times \text{Hypergeom}\left(\left[1, \frac{q}{2}+\frac{1}{2}\right], \left[2, \frac{3}{2}, \frac{3}{2}+\frac{q}{2}\right], -\frac{\alpha^2 R^2}{R_0^2}\right), \qquad (9)$$

$$\int |\psi_2|^2 R^q dR$$

$$= 2^{(2-q)} \left(\frac{\mu}{R_0}\right)^{-q} \frac{\mu^2}{R_0} e^{2\mu}$$

$$\times \sin\alpha^2 \left[\frac{1}{(q-2)(q-1)q(q+1)} \left(\frac{\mu R}{R_0}\right)^{-q/2}\right.$$

$$\times e^{-\mu R/R_0} R^q 2^{q/2} \left(\frac{\mu}{R_0}\right)^q$$

$$\times WM\left(\frac{q}{2}, \frac{q}{2}+\frac{1}{2}, \frac{2\mu R}{R_0}\right)$$

$$+ \frac{2^{q/2-3}}{(q-2)(q-1)q\mu^3} \left(\frac{\mu R}{R_0}\right)^{-q/2}$$

$$\times e^{-\mu R/R_0} R_0^3 R^{q-3} \left(\frac{\mu}{R_0}\right)^q$$

$$\times \left(4\mu^2 \frac{R^2}{R_0^2} + \frac{2q\mu R}{R_0} - q + q^2\right)$$

$$\left.\times WM\left(\frac{q}{2}+1, \frac{q}{2}+\frac{1}{2}, \frac{2\mu R}{R_0}\right)\right], \qquad (10)$$

$$\int \frac{1}{R}\left|\frac{d\psi_1}{dR}\right|^2 dR = \frac{1}{8R_0^2 R^4}\left[-2\alpha^2 R^2 + 2R_0\alpha R\sin\left(\frac{2\alpha R}{R_0}\right)\right.$$

$$\left.-R_0^2 + R_0^2\cos\left(\frac{2\alpha R}{R_0}\right)\right], \qquad (11)$$

$$\int \frac{1}{R}\left|\frac{d\psi_2}{dR}\right|^2 dR$$

$$= \frac{\sin^2\alpha}{120 R_0^4 R^5}\left[63R_0^4\mu R e^{(-2\mu(-R_0+R)/R_0)}\right.$$

$$- 2R_0^3\mu^2 R^2 e^{(-2\mu(-R_0+R)/R_0)}$$

$$+ 2R_0^2\mu^3 R^3 e^{(-2\mu(-R_0+R)/R_0)}$$

$$- 4R_0\mu^4 R^4 e^{(-2\mu(R_0+R)/R_0)}$$

$$+ 54R_0^5 e^{(-2\mu(-R_0+R)/R_0)}$$

$$\left.+8\mu^5 R^5 e^{2\mu} Ei\left(1, \frac{2\mu R}{R_0}\right)\right], \qquad (12)$$

$$\int \left|\frac{d^2\psi_1}{dR^2}\right|^2 R dR = \frac{\alpha^4}{2R_0^4}\left[\log\left(\frac{\alpha R}{R_0}\right) - Ci\left(\frac{2\alpha R}{R_0}\right)\right]$$

$$- \frac{\alpha^2}{R_0^2 R^2}\cos\left(\frac{2\alpha R}{R_0}\right) + \frac{\alpha}{R_0 R^3}\sin\left(\frac{2\alpha R}{R_0}\right)$$

$$- \frac{1}{2R^4} + \frac{1}{2R^4}\cos\left(\frac{2\alpha R}{R_0}\right), \qquad (13)$$

$$\int \left|\frac{d^2\psi_2}{dR^2}\right|^2 R dR$$

$$= \frac{\sin^2\alpha}{32R_0^4 R^5} e^{2\mu}\left[-90R_0^5 e^{(-2\mu R/R_0)}\right.$$

$$- 135R_0^4\mu R e^{(-2\mu R/R_0)}$$

$$- 86R_0^3\mu^2 R^2 e^{(-2\mu R/R_0)}$$

$$- 10R_0^2\mu^3 R^3 e^{(-2\mu R/R_0)}$$

$$- 12R_0\mu^4 R^4 e^{(-2\mu R/R_0)}$$

$$\left.+24\mu^5 R^5 Ei\left(1, \frac{2\mu R}{R_0}\right)\right], \qquad (14)$$

$$\int \left|\frac{d^3\psi_1}{dR^3}\right|^2 R dR$$

$$= \alpha^6\left[\frac{1}{2R_0^6}\log\left(\frac{\alpha R}{R_0}\right) - \frac{1}{2R_0^6}Ci\left(\frac{2\alpha R}{R_0}\right)\right]$$

$$+ \frac{\alpha^5}{2R_0^5 R}\sin\left(\frac{2\alpha R}{R_0}\right)$$

$$+ \alpha^4\left[\frac{5}{4R_0^4 R^2}\cos\left(\frac{2\alpha R}{R_0}\right) + \frac{3}{4R_0^4 R^2}\right]$$

$$- \frac{4\alpha^3}{R_0^3 R^3}\sin\left(\frac{2\alpha R}{R_0}\right) - \frac{6\alpha^2}{R_0^2 R^4}\cos\left(\frac{2\alpha R}{R_0}\right)$$

$$+ \frac{6\alpha}{R_0 R^5}\sin\left(\frac{2\alpha R}{R_0}\right) + \frac{3}{R^6}\cos\left(\frac{2\alpha R}{R_0}\right) - \frac{3}{R^6}, \qquad (15)$$

$$\int \left|\frac{d^3\psi_2}{dR^3}\right|^2 R dR R$$

$$= \frac{\sin^2\alpha}{64R_0^6 R^7} e^{2\mu}$$

$$\times \left[40\mu^7 R^7 Ei\left(1, \frac{2\mu R}{R_0}\right) + 1575 R_0^7 e^{-(2\mu R/R_0)} \right.$$
$$+ 2625 R_0^6 \mu R e^{-(2\mu R/R_0)} + 2082 R_0^5 \mu^2 R^2 e^{-(2\mu R/R_0)}$$
$$+ 999 R_0^4 \mu^3 R^3 e^{-(2\mu R/R_0)} + 246 R_0^3 \mu^4 R^4 e^{-(2\mu R/R_0)}$$
$$\left. + 42 R_0^2 \mu^5 R^5 e^{-(2\mu R/R_0)} - 20 R_0 \mu^6 R^6 e^{-(2\mu R/R_0)} \right],$$

$$(16)$$

$$\int \frac{1}{R} \left| \frac{d\psi_1}{dR} \right| \left| \frac{d^3 \psi_1}{dR^3} \right| dR$$
$$= \frac{1}{8 R_0^4 R^6} \left[-4 R_0^4 - 3\alpha^2 R_0^2 R^2 + 2\alpha^4 R^4 \right.$$
$$- 2\alpha^3 R_0 R^3 \sin\left(\frac{2\alpha R}{R_0}\right) + 8\alpha R_0^3 R \sin\left(\frac{2\alpha R}{R_0}\right)$$
$$\left. - 5\alpha^2 R_0^2 R^2 \cos\left(\frac{2\alpha R}{R_0}\right) + 4 R_0^4 \cos\left(\frac{2\alpha R}{R_0}\right) \right],$$

$$(17)$$

$$\int \frac{1}{R} \left| \frac{d\psi_2}{dR} \right| \left| \frac{d^3 \psi_2}{dR^3} \right| dR$$
$$= \frac{\sin^2\alpha}{240 R_0^6 R^7 e^{(2\mu R/R_0)}} e^{2\mu}$$
$$\times \left[975 R_0^6 \mu R + 474 R_0^5 \mu^2 R^2 \right.$$
$$+ 123 R_0^4 \mu^3 R^3 - 2 R_0^3 \mu^4 R^4$$
$$+ 2 R_0^2 \mu^5 R^5 - 4 R_0 \mu^6 R^6$$
$$+ 8 \mu^7 R^7 e^{(2\mu R/R_0)} Ei\left(1, \frac{2\mu R}{R_0}\right)$$
$$\left. + 675 R_0^7 \right],$$

$$(18)$$

$$\left| \frac{d^2 \psi}{dR^2} \right| = \frac{\sin\alpha}{4 R^5 (R_0/R)^{3/2}} e^\mu e^{-(\mu R/R_0)}$$
$$\times \left(15 R_0^2 + 12 R_0 \mu R + 4\mu^2 R^2 \right),$$

$$(19)$$

where ψ_1 and ψ_2 are given in (26). α, R_0, and μ are the three variational parameters present in the fundamental modal solution.

$Ei(z)$ is exponential integral given by [23] as follows:

$$Ei(1, z) = \int_z^\infty \frac{e^{-t}}{t} dt. \qquad (20)$$

$Ci(z)$ is the cosine integral function, defined by [23], as follows:

$$Ci(z) = \chi + \ln(z) + \int_0^z \frac{\cos t - 1}{t} dt, \qquad (21)$$

where χ is Euler's constant 0.5772.

$WM(k, m, z)$ are the Whittaker functions which are solutions to the Whittaker differential equation [23].

$Hypergeom(n, d, z)$ is the generalized hypergeometric function $F(n, d, z)$, where [23]

$$F(n, d, z) = \sum_{k=0}^\infty \frac{C_{n,k}}{C_{d,k}} \cdot \frac{z^k}{k!}, \qquad (22)$$

with,

$$C_{v,k} = \prod_{j=1}^v \frac{\Gamma(v_j + k)}{\Gamma(v_j)}, \qquad (23)$$

where $\Gamma(a)$ is the gamma function [23].

2.2. *Basic Formulations.* The refractive index profile for a weakly guiding fiber is given by

$$n_{f1}^2 = n_2^2 + \left(n_1^2 - n_2^2\right) f_1, \quad \text{for } 0 \le R \le S_0,$$
$$n_{f2}^2 = n_2^2 + \left(n_1^2 - n_2^2\right) f_2, \quad \text{for } S_0 \le R \le 1, \qquad (24)$$
$$n_{f3}^2 = n_2^2 + \left(n_1^2 - n_2^2\right) f_3, \quad \text{for } R > 1,$$

where the normalized profile functions for the trapezoidal and triangular index profiles f_i $(i = 1, 2, 3)$ are given by

$$f_1 = 1,$$
$$f_2 = \frac{1 - R}{1 - S_0}, \qquad (25)$$
$$f_3 = 0.$$

Here, S_0 is the aspect ratio, R is the normalized radius $(= r/a)$, a is the core radius, r is the actual radius of the optical fiber, and n_1 and n_2 are, respectively, the refractive indices of the core axis and cladding.

For the present study, the following approximations for the fundamental mode as the trial field have been proposed:

$$\psi_1 = \frac{\sin(\alpha R/R_0)}{R} \quad \text{for } R \le R_0,$$
$$\psi_2 = \left(\frac{\sin(\alpha)}{R}\right) e^{\mu(1-(R/R_0))} \sqrt{\frac{R_0}{R}} \quad \text{for } R > R_0, \qquad (26)$$

where α, R_0, and μ are the three variational parameters present in the fundamental modal solution.

To employ variational technique, first the scalar variational expression for the propagation constant β as given by

(27) has been considered and is shown in equations through (28) to (30) as follows:

$$\beta^2 = \frac{k^2 \int_0^\infty n^2(R)|\psi(R)|^2 RdR - (1/a^2)\langle\psi'^2\rangle}{\langle\psi^2\rangle}, \quad (27)$$

$$\beta^2 = \frac{1}{\langle\psi^2\rangle}\left[\int_0^{R_0} k^2 n_{f1}^2|\psi_1|^2 RdR + \int_{R_0}^{S_0} k^2 n_{f1}^2|\psi_2|^2 RdR\right.$$
$$+ \int_{S_0}^1 k^2 n_{f2}^2|\psi_2|^2 RdR + \int_1^\infty k^2 n_{f3}^2|\psi_2|^2 RdR$$
$$\left. - \frac{1}{a^2}\langle\psi'^2\rangle\right], \quad (28)$$

for $R_0 < 1$ and $R_0 < S_0$,

$$\beta^2 = \frac{1}{\langle\psi^2\rangle}\left[\int_0^{S_0} k^2 n_{f1}^2|\psi_1|^2 RdR + \int_{S_0}^{R_0} k^2 n_{f2}^2|\psi_1|^2 RdR\right.$$
$$+ \int_{R_0}^1 k^2 n_{f2}^2|\psi_2|^2 RdR + \int_1^\infty k^2 n_{f3}^2|\psi_2|^2 RdR$$
$$\left. - \frac{1}{a^2}\langle\psi'^2\rangle\right], \quad (29)$$

for $R_0 < 1$ and $R_0 > S_0$,

$$\beta^2 = \frac{1}{\langle\psi^2\rangle}\left[\int_0^{S_0} k^2 n_{f1}^2|\psi_1|^2 RdR\right.$$
$$+ \int_{S_0}^1 k^2 n_{f2}^2|\psi_1|^2 RdR + \int_1^{R_0} k^2 n_{f3}^2|\psi_1|^2 RdR$$
$$\left. + \int_{R_0}^\infty k^2 n_{f3}^2|\psi_2|^2 RdR - \frac{1}{a^2}\langle\psi'^2\rangle\right], \quad (30)$$

for $R_0 > 1$, where

$$\langle\psi^2\rangle = \int_0^{R_0}|\psi_1|^2 RdR + \int_{R_0}^\infty|\psi_2|^2 RdR,$$
$$\langle\psi'^2\rangle = \int_0^{R_0}\left|\frac{d\psi_1}{dR}\right|^2 RdR + \int_{R_0}^\infty\left|\frac{d\psi_2}{dR}\right|^2 RdR. \quad (31)$$

Now, the core parameter U is given by

$$U^2 = a^2\left(k^2 n_1^2 - \beta^2\right). \quad (32)$$

Now for a fixed value of normalized frequency, the core parameter U is minimized with respect to the variational parameters α, R_0, and μ. Once the optimized values of these three parameters are obtained, the propagation constant and other design parameters can be obtained as explained in the next section.

2.3. Splice Loss.
For small angular misalignment (θ) at the splice of two optical fibers, following Hosain et al. [11], the well-known overlap integral can be represented as

$$C_a(p) = \int_0^{2\pi}\int_0^\infty d\phi R\,dR|\psi(R)|^2 \exp(ipR\cos\phi), \quad (33)$$

where $p = akn\theta$, n being refractive index of the index matching fluid joining the fibers and θ being the angular misalignment.

The transmission coefficient $T_a(p)$ at the splice with angular mismatch can then be expressed as

$$T_a(p) = \left|\frac{C_a(p)}{C_a(0)}\right|^2. \quad (34)$$

Expanding the exponential term, (33) can be written as

$$C_a(p) = 2\pi\sum_{n=0}^{n=\infty}\frac{(-p^2/4)^n}{(n!)^2}\int_0^\infty|\psi(R)|^2 R^{2n+1}dR, \quad (35a)$$

and from (33), $C_a(0)$ is given by

$$C_a(0) = 2\pi\int_0^\infty R\,dR|\psi(R)|^2. \quad (35b)$$

According to Hosain et al. [11], only the first four terms in (35a) are enough to obtain sufficient accuracy for misalignment up to 1^0, which corresponds to $p \approx 0.8$ for an optical fiber with $n = 1.5$ and $a = 4\,\mu m$ working at a wavelength $\lambda = 0.8\,\mu m$. Here, up to the fifth term of (35a) have been calculated and the required expressions are given in (9)-(10).

The transmission coefficient $T_t(\Delta)$ at the splice for a transverse offset d is expressed as follows:

$$T_t(\Delta) = \left|\frac{C_t(\Delta)}{C_t(0)}\right|^2, \quad (36)$$

where $\Delta = d/a$ is the normalized transverse offset, and in practice for $\Delta \leq 0.8$, one can approximately write [11]

$$\frac{C_t(\Delta)}{C_t(0)} = 1 - \frac{B_1}{B_0}\left(\frac{\Delta}{2}\right)^2 + \frac{B_2}{B_0}\left(\frac{\Delta}{2}\right)^4 - \frac{B_3}{B_0}\left(\frac{\Delta}{2}\right)^6, \quad (37)$$

where

$$B_0 = \langle\psi^2\rangle, \quad (38)$$

$$B_1 = \langle\psi'^2\rangle, \quad (39)$$

$$B_2 = \frac{1}{4}\left(\int_0^\infty\left|\frac{d^2\psi}{dR^2}\right|^2 RdR + \int_0^\infty\left|\frac{d\psi}{dR}\right|^2\frac{dR}{R}\right), \quad (40)$$

$$B_3 = \frac{1}{36}\int_0^\infty\left|\frac{d^3\psi}{dR^3}\right|^2 RdR - \frac{1}{12}\int_0^\infty\frac{d\psi}{dR}\frac{d^3\psi}{dR^3}\frac{dR}{R}$$
$$- \frac{1}{24}\left(\frac{d^2\psi}{dR^2}\right)_{R=0}. \quad (41)$$

Integrals given in (38)–(41) can be evaluated by using the expression of fundamental modal field given by (26). Hence, (34) and (36) can be evaluated with the help of (35a), (35b), and (37).

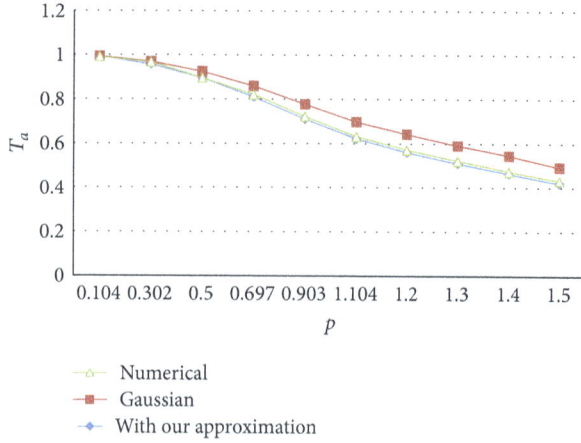

FIGURE 1: Variation of power transmission coefficients T_a with the normalized angular offset p for splicing of two identical single mode triangular index fibers with $V = 2.7$ (exact numerical results [10, 13]; results by our approximation; results based on Gaussian approximation [10, 13]).

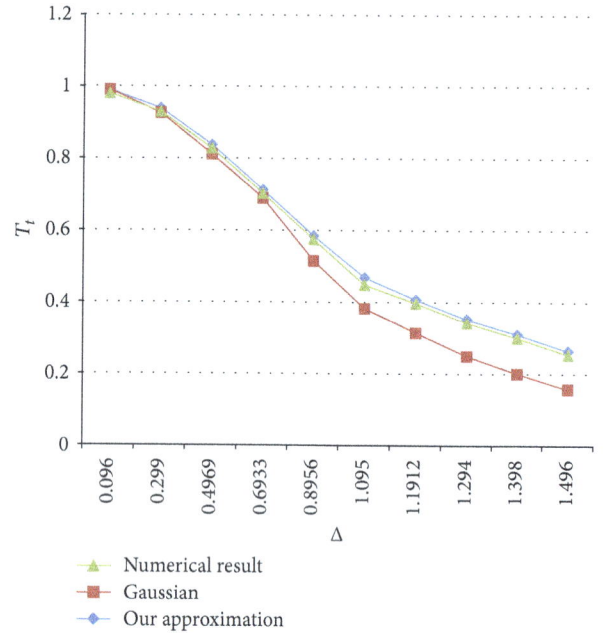

FIGURE 2: Variation of power transmission coefficients T_t with the normalized transverse offset Δ for splicing of two identical single mode triangular index fibers with $V = 2.7$ (exact numerical results [10, 13]; results by our approximation; results based on Gaussian approximation [10, 13]).

2.4. Evaluation of Integrals. Evaluation of integrals to determine propagation constant and splice loss is presented in (1)–(23). Substituting (1)–(6) into (28), (29), and (30), analytical expression of propagation constant β can be obtained with the help of (26) and (27). The transmission coefficient $T_a(p)$ (34) at the splice with angular mismatch can be obtained by substituting (9)-(10) into (35a) and (35b). Using (31), (1), (2), (5), and (6), (38) can be obtained. Equations (40) and (41) can be evaluated using (11). Once (38)–(41) are evaluated, $C_t(\Delta)/C_t(0)$ (see (37)) can be calculated. Then, the analytical expression of transmission coefficient $T_t(\Delta)$ at the splice between two identical optical fibers having a transverse offset can be evaluated using (36).

3. Results and Discussions

Detailed comparison between the proposed formulation and available exact numerical results [10, 13] has been carried out in terms of accuracy assessment. It has been justified by many authors [1–3] that two-parameter approximations are more accurate than single-parameter approximation. The proposed approximation of fundamental field involving three optimizing parameters incorporates more flexibility to modify the fundamental modal solution of optical fibers having different specifications. Optimized values of these parameters for different normalized frequencies are given in Tables 1 and 2 for a particular specification of optical fiber having trapezoidal and triangular index profiles, respectively. Values for other normalized frequencies having different specification of optical fiber can also be obtained by using Nelder-Mead method of nonlinear unconstrained minimization.

In order to verify the feasibility of the proposed approximation, the outcomes of the proposed study have been compared with the earlier reported numerical results [10, 13]. In the present study, $S_0 = 0.25$ and $V = 2.4$ are considered for trapezoidal index profile, which corresponds to a typical

TABLE 1: Values of optimizing parameters with different normalized frequencies for trapezoidal index profile.

α	R_0	μ	V
1.839506	1.487480	0.006634	1.6000
1.854451	1.152568	0.040590	1.8000
1.878781	1.006541	0.097612	2.0000
1.908874	0.923490	0.171137	2.2000
1.941083	0.867514	0.253638	2.4000
1.954907	0.814689	0.289126	2.6000
1.987346	0.781444	0.277328	2.8000
1.967303	0.722072	0.256257	3.0000
2.207612	0.797867	0.283156	3.2000
1.930533	0.637208	0.226140	3.4000

dispersion shifted silica fiber with $a = 3.2\,\mu$m, $\delta = (n_1{}^2 - n_2{}^2)/2n_1{}^2 = 0.008$, and zero dispersion wavelength at $1.55\,\mu$m [12]. For triangular index profile, $V = 2.7$ and $S_0 = 0$ have been chosen, taking $a = 3.5\,\mu$m, $\delta = 0.008$, and zero dispersion wavelength at $1.5\,\mu$m [12].

For evaluation of splice loss, the applicability of the proposed formulations in case of power transmission coefficients $T_a(p)$ and $T_t(\Delta)$ at splices between two identical optical fibers has been considered. Gaussian approximation gives accurate result for the evaluation of transmission coefficient only in the region near the cutoff of single mode operation, but it leads to considerable error throughout the single mode region [11]. The variation of T_a with p and T_t with Δ, in case of splicing of two identical triangular index fibers, has been plotted in

Splice Loss of Graded-Index Fibers: Accurate Semianalytical Descriptions Using Nelder-Mead Nonlinear Unconstrained
Optimization with Three-Parameter Fundamental Modal Field

7

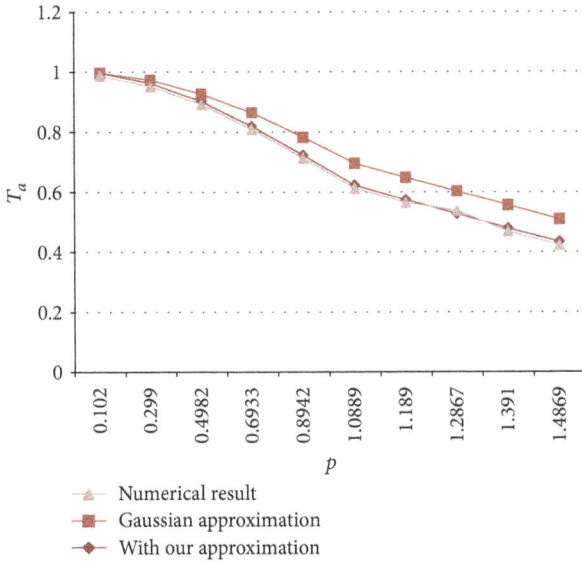

FIGURE 3: Variation of power transmission coefficients T_a with the normalized angular offset p for splicing of two identical single mode trapezoidal index fibers with $V = 2.4$ (exact numerical results [10, 13]; results by our approximation; results based on Gaussian approximation [10, 13]).

TABLE 2: Values of optimizing parameters with different normalized frequencies for triangular index profile.

α	R_0	μ	V
1.836461	1.962401	−0.000309	1.7000
1.843169	1.323589	0.014877	1.9000
1.858852	1.102224	0.050720	2.1000
1.881018	0.986778	0.103042	2.3000
1.907224	0.914502	0.167036	2.5000
1.934891	0.862840	0.237307	2.7000
1.954757	0.817735	0.290207	2.9000
1.947828	0.764272	0.271233	3.1000
1.941593	0.718887	0.255127	3.3000
2.150373	0.784010	0.278234	3.5000

4. Conclusions

An accurate three-parameter approximation of fundamental modal field solution of an optical fiber has been presented, which can effectively be used to estimate the power transmission coefficients in case of splicing of two identical single mode graded-index fibers in presence of both transverse and angular misalignments. Taking trapezoidal and triangular index fibers as examples, it has been shown that the results obtained with our function are excellently matching with the exact available and numerical results [10, 13]. Besides providing values of optimizing parameters involved in the approximate field obtained by Nelder-Mead method of nonlinear unconstrained minimization, all related simplified analytical expressions have also been presented, which can be used directly by optical fiber designer while predicting splice losses of an optical fiber, having triangular and trapezoidal index profiles for a wide range of normalized frequencies. The salient features of the proposed solution are easy computation on an ordinary personal computer and a robust algorithm for nonlinear unconstrained optimization being applied in an optical fiber having triangular and trapezoidal index profiles.

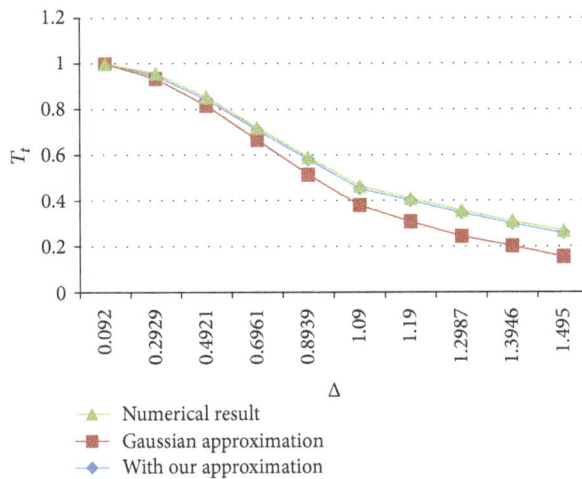

FIGURE 4: Variation of power transmission coefficients T_t with the normalized transverse offset Δ for splicing of two identical single mode trapezoidal index fibers with $V = 2.4$ (exact numerical results [10, 13]; results by our approximation; results based on Gaussian approximation [10, 13]).

References

[1] R. Tewari, S. I. Hosain, and K. Thyagarajan, "Scalar variational analysis of single mode fibers with Gaussian and smoothed-out profiles," *Optics Communications*, vol. 48, no. 3, pp. 176–180, 1983.

[2] A. Sharma, S. I. Hosain, and A. K. Ghatak, "The fundamental mode of graded-index fibres: simple and accurate variational methods," *Optical and Quantum Electronics*, vol. 14, no. 1, pp. 7–15, 1982.

[3] A. Sharma and A. K. Ghatak, "A variational analysis of single mode graded-index fibers," *Optics Communications*, vol. 36, no. 1, pp. 22–24, 1981.

[4] A. Ankiewicz and G.-D. Peng, "Generalized Gaussian approximation for single-mode fibers," *Journal of Lightwave Technology*, vol. 10, no. 1, pp. 22–27, 1992.

[5] S. Chieh, W.-H. Tsai, and M.-S. Wu, "Extended Gaussian approximation for single-mode graded-index fibers," *Journal of Lightwave Technology*, vol. 12, no. 3, pp. 392–395, 1994.

Figures 1 and 2, respectively. Similarly, the variations of these power transmission coefficients for the case of splicing of two identical trapezoidal index fibers are illustrated in Figures 3 and 4. For the practical range of p and Δ, the results obtained by the proposed approximation are identically matching with the exact available and numerical results [10, 13].

[6] G. De Angelis, G. Panariello, and A. Scaglione, "Variational method to approximate the field of weakly guiding optical fibers by Laguerre-Gauss/Bessel expansion," *Journal of Lightwave Technology*, vol. 17, no. 12, pp. 2665–2674, 1999.

[7] F. Chiadini, G. Panariello, and A. Scaglione, "Variational analysis of matched-clad optical fibers," *Journal of Lightwave Technology*, vol. 21, no. 1, pp. 96–105, 2003.

[8] Q. Cao and S. Chi, "Approximate analytical description for fundamental-mode fields of graded-index fibers: beyond the Gaussian approximation," *Journal of Lightwave Technology*, vol. 19, no. 1, pp. 54–59, 2001.

[9] A. K. Ghatak, R. Srivastava, I. F. Faria, K. Thyagarajan, and R. Tiwari, "Accurate method for characterising single-mode fibres: theory and experiment," *Electronics Letters*, vol. 19, no. 3, pp. 97–99, 1983.

[10] J. P. Meunier and S. I. Hosain, "An efficient model for splice loss evaluation in single-mode graded-index fibers," *Journal of Lightwave Technology*, vol. 9, no. 11, pp. 1457–1463, 1991.

[11] S. I. Hosain, A. Sharma, and A. K. Ghatak, "Splice-loss evaluation for single-mode graded-index fibers," *Applied Optics*, vol. 21, no. 15, pp. 2716–2720, 1982.

[12] U. C. Paek, "Dispersionless single-mode fibers with trapezoidal-index profiles in the wavelength region near 1.5 μm," *Applied Optics*, vol. 22, no. 15, pp. 2363–2369, 1983.

[13] K. Kawano and T. Kitoh, *Introduction to Optical Waveguide Analysis*, John Willy and Sons, New York, NY, USA, 2001.

[14] J. C. Lagarias, J. A. Reeds, M. H. Wright, and P. E. Wright, "Convergence properties of the Nelder-Mead simplex method in low dimensions," *SIAM Journal on Optimization*, vol. 9, no. 1, pp. 112–147, 1998.

[15] J. M. Parkinson and D. Hutchinson, "An investigation into the efficiency of variants on the simplex method," in *Numerical Methods for Nonlinear Optimization*, F. A. Lootsma, Ed., pp. 115–135, Academic Press, New York, NY, USA, 1972.

[16] M. H. Wright, "Direct search methods: once scorned, now respectable," in *Proceedings of the Dundee Biennial Conference in Numerical Analysis*, D. F. Griffiths and G. A. Watson, Eds., pp. 191–208, Addison Wesley; Longman, Harlow, UK, 1996.

[17] J. A. Nelder and R. Mead, "A simplex method for function minimization," *Computer Journal*, vol. 7, pp. 308–313, 1965.

[18] T. H. Rowan, *Functional stability analysis of numerical algorithms [Ph.D. thesis]*, University of Texas, Austin, Tex, USA, 1990.

[19] S. Singer and S. Singer, "Complexity analysis of Nelder-Mead search iterations," in *Proceedings of the 1st Conference on Applied Mathematics and Computation, Dubrovnik, Croatia, 1999*, M. Rogina, V. Hari, N. Limić, and Z. Tutek, Eds., pp. 185–196, PMF-Matematički odjel, Zagreb, 2001.

[20] S. Singer and S. Singer, "Efficient implementation of the Nelder-Mead search algorithm," *Applied Numerical Analysis and Computational Mathematics*, vol. 1, no. 3, pp. 524–534, 2004.

[21] R. Roychoudhury and A. Roychoudhury, "Accurate semi analytical model of an optical fiber having Kerr nonlinearity using a robust nonlinear unconstrained optimization method," *Optics Communications*, vol. 284, no. 4, pp. 1038–1044, 2011.

[22] I. S. Gradshteyn and I. M. Ryzhik, *Table of Integrals, Series and Products*, Academic Press, New York, NY, USA, 1980.

[23] M. Abramowitz and I. A. Stegun, *Handbook of Mathematical Functions*, Dover, New York, NY, USA, 1981.

Optimization Efficiency of Monte Carlo Simulation Tool for Evanescent Wave Spectroscopy Fiber-Optic Probe

Daniel Khankin,[1] Shaul Mordechai,[2] and Shlomo Mark[1, 3]

[1] Software Engineering Department, Shamoon College of Engineering (SCE), 84100 Beer Sheva, Israel
[2] Department of Physics, Ben-Gurion University, 84105 Beer Sheva, Israel
[3] Negev Monte Carlo Research Center, Shamoon College of Engineering (SCE), 84100 Beer Sheva, Israel

Correspondence should be addressed to Shlomo Mark, marks@sce.ac.il

Academic Editor: Yeshoshua Kalisky

In a previous work, we described the simulation tool (FOPS 3D) (Khankin et al., 2001) which can simulate the full three-dimensional geometrical structure of a fiber and the propagation of a light beam sent through it. In this paper we are focusing on three major points: the first concerns the improvements made with respect to the simulation tool and the second, optimizations implemented with respect to the calculations' efficiency. Finally, the major research improvement from our previous works is the simulation results of the optimal absorbance value, as a function of bending angle for a given uncladded part diameter, that are presented; it is suggested that fiber-bending may improve the efficiency of recording the relevant measurements. This is the third iteration of the FOPS development process (Mann et al., 2009) which was significantly optimized by decreasing memory usage and increasing CPU utilization.

1. Introduction

The evanescent wave spectroscopy technique, generally used in the IR range, is useful for inspecting materials and examining their properties, as well as for establishing biomedical diagnoses [1]. The diagnosis technique is based on a phenomenon called Attenuated Total Reflection (ATR) [2], in which incident rays are completely reflected within the medium, leaving evanescent waves on the interface between the medium and the adjacent sample. As the number of reflections occurring in the sensing fiber increases, a greater number of evanescent waves are created. Consequently, use of this spectroscopic method requires detailed planning, since fiber wave spectroscopy intensifies as fiber absorbance increases.

Two primary methods can be used, the first involves tapering the untapped part of the fiber and the second, bending the fiber about its untapped part. These two actions make it more difficult for the light beam to propagate, causing more hits in the untapped part and thus creating additional evanescent waves, which in turn increase the absorption intensity. The first method was investigated experimentally by [3] and by [1]; these papers provided correlations between the absorbance and the thickness of the untapped fiber section, [1] the optimal width d of the untapped part. In our previous work [4], we computed and investigated the relation between the diameter of the uncladded part and the resultant absorption value, using an in-house Monte Carlo simulation tool. In this work, we investigate an additional method of determining the optimal bending angle of a fiber with a constant width.

In a previous work, the Monte Carlo simulation tool (FOPS 3D) was described [5]; this tool can simulate the full three-dimensional geometrical structure of the fiber and the propagation of a light beam through it. Evanescent waves are formed when waves travelling through a medium undergo total internal reflections, on which the Attenuated Total Reflectance (ATR) sampling technique is based [1, 6]. In fact, evanescent waves decrease exponentially as they propagate further into the sample, and optical fibers, which are transparent in the mid-IR range, are used as the ATR elements. As the number of reflections causing the evanescent fields increases, the measurement becomes more efficient. Thus, the curvature of the fiber and the diameter of its uncladded part, which is in contact with the sample, both contribute to the efficiency of the measurement.

2. System Description

The user may define several properties for the simulation system, including the simulated fiber's length, radius, the radius of the uncladded part, reflection coefficients, and bending angle (see Figure 1). Next, the simulated light beam is propagated through the fiber, undergoes an emission process, and hits the fiber medium. The emission source, which emits rays according to a radial Gaussian distribution (which is both uniform and isotropous), is simulated with Monte Carlo techniques.

The simulation history is defined by a light beam that hits the uncladded part and successfully travels through the fiber up to the light sensor. The number of successful histories is an estimator of the efficiency of the simulated fiber shape. The unbiased mean value of hits in successful histories is used for calculating the Fresnel transmission coefficient [6], which is the absorption property of the simulated shape.

The simulation tool, as it is virtual, has the flexibility of freely bending the fiber, thus providing the possibility of creating a fiber folded in any possible curvature. In turn, folding the fiber increases the number of beam hits in the uncladded part by slowing down the beam's propagation. A second feature of the simulation tool is the possibility of adjusting the radius of the uncladded part alone, specifically decreasing its radius relative to the rest of the fiber's radius; this affects the number of beam hits in the uncladded section. In addition, the fiber maybe deformed and clay modeled, providing the possibility of creating alternative geometrical shapes and inspecting their efficiency as ATR elements [1, 3, 4].

The Monte Carlo approach of simulating physical phenomena is based on the creation of a large sample of random occurrences, used in order to reconstruct the dynamics of a particular system. The simulation tool can then provide, as a system output, the utilization of a certain geometrical fiber configuration. The simulation results include estimates of the probability of rays successfully passing through a fiber in a particular, user-defined geometry.

3. Physical Background

Fiber evanescent wave spectroscopy, primarily used with an IR light source, consists of emitting rays into a flexible optical fiber. The emitted energy is passed on to the distal end of the fiber and into a Fourier transform infrared spectroscopy (FTIR) detector. In this section, we will briefly define the main physical phenomena modeled by the simulation algorithm.

The Gaussian distribution for a beam waist is given by [7]:

$$I(r) = \frac{I_0 e^{-8r^2}}{r_0^2}, \qquad (1)$$

where r_0 is the radius of the laser aperture and I_0 is the irradiance at the center of the aperture.

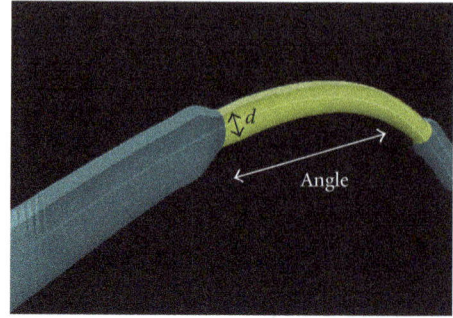

FIGURE 1: The simulated fiber.

The incidence angles of the rays must be less than $\theta_{cl} = \sin^{-1}(\sqrt{n_1^2 - n_{cl}^2})$, where θ_{cl} is the critical angle of the cladded section, n_1 is the refractive index of the fiber core, and n_{cl} is the refractive index of the cladding, and is sampled from the angular distribution given by [8]

$$p(\theta) \propto f(x) = \begin{cases} \dfrac{\sin\theta}{\cos^3\theta}, & \theta < \theta_{cri}, \\ 0, & \theta \geq \theta_{cri}. \end{cases} \qquad (2)$$

According to the Beer-Lambert Law, the transmitted power $P(z)$ along a fiber is given by

$$P(z) = p_0 e^{-\gamma z}, \qquad (3)$$

where p_0 is the power at the input face of the fiber, and γ is the effective evanescent absorption coefficient (which varies for different orientations of the ray) [9].

In the Monte Carlo approach applied in the simulation, Ruddy's equation [9] is used:

$$\gamma = NT, \qquad (4)$$

where N is the number of reflections per unit length, and T is the Fresnel transmission coefficient for an evanescent wave, defined for any angle θ_z relative to the core axis.

In the Monte Carlo simulation, rays are fired into the fiber according to the radial and angular distributions of (1) and (2), respectively.

4. Development Process

The current simulation tool belongs to the family of scientific software [10, 11], softwares that require a knowledgeable user, that appreciates both its input and its output. In addition, successful scientific software development requires a knowledgeable tester, who should understand the mathematical as well as the scientific models used in the software. In third iteration the simulation tool had to be transferred from a 32-bit environment to a 64-bit operating system and, correspondingly, a 64-bit Java environment. Though from an API perspective the environments are similar, this transfer required taking some aspects into account. For instance, the data structures in a 64-bit environment require more memory than required in the 32-bit environment. Taking the above specifications into account, the Agile [12, 13] software development methodology was chosen, as it is suited for small teams with constantly changing requirements. However, the changes and optimizations to the software are not trivial and should be carefully considered before implementation. Therefore, test-driven development (TDD) methodology should be integrated into the software development process.

Test-driven development is a software development approach based on "test-first development", according to which tests should be written before coding. Since the development process is broken into small units, it is very easy to adapt it to black-box testing. This approach tests all possible combinations of end-user actions. Black-box testing does not require knowledge of code and is intended to simulate the end-user experience of the final product. These tests determine appropriately correct output for valid or invalid input. The Java platform was chosen for development since it is efficient time wise with regards to the development of existing data structures and helps avoid memory management issues.

The simulation tool was initially modeled and designed by Unified Modeling Language (UML). The simulation's architectural blueprints were visualized by UML elements. Objects were described by object modeling techniques and information flow by data flow diagrams and entity relationship diagrams. Entity relationship diagrams represent abstract and conceptual data elements; for example, the simulation tool uses the fiber structure as the conceptual data. In order to describe the data flow between object entities, as well as modifications on data structures, data flow diagrams were created. Data flow diagrams help visualize data processing in an information system. FOPS 3D is computation-intensive software, and for this reason, more tests were conducted to ensure high level of reliability. Finally, it was verified that the results were in accordance with previous experiments and simulations.

5. Optimization

Extensive optimization of the tool was performed. The main aspect optimized was memory usage, which was especially high after the transfer to a 64-bit JVM. As the system was first ported to the new architecture, memory usage was extensive, (around 600 MB). In previous versions, memory cleaning was mostly performed by the JVM garbage collector. However, with the advances in CPU speed and architecture, the JVM's garbage collector was too slow to keep memory usage at a consistent level. Data structure objects were not deleted quickly enough and memory was not retrieved at a sufficient pace, while the simulation continued to run and demand memory space allocations. To solve this problem, an inner agent was introduced, with its own cleaning policy. The agent's primary concern was to clear data structures by clearing all references to it and marking it for garbage collection. Meanwhile, the agent replaced the existing data structures with new, smaller sized structures, since in Java clearing only the data structure will not reduce its memory size. Furthermore, the size of data objects was reduced as much

as possible. In addition, some redundancy was found with respect to data objects, especially in data structures regarding animation of recurring shapes, like rays (consisting of points). Such data structures were adapted to a limit of containing data objects, so they will not hold too many objects.

After implementing the above-mentioned changes, memory usage declined to 230 MB. Of course, there is always a tradeoff between resource usage and accuracy of calculation. In this case, reducing memory usage allowed for an increase in calculation accuracy. The latter depends on the number of fiber parts constituting a single fiber, where a greater number of fiber parts results in increased accuracy. Thus, the reduction in memory usage allowed for a significant increase in the number of fiber parts represented. Moreover, simplifying the data objects led to significant improvement in the overall simulation runtime. The simulation engine is now able to process both the fiber data structures and ray beam structures more quickly. In other words, calculations of the ray beam's advancement and its collision with the fiber medium are more accurate, and the error tolerance of the calculations is of a much smaller order. The running time for the previous version, at 1,000,000 histories, was days, while the running time for the new version (at the same number of histories) was around hours. Overall, there was an approximate 43% improvement in running time and 62% improvement of memory usage.

6. Results and Discussion

The verification phase in the software development process of the new version of the FOPS 3D simulation tool consisted of comparing the compatibility of its results with those of a previous work [5]. The verification case was a fiber (indices of refraction: $n_1 = 1.46$, $n_2 = 1.34$, and $n_{cl} = 1.45$) 100 mm long and 0.9 mm thick, with a flattened section of 35 mm and conical tapered sections with a 6° slope, where the thickness of the flattened section varied between 0.32 mm to 0.19 mm. The results showed an exact match between the old and new results [5], which allowed for validating the new version of FOPS 3D so that other conditions could be investigated.

As in the previous study, in order to seek the optimal width of the uncladded section, the relative absorbance was simulated with respect to the following fiber properties: fiber length 100 mm, diameter 0.9 mm, and refractive indices $n_1 = 1.457$ and $n_2 = 1.357$. As opposed to the previous simulation, where the midsection was a **flattened section**, in this simulation the midsection was a **circular** section with a diameter of 0.2–0.36 mm, length of 50 mm, and conical tapered sections with a 6° slope.

In Figure 2, the simulation results for the given fiber properties (in the case of no bending angle) are shown. These results were presented in our previous works [4, 5] and were actually used to validate all three versions of FOPS 3D and were also used as acceptance tests.

As seen in Figure 2, the results agree with those presented earlier [4] (where a tapped midsection was used), with a slight difference, probably because of the different midsections used in the simulation. As shown in Figure 2, the absorbance increases as thickness of the narrowed section

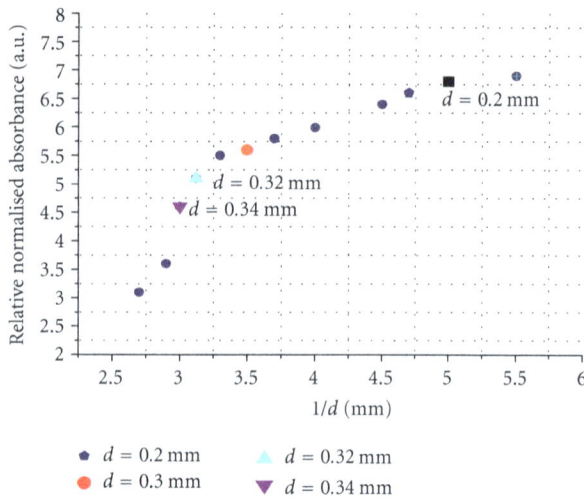

FIGURE 2: Average relative absorbance for a flattened fiber with a narrowed midsection.

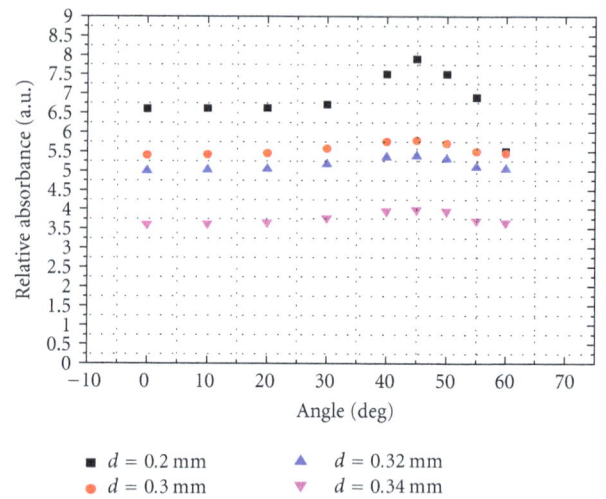

FIGURE 3: Average relative absorbance of bending angle for a given uncladded part diameter.

decreases, due to the larger number of reflections that takes place within the narrower section. In other words, these results clearly indicate that even for a circular midsection, using thinner tapered midsection radii may increase the absorbance intensity dramatically and improve the spectral signal-to-noise ratio.

In this work, the goal was to explore the effect of fiber bending on the fiber effectiveness. Accordingly, the bending angles for the untapped midsection were changed, while the midsection itself was kept constant, in order to see determine the optimal absorbance value. In other words, the relative absorbance for different bending angles was measured, while defining the uncladded section at the constant that was shown in Figure 2.

The results of the simulation are shown in Figure 3. Clearly, as the fiber is bent, the relative absorbance increases, for the reasons mentioned earlier regarding fiber bending, which enables the light beam to hit the uncladded section more often. In Figure 3, the results for four different cases are shown. The first case is the optimal one, as was found in Figure 2, that is, a circular midsection with a diameter of 0.2 mm; the other three cases show results for additional (arbitrary) diameter values ($d = 0.3, 0.32, 0.34$ mm). For the optimal case, this goes on as far as a certain optimal bending angle (around 45°), where the efficiency increases almost 20% over that of the comparable unbent case, after which the relative absorbance decreases by 20% as the bending angle grows. Other widths below the optimal value were tested as well, and the same pattern was observed: an increase until the optimal angle (around 45°) and a decrease following it; however, the difference in the efficiency of the optimal angle is no more than 5%. Figure 3 reveals the most dramatic changes in efficiency that soccur when the fiber bending occurs mostly in the optimal diameter. Thus, both thickness and bending angle are two complementary parameters with which optimal efficiency can be achieved.

7. Conclusions

As predicted, it indeed is beneficial to bend the fiber so that the ray beam advances more slowly and hits the fiber medium a greater number of times, in order to increase the efficiency of spectroscopy with fiber optics. More hits on the fiber medium transfer additional energy into the tissue or sample tested, thus increasing the accuracy of the method. The most significant finding of this work is that there are two complementary parameters determining optimal fiber efficiency: the thickness of the midsection and the bending angle, which is optimally around 45°. The relation between these two parameters is that the efficiency achieved by bending at the optimal angle is much more significant when the midsection is at its optimal width.

The simulation tool still requires further optimization at the algorithmic level. With the recent advances of multicore processors and parallel programming languages, FOPS 3D may be adapted for parallel execution in order to increase the simulation processing speed. This, along with the use of grids or genetic algorithms, can further improve the simulation tool, enabling it to find the optimal bending level once the optimal width for given fiber properties is determined.

References

[1] Y. Raichlin, L. Fel, and A. Katzir, "Evanescent-wave infrared spectroscopy with flattened fibers as sensing elements," *Optics Letters*, vol. 28, no. 23, pp. 2297–2299, 2003.

[2] N. J. Harrick, *Internal Reflection Spectroscopy*, Harrick Scientific Corporation, New York, NY, USA, 1979.

[3] D. Gupta, C. D. Singh, and A. Sharma, "Fiber optic evanescent field absorption sensor: effect of launching condition and the geometry of the sensing region," *Optical Engineering*, vol. 33, no. 6, pp. 1864–1868, 1994.

[4] M. P. Mann, S. Mark, Y. Raichlin, A. Katzir, and S. Mordechai, "Optimization of fiber-optic evanescent wave spectroscopy: a Monte Carlo approach," *Applied Spectroscopy*, vol. 63, no. 9, pp. 1057–1061, 2009.

[5] D. Khankin, S. Mark, and S. Mordechai, "Monte Carlo Simulation Tool of Evanescent Waves Spectroscopy Fiber-Optic Probe for Medical Applications (FOPS 3D)," InTech, 2001.

[6] A. Messica, A. Greenstein, and A. Katzir, "Theory of fiber-optic, evanescent-wave spectroscopy and sensors," *Applied Optics*, vol. 35, no. 13, pp. 2274–2284, 1996.

[7] E. Hecht, *Optics*, Addison-Wesley, 2002.

[8] W. Snyder and J. D. Love, *Optical Waveguide Theory*, Chapman and Hall, London, UK, 1983.

[9] V. Ruddy, B. D. MacCraith, and J. A. Murphy, "Evanescent wave absorption spectroscopy using multimode fibers," *Journal of Applied Physics*, vol. 67, no. 10, pp. 6070–6074, 1990.

[10] D. Khankin, A. D. Solomon, Y. Shpungin, Y. Shtoland, and S. Mark, "A Monte Carlo package for optimizing fiber-optic evanescent wave spectroscopy as a test case for scientific programming," in *Proceedings of the The International Symposium on Stochastic Models in Reliability Engineering, Life Sciences and Operations Management (SMRLO '10)*, Beer Sheva, Israel, 2010.

[11] I. Sommerville, *Software Engineering*, Addison-Wesley, New York, NY, USA, 7th edition, 2004.

[12] A. Cockburn, *Agile Software Development*, Addison-Wesley, Boston, Mass, USA, 2002.

[13] J. Highsmith and A. Cockburn, "Agile software development: the business of innovation," *Computer*, vol. 34, no. 9, pp. 120–122, 2001.

Cutting Properties of Austenitic Stainless Steel by Using Laser Cutting Process without Assist Gas

Hitoshi Ozaki, Yosuke Koike, Hiroshi Kawakami, and Jippei Suzuki

Graduate School of Engineering, Mie University, 1577 Kurima-machiya, Tsu, Mie 514-8507, Japan

Correspondence should be addressed to Hitoshi Ozaki, ozaki@mach.mie-u.ac.jp

Academic Editor: Augusto Belendez

Recently, laser cutting is used in many industries. Generally, in laser cutting of metallic materials, suitable assist gas and its nozzle are needed to remove the molten metal. However, because of the gas nozzle should be set closer to the surface of a workpiece, existence of the nozzle seems to prevent laser cutting from being used flexible. Therefore, the new cutting process, Assist Gas Free laser cutting or AGF laser cutting, has been developed. In this process, the pressure at the bottom side of a workpiece is reduced by a vacuum pump, and the molten metal can be removed by the air flow caused by the pressure difference between both sides of the specimen. In this study, cutting properties of austenitic stainless steel by using AGF laser cutting with 2 kW CO_2 laser were investigated. Laser power and cutting speed were varied in order to study the effect of these parameters on cutting properties. As a result, austenitic stainless steel could be cut with dross-free by AGF laser cutting. When laser power was 2.0 kW, cutting speed could be increased up to 100 mm/s, and kerf width at specimen surface was 0.28 mm.

1. Introduction

Laser cutting is one of the thermal cutting processes such as gas cutting and plasma cutting [1]. In laser cutting, as a focused laser beam melts or vapors a workpiece locally, it has the following features. First, width of cut kerf is narrower than that of the conventional cutting process. Second, laser cutting can achieve precision cut with small distortion. Therefore, laser cutting is applied in many industries such as automobile, aerospace, shipbuilding, and so on [2].

In laser cutting of metallic materials, molten metal is usually removed by using suitable assist gas at high pressure as shown in Figure 1 [3]. Thus, the special gas nozzle is needed to provide the assist gas such as oxygen, nitrogen, air, and so on. However, because of the gas nozzle should be moved closer to the position about 1 mm from the surface of a workpiece, it's thought that existence of the nozzle causes lack of flexibility of laser cutting. For example, when the laser cutting of odd-shapes material pressed into

shapes is intended, the gas nozzle may touch the material. Additionally, in the remote laser cutting studied in these days [4–6], gas nozzle using itself is difficult. From this viewpoint, it's expected that the application range of laser cutting will expand by realizing without assist gas.

For this reason, the new cutting process, Assist Gas Free laser cutting, hereafter, called as AGF laser cutting, has been developed and investigated about cutting properties in our laboratory [7]. In this process, as shown in Figure 2, the pressure at the bottom side of a workpiece is reduced by a vacuum pump, and the molten metal can be removed by the air flow caused by the pressure difference between top and bottom sides of the specimen. In addition, the pressure difference is important for this process, since it corresponds to the assist gas pressure in conventional laser cutting process.

In the present study, cutting properties of austenitic stainless steel by using AGF laser cutting were investigated. At first, laser power and cutting speed were varied in order

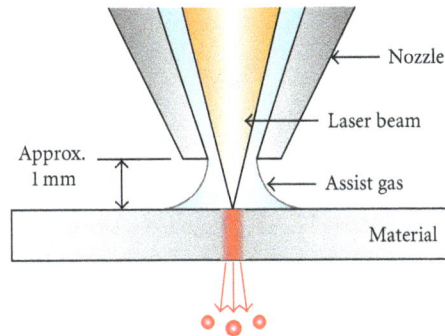

FIGURE 1: Schematic of conventional laser cutting process of metallic materials with assist gas.

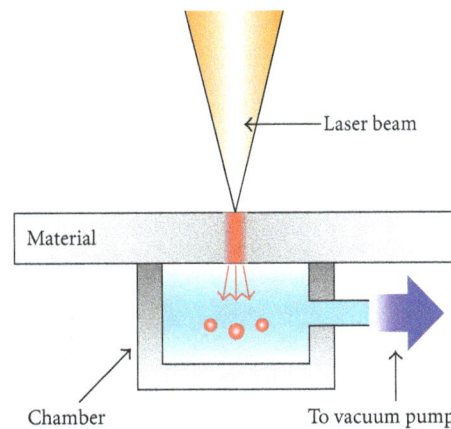

FIGURE 2: Schematic of Assist Gas Free laser cutting process.

to study the effect of these parameters on cutting properties. Moreover, effect of cutting condition on pressure in chamber during cutting process was discussed.

2. Experimental Procedure

2.1. Materials Used. As materials, austenitic stainless steel, JIS SUS304, was used for AGF laser cutting. Thickness of the SUS304 plate was 1 mm in this study. Width of the specimen was 39 mm, and length was 300 mm. Chemical compositions of the material used were shown in Table 1.

2.2. Procedure of AGF Laser Cutting Process. A schematic of AGF laser cutting equipment using a 2 kW CO_2 laser was shown in Figure 3.

A chamber connected to a vacuum pump was mounted on an NC work table. After a sheet material was set on top side of the chamber, the pressure in the chamber was reduced by the vacuum pump. When inside of the chamber became near vacuum state, AGF laser cutting experiment was carried out. Laser beam melts the material locally; the molten metal can be removed by the pressure difference between both sides of the sheet.

2.3. Process Parameters for AGF Laser Cutting. Process parameters for AGF laser cutting were shown in Table 2.

A 2 kW CO_2 laser of continuous wave with circular polarization was used. In this study, laser power was varied from 1.0 to 2.0 kW. Focal length of lens was 127 mm, and focal position was set at the surface of the SUS304 plate. Piercing time was 1 s every time. Cutting speed was varied from 10 to 110 mm/s. Shielding gas of nitrogen was used to protect the focusing lens. Cut length 100 mm was controlled by the NC program.

3. Experimental Results and Discussions

3.1. Process Window of AGF Laser Cutting. Process window of AGF laser cutting is shown in Figure 4.

The condition which could be cut with all length of 100 mm is shown as circles, and that which could not be cut is shown as crosses. When laser power was increased, the range of cutting speed which could be cut was expanded. This is because heat input to the SUS304 plate is augmented by the increment of laser power at higher cutting speeds.

From this result, critical cutting speed V_c for each laser power was found and shown in Table 3.

V_c was 50 mm/s at laser power 1.0 kW, and 100 mm/s at 2.0 kW. When laser power was increased, the critical cutting speed V_c was also increased, since sufficient heating was applied to the material in spite of high cutting speed. Calculated heat input at the critical cutting speed with laser power of 1.0 kW is 20 J/mm, 1.5 kW is 19 J/mm and 2.0 kW is

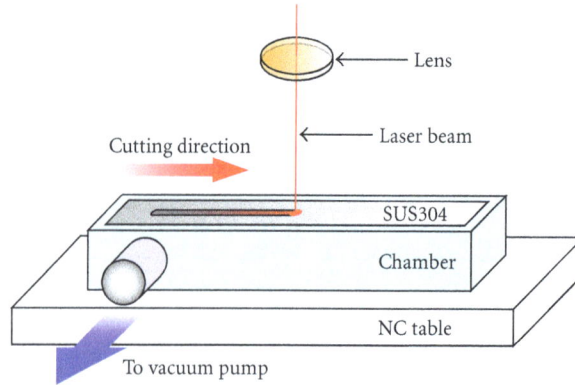

FIGURE 3: Schematic of AGF laser cutting equipment.

TABLE 1: Chemical compositions of JIS SUS304, standard value, mass %.

C	Si	Mn	P	S	Ni	Cr
≦0.08	≦1.00	≦2.00	≦0.045	≦0.030	8.00 ~ 10.50	18.00 ~ 20.00

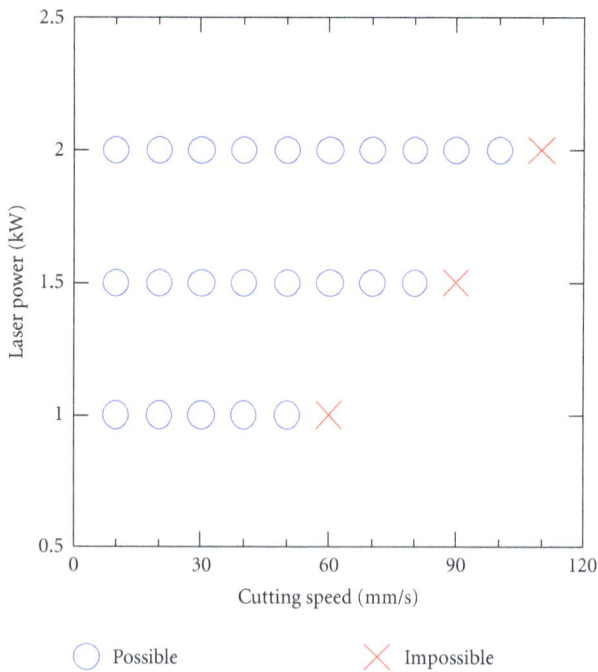

○ Possible ✕ Impossible

FIGURE 4: Process window of AGF laser cutting with different laser power and cutting speed.

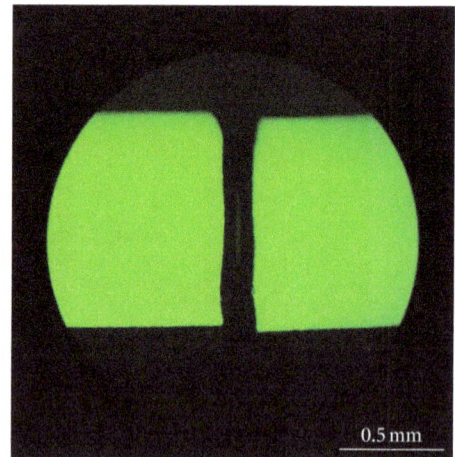

FIGURE 5: Macro cross-section of specimen after AGF laser cutting with laser power of 1.0 kW and cutting speed of 50 mm/s.

TABLE 2: Process parameters for AGF laser cutting.

Laser type	CO_2 laser (CW)
Polarization of laser	Circular polarization
Laser power	1.0 ~ 2.0 kW
Focal length of lens	127 mm
Defocused distance	±0 mm
Piercing time	1 s
Cutting speed	10 ~ 110 mm/s
Shielding gas	N_2 : 15 L/min
Cut length	100 mm

20 J/mm, respectively. Therefore, the heat input of 19 J/mm, is minimum required for AGF laser cutting of the SUS304 plate of 1 mm thickness.

3.2. *Cross-Sectional Shapes of Kerf.* Macro cross-section of kerf with laser power of 1.0 kW and cutting speed of 50 mm/s is shown in Figure 5.

Cut kerf was formed and seen at the center of the picture. In order to discuss cross-sectional shapes of kerf, its profiles were measured with all cut specimens. The cross-sectional shapes of kerf at different laser power and cutting speed are shown in Figure 6.

In this figure, the upper side is corresponded to the laser irradiated surface in each graph. In all cutting conditions, cut kerfs are formed perpendicular to the surface of the plate.

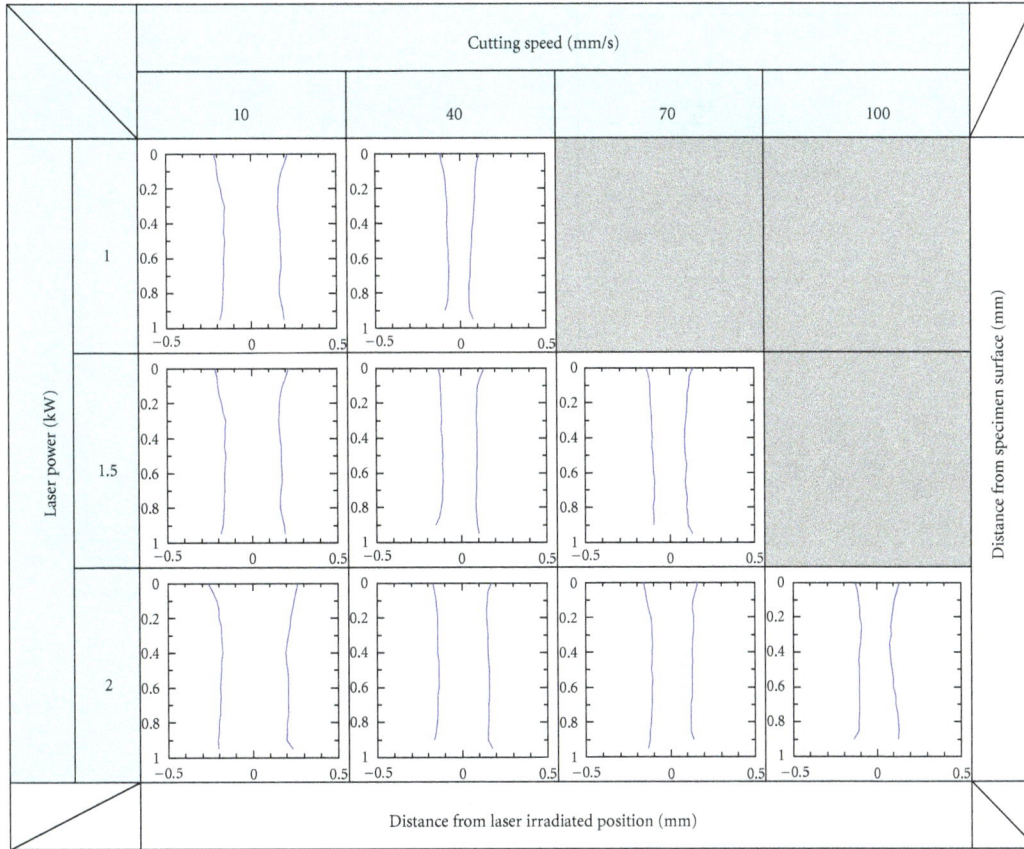

FIGURE 6: Cross-sectional shapes of kerf at different laser power and cutting speed.

TABLE 3: Critical cutting speed V_c of AGF laser cutting for each laser power.

Laser power	V_c
1.0 kW	50 mm/s
1.5 kW	80 mm/s
2.0 kW	100 mm/s

power. The width was measured at the specimen surface of the laser irradiated side.

At same cutting speed, when laser power was increased, kerf width was also increased, since the heat input was augmented. By contrast, when cutting speed was increased, kerf width was decreased, as the heat input was also decreased. Maximum kerf width was 0.56 mm at laser power 2.0 kW and cutting speed 10 mm/s. Minimum was 0.25 mm at laser power 1.0 kW and cutting speed 50 mm/s.

3.4. The Effect of Cutting Speed on Removed Area of Kerf. Likewise, Figure 9 shows the effect of cutting speed on removed area of kerf. The removed area of kerf means the cross-sectional area of kerf.

At same cutting speed, when laser power was increased, removed area of kerf was also increased. By contrast, when cutting speed was increased, removed area of kerf was decreased. As mentioned above, this is because the influence of increasing and decreasing of the heat input. Maximum removed kerf area was 0.37 mm^2 at laser power 2.0 kW and cutting speed 10 mm/s. Minimum was 0.11 mm^2 at laser power 1.0 kW and cutting speed 50 mm/s.

3.5. The Effect of Heat Input on Kerf Shapes. Laser power and cutting speed can be treated as one parameter of the heat input. Thus, the kerf width and the removed area of

Furthermore, the kerf shapes are almost the same though laser power or cutting speed is varied. These facts were confirmed by measuring kerf taper. In this study, the taper was defined as one-half of the difference between upper and lower width of kerf, as mentioned in the literature [3].

Figure 7 shows the effect of cutting speed on kerf taper at different laser power. Maximum of the taper was about 0.06 mm at laser power 2.0 kW and cutting speed 10 mm/s. Hence, it was found that formation of the taper was suppressed, because of the taper was below 0.06 mm in all cutting conditions.

However, as shown in Figure 6, kerf width and area are different depending on the cutting condition. Therefore, they are quantitatively evaluated in the following section.

3.3. The Effect of Cutting Speed on Kerf Width. Figure 8 shows the effect of cutting speed on kerf width at different laser

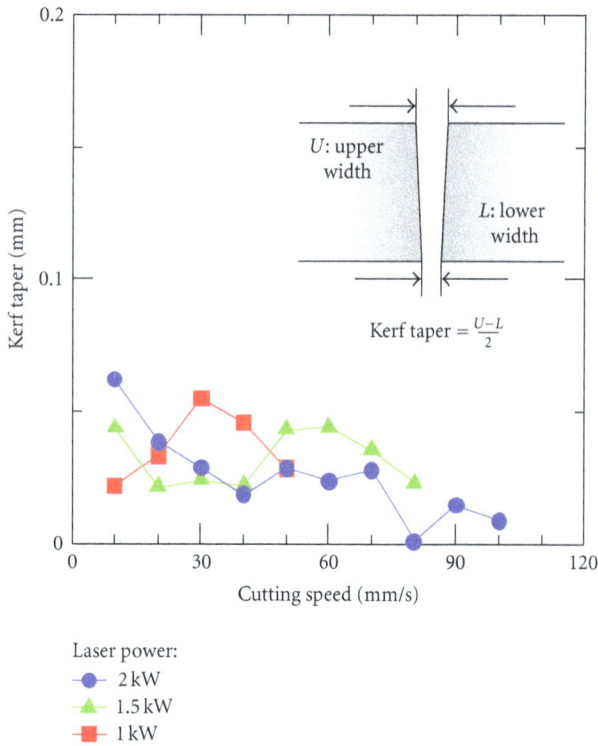

FIGURE 7: Effect of cutting speed on kerf taper at different laser power.

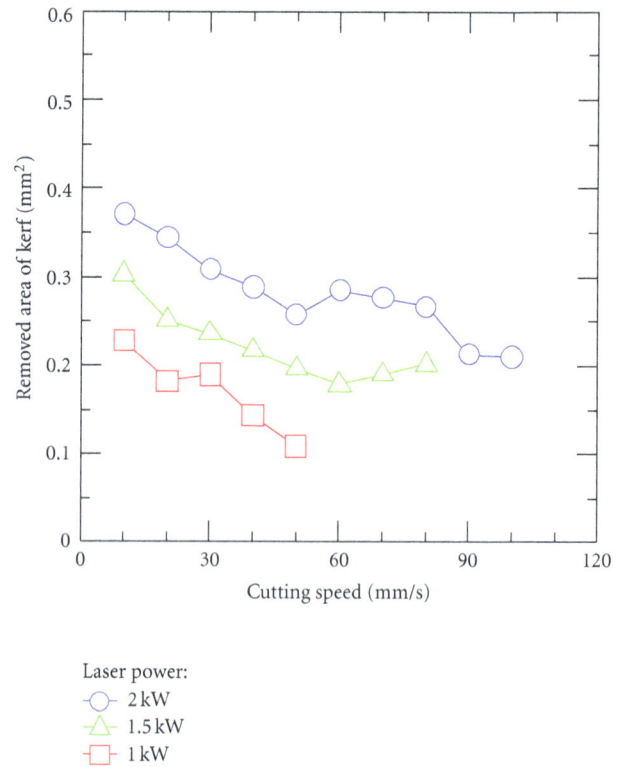

FIGURE 9: Effect of cutting speed on removed area of kerf at different laser power.

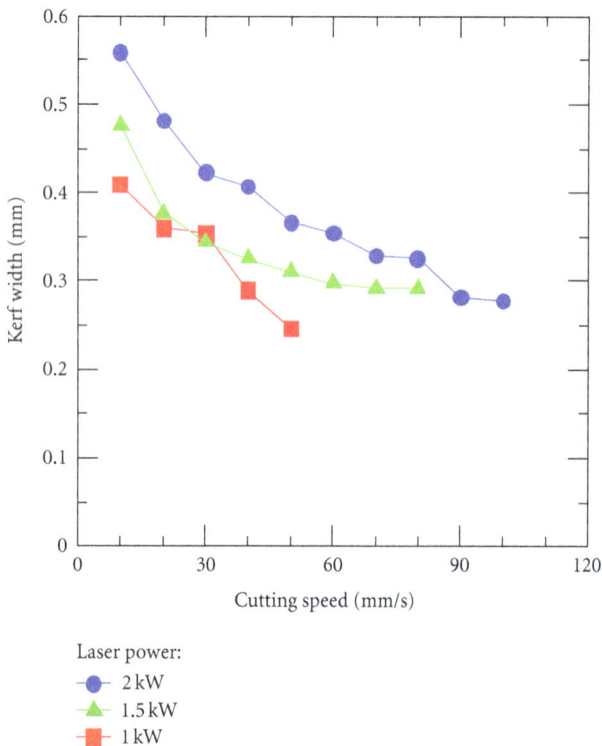

FIGURE 8: Effect of cutting speed on kerf width at different laser power.

kerf were organized by the heat input. Figure 10 shows the effect of the heat input on kerf shapes. Figure 10(a) shows the relationship between the heat input and the kerf width at the specimen surface, and Figure 10(b) shows the heat input and the removed area of kerf.

When heat input was increased, both kerf width and removed kerf area were also increased. As shown in Figure 10(a), the kerf width is roughly corresponding at the low heat input. In contrast, the kerf width of 2.0 kW laser power is wider than that of the other laser powers at the high heat input. Beam transverse mode of the CO_2 laser used is TEM_{00} under the laser power of 1.0 kW, after that, the mode turns into TEM_{01}^* as increment of the laser power. Hence, it's considered that the kerf width was affected by the laser beam mode.

As shown in Figure 10(b), the removed area of kerf is different in three laser powers regardless of the same heat input. This is because the various kerf shapes were present as shown in Figure 6. At the low laser power like 1.0 kW, the kerf shape was near inverted trapezoid. When the laser power increased, the kerf shape became rectangle, since the lower part of the plate was heated well.

3.6. Appearance of Cut Surface. In order to evaluate cutting quality, the appearance of cut surface was observed. Figure 11 shows the appearance of cut surface at different laser power and cutting speed.

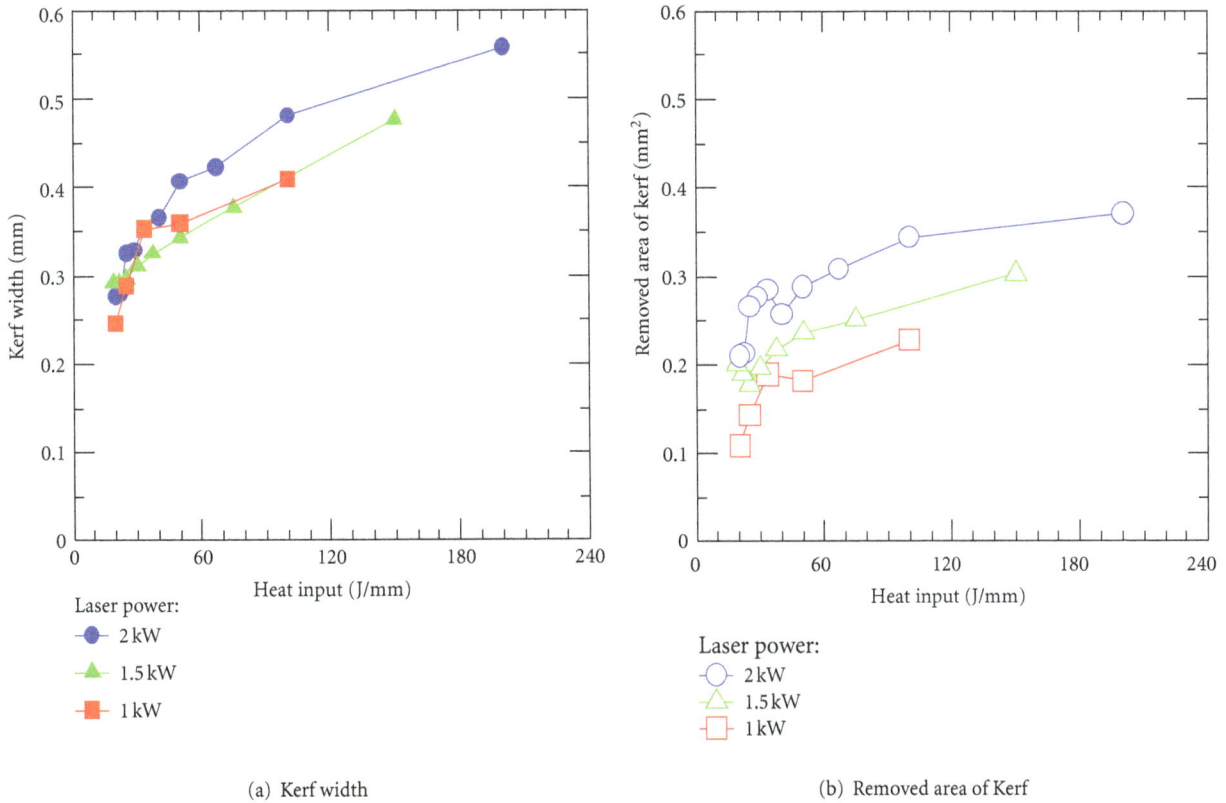

(a) Kerf width

(b) Removed area of Kerf

FIGURE 10: Effect of heat input on kerf shapes.

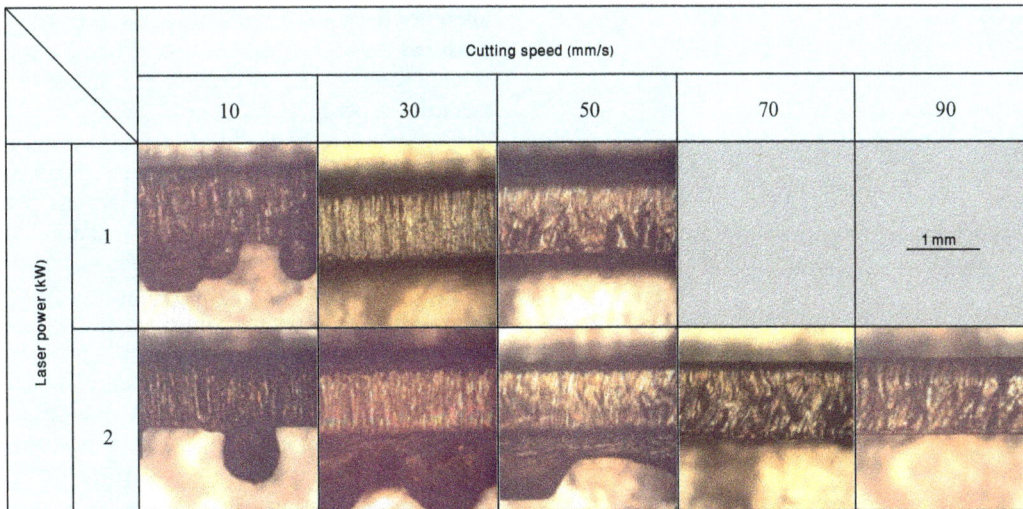

FIGURE 11: Appearance of cut surface at different laser power and cutting speed.

With low cutting speed such as 10 mm/s, the formation of dross was confirmed. When cutting speed was decreased, because the amount of the molten metal was increased, part of the molten metal was not removed from the kerf and remained at the bottom surface of the plate as the dross. However, when cutting speed was increased, dross-free cutting was achieved. In the high cutting speed, since the amount of molten metal is few, the metal was ejected smoothly. Therefore, it's thought that high-speed cutting is effective in dross-free cutting.

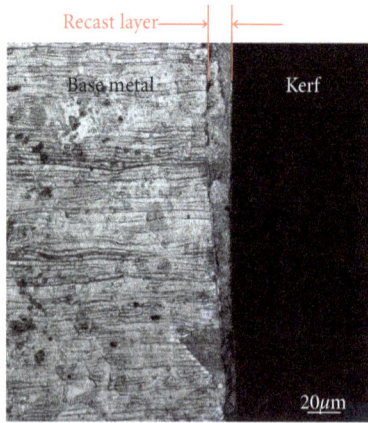

FIGURE 12: Microstructure of cut surface with laser power of 2.0 kW and cutting speed of 100 mm/s.

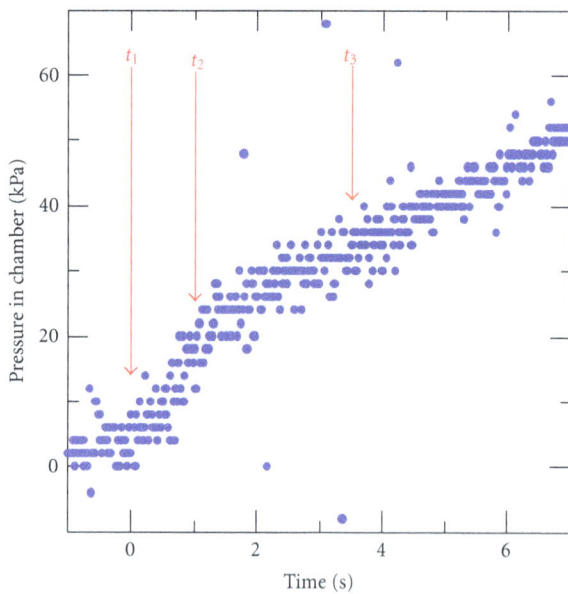

FIGURE 13: Change of pressure in chamber during AGF laser cutting with laser power of 2.0 kW and cutting speed of 40 mm/s.

In addition, cut surface was observed in detail after etching the kerf cross-section with hydrochloric acid. Figure 12 shows microstructure of the cut surface near the center of the plate thickness with laser power of 2.0 kW and cutting speed of 100 mm/s.

An approximately $10\,\mu m$ thin layer of different microstructure from the base metal was observed on the cut surface. This layer may be formed by solidification of the molten metal which wasn't removed, that is, recast layer. Similar thin layer was also confirmed at other cutting conditions.

3.7. Effect of Cutting Condition on Pressure in Chamber during AGF Laser Cutting. As mentioned above, in AGF laser cutting, the molten metal can be removed by the air flow caused by the pressure difference between both sides of the

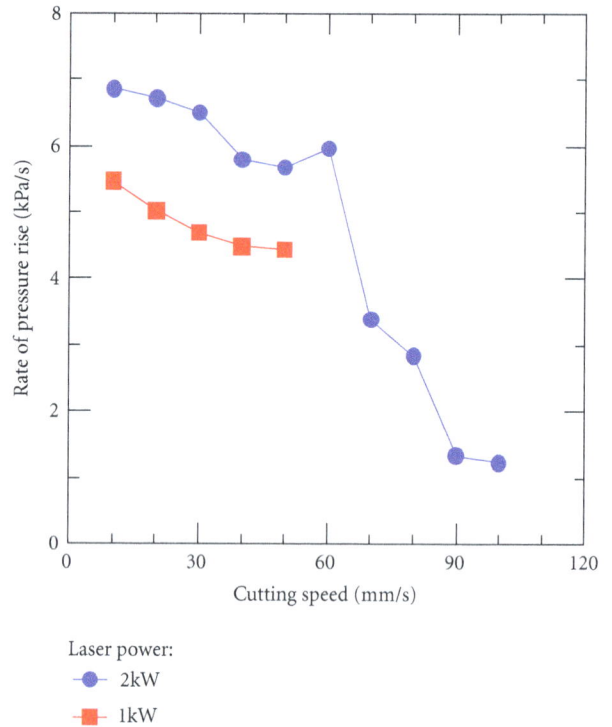

FIGURE 14: Effect of cutting speed on rate of pressure rise during AGF laser cutting at different laser power.

specimen. However, as cutting progresses, air flows into the chamber from the cut kerf. In other words, the pressure difference decreases. Thus, the pressure in the chamber was measured during cutting process by using a pressure sensor. As an example, the change of the pressure during AGF laser cutting with laser power of 2.0 kW and cutting speed of 40 mm/s was shown in Figure 13, where, start time of piercing was defined as "t_1", start time of cutting as "t_2", and end time of cutting as "t_3." Before t_1, the pressure in the chamber was about 0 kPa, or near vacuum state. From t_1 to t_2, because of the piercing process had been undergone, the pressure rose. From t_2 to t_3, cutting process had been in progress, the pressure continued to increase for further inflow of atmosphere.

In order to discuss pressure change during AGF laser cutting process, between t_2 and t_3, it's defined as rate of pressure rise. The rate was calculated from the slope of the line obtained by least squares approximation from t_2 to t_3 in each cutting condition. Figure 14 shows the effect of cutting condition on the rate of pressure rise during AGF laser cutting at different laser power.

At same cutting speed, when laser power was increased, the rate of pressure rise was also increased, since the kerf width was expanded as shown in Figure 8. By contrast, when cutting speed was increased, the pressure rise rate was decreased, as the kerf width was also decreased as shown in Figure 8.

Therefore, AGF laser cutting by low heat input is desirable, because narrow kerf width and dross-free cutting can be obtained as above. In addition, since the rate of pressure rise during cutting is low, longer cutting length can be ensured under low-heat-input cutting.

4. Conclusions

The present study is focused on cutting properties of austenitic stainless steel by using AGF laser cutting process. The following conclusions can be drawn.

(1) Increase in the laser power led to increase of the critical cutting speed. With the laser power of 2.0 kW, the critical cutting speed was 100 mm/s.

(2) The cross-sectional shapes of kerf were almost the same though the laser power or the cutting speed was varied. The kerf taper was below 0.06 mm in all cutting conditions.

(3) When the heat input was increased, the kerf width and the removed area of kerf were also increased.

(4) When the cutting speed was high, the dross-free cutting was achieved. The thin recast layer, about 10 μm, was observed on the cut surface.

(5) When the laser power was decreased or the cutting speed was increased, the rate of pressure rise in the chamber during cutting process was decreased.

References

[1] Japan Welding Society, *New Edition Advanced Theory of Welding and Joining Technology*, Sanpo Publications, 2011.

[2] T. Arai, M. Kutsuna, and I. Miyamoto, *Laser Cutting Process*, Machinist Publishing, 1994.

[3] M. Kanaoka, *Practical Business of Laser Materials Processing—Vital Point of Work and Measure Against Trouble*, Nikkan Kogyo Shimbun, 2007.

[4] M. F. Zaeh, J. Moesl, J. Musiol, and F. Oefele, "Material processing with remote technology—revolution or evolution?" *Physics Procedia*, vol. 5, pp. 19–33, 2010.

[5] J. Hatwig, G. Reinhart, and M. F. Zaeh, "Automated task planning for industrial robots and laser scanners for remote laser beam welding and cutting," *Production Engineering*, vol. 4, no. 4, pp. 327–332, 2010.

[6] H. Shamoto, T. Ikeda, and A. Tsuboi, "Development of remote cutting technology by high quality solid state laser," in *Reports on The 406th Topical Meeting of The Laser Society of Japan*, pp. 17–21, 2010.

[7] J. Suzuki, *Japanese Patent Publication*, 2007-190590.

4

Dynamics of 1.55 μm Wavelength Single-Mode Vertical-Cavity Surface-Emitting Laser Output under External Optical Injection

Kyong Hon Kim, Seoung Hun Lee, and Vijay Manohar Deshmukh

Department of Physics, Inha University, Incheon 402-751, Republic of Korea

Correspondence should be addressed to Kyong Hon Kim, kyongh@inha.ac.kr

Academic Editor: Zoran Ikonic

We review the temporal dynamics of the laser output spectrum and polarization state of 1.55 μm wavelength single-mode (SM) vertical-cavity surface-emitting lasers (VCSELs) induced by external optical beam injection. Injection of an external continuous-wave laser beam to a gain-switched SM VCSEL near the resonance wavelength corresponding to its main polarization-mode output was critical for improvement of its laser pulse generation characteristics, such as pulse timing-jitter reduction, linewidth narrowing, pulse amplitude enhancement, and pulse width shortening. Pulse injection of pulse width shorter than the cavity photon lifetime into the SM VCSEL in the orthogonal polarization direction with respect to its main polarization mode caused temporal delay of the polarization recovery after polarization switching (PS), and its delay was found to be the minimum at an optimized bias current. Polarization-mode bistability was observed even in the laser output of an SM VCSEL of a standard circularly cylindrical shape and used for all-optical flip-flop operations with set and reset injection pulses of very low pulse energy of order of the 3.5~4.5 fJ.

1. Introduction

Dynamical laser output of the vertical-cavity surface-emitting lasers (VCSELs) under injection of an external laser beam has been investigated widely for potential application to pulse timing-jitter reduction, polarization switching, all-optical flip-flop operation, and long-distance fiber transmissions [1–15]. VCSELs are known to be low power-consuming optical signal sources compared to the existing edge-emitting laser diodes (LDs) and to be potentially useful all-optical logic gate devices in a two-dimensional array. Beside the conventional signal source application of the stand-alone VCSELs for relatively short-distance optical communications or interconnects, the injection locking of an external optical beam to the VCSELs allows new application areas of the VCSELs possible and improves fiber transmission properties of the VCSELs' output. Weak continuous-wave (CW) laser beam injection to a gain-switched VCSEL lowered its pulse timing jitter [2, 3]. Laser pulse beam injection of an

orthogonal or circular polarization into a VCSEL caused a high-speed polarization switching (PS) [4–7] or spin-induced polarization oscillation [8, 9]. The PS and induced bistability mechanism of the VCSEL output under a laser beam injection of orthogonal polarization to the VCSEL's main polarization mode were used for demonstration of all-optical flip-flop operation and optical buffer memory application [10–13]. An external laser beam injection to high-speed direct-modulated VCSELs reduced the frequency chirp, enhanced the small-signal modulation bandwidth, and thus helped extension of the transmission distance of the VCSEL signals over long fiber spans [14–16].

In this paper, we will first review what optimum external beam injection condition required for the gain-switched VCSELs to have the minimized timing jitter state. Then, we discuss how the polarization-switching dynamics vary with external optical beam injection conditions. Finally, we introduce how the PS dynamics can be used for high-speed all-optical flip-flop application.

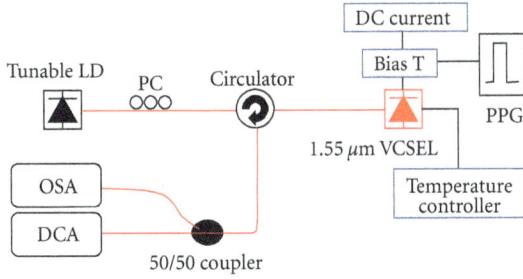

FIGURE 1: Experimental setup used for the timing-jitter reduction of the gain-switched VCSEL pulses with a laser beam injection; PPG: pulse pattern generator; PC: polarization controller, OSA: optical spectrum analyzer; DCA: digital communication analyzer.

2. Timing-Jitter Reduction of Gain-Switched Single-Mode VCSELs

Gain-switched or mode-locked optical pulses of low timing jitter are very useful for high-speed optical time-division multiplexed communications and all-optical signal processing. It has been demonstrated previously that the timing jitter of gain-switched semiconductor lasers, such as Fabry-Perot (FP) LDs and distributed feedback (DFB) LDs, could be reduced by an external laser beam injection [17, 18]. However, it was observed that a trade-off relationship existed between the jitter reduction and pulse width shortening of the gain-switched LDs.

Recently, timing-jitter reduction as well as pulse width shortening, pulse amplitude enhancement, spectral linewidth narrowing, and pulse amplitude increase of gain-switched pulses from a 1.55 μm wavelength single-mode VCSEL was observed experimentally with an optimized injection laser beam wavelength [3]. Figure 1 shows the experimental setup used for timing-jitter reduction of the gain-switched VCSEL pulses with a tunable laser beam injection. The VCSEL was a commercially available single-longitudinal- and transverse-mode (SM) VCSEL based on monolithically grown InAlGaAs-InGaAs quantum well layers emitting a laser beam at 1.55 μm wavelength. It was packaged into a transistor-outlined-can (TO-CAN) with a single-mode fiber (SMF) pigtail and with a thermoelectric cooler (TEC) inside. This TO-CAN type-packaged VCSEL had a threshold current of about 2.1 mA at 21.4°C. The VCSEL was gain switched by rectangular electric pulses of 400 ps pulse width and 280 mV amplitude at 1.25 GHz repetition rate with a DC bias current of 3 mA, while its temperature was kept at 21.4°C with the TEC control.

Figure 2(a) shows measured optical spectra of the free-running continuous-wave (CW) and gain-switched VCSEL's outputs with an optical spectrum analyzer (OSA). The free-running CW spectrum illustrates the linearly polarized main mode at 1,553.86 nm with a 3 dB linewidth below the OSA resolution limit of 0.06 nm, while the side mode of orthogonal polarization appears at a shifted wavelength of 0.35 nm from the main mode with its intensity suppressed relatively below 32 dB from the main peak. The gain switching caused spectral broadening of the two polarization-mode outputs

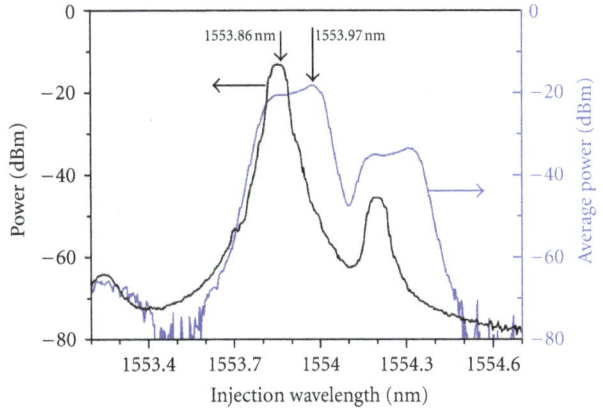

(a)

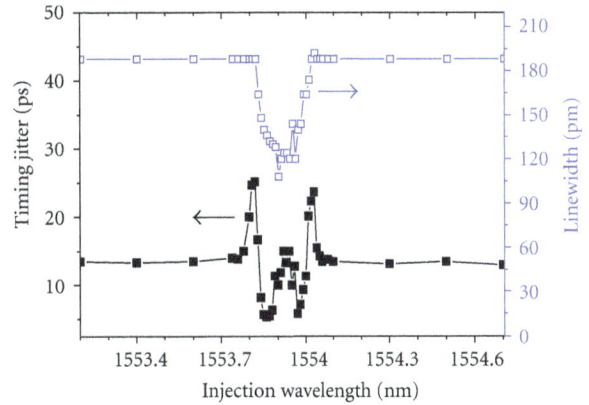

(b)

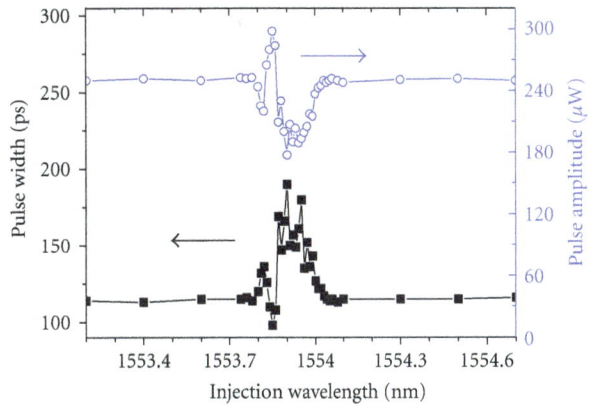

(c)

FIGURE 2: (a) Measured optical spectra of the VCSEL in a free-running continuous-wave (black line with the Y-axis scale on the left-hand side) and gain-switched operations (blue line with the Y-axis scale on the right-hand side); measured (b) timing jitter and linewidth; (c) pulse width and pulse amplitude of the gain-switched VCSEL pulses under an external laser beam injection.

probably due to the gain-modulated spectral chirping inside the VCSEL. Then, a CW optical beam from a tunable laser source was coupled into the VCSEL via an optical circulator and polarization controller (PC) in the same polarization direction as that of the main mode. Figures 2(b) and 2(c) show the measured root-mean-square (rms) timing jitter,

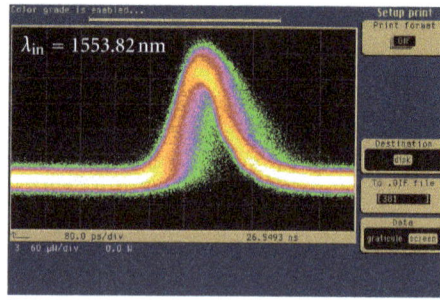
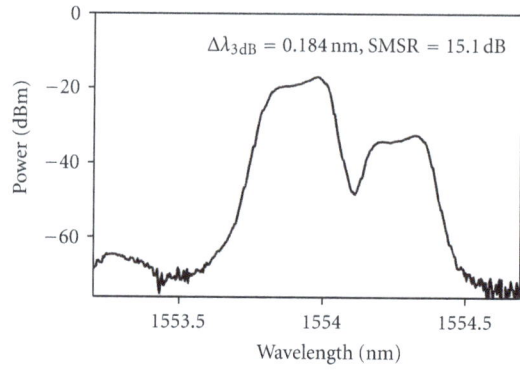

(a)

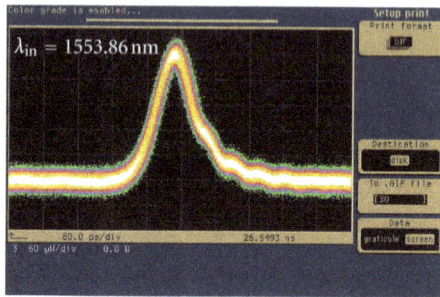
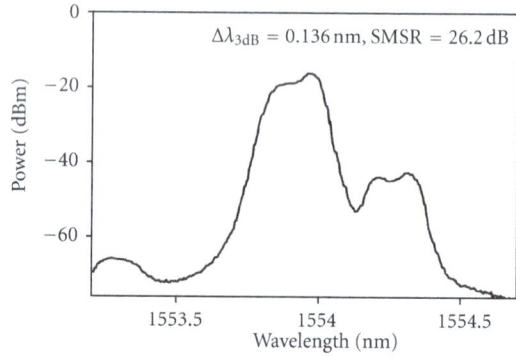

(b)

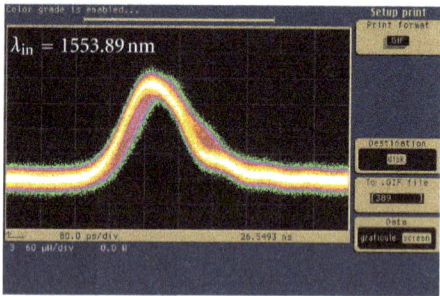
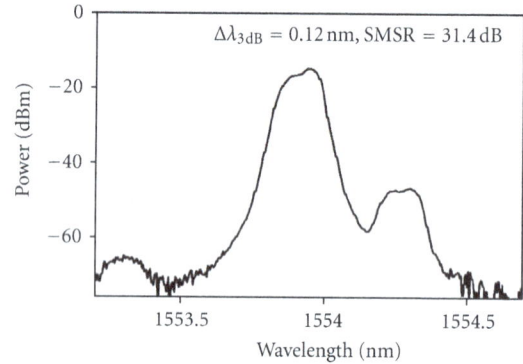

(c)

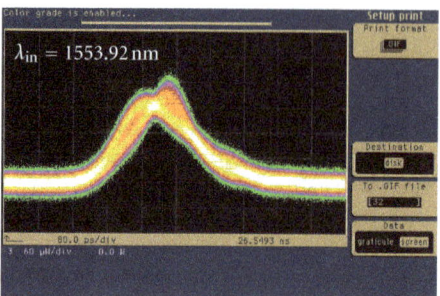
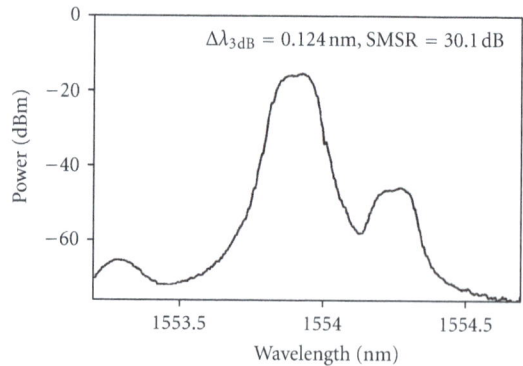

(d)

FIGURE 3: The measured pulse shapes and optical spectra of the gain-switched pulses with external beam injection of various detuned wavelengths.

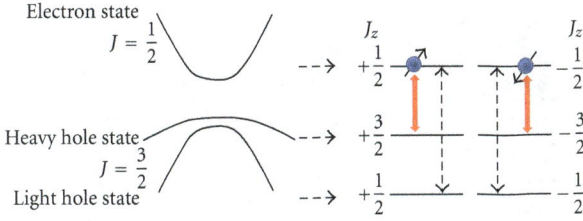

FIGURE 4: The band structure of semiconductor quantum wells and allowed transitions [19].

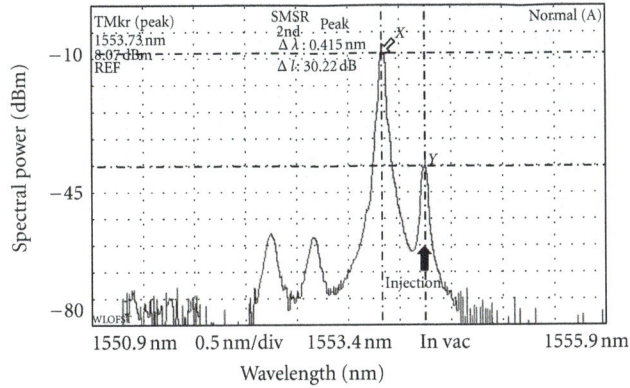

FIGURE 5: Observed output spectrum of a single-mode 1.5 μm wavelength VCSEL.

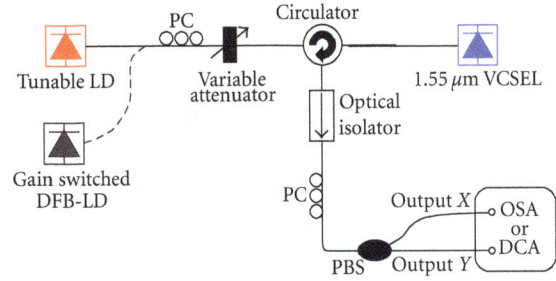

FIGURE 6: Experimental setup used for study of PS dynamics of a single-mode 1.5 μm wavelength VCSEL. PC: polarization controller, PBS: polarization beam splitter.

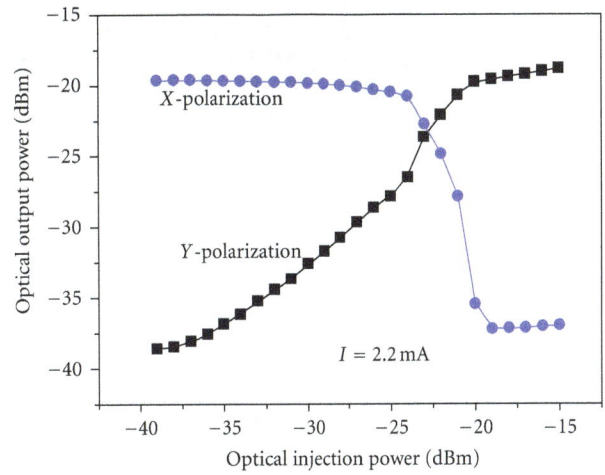

(a)

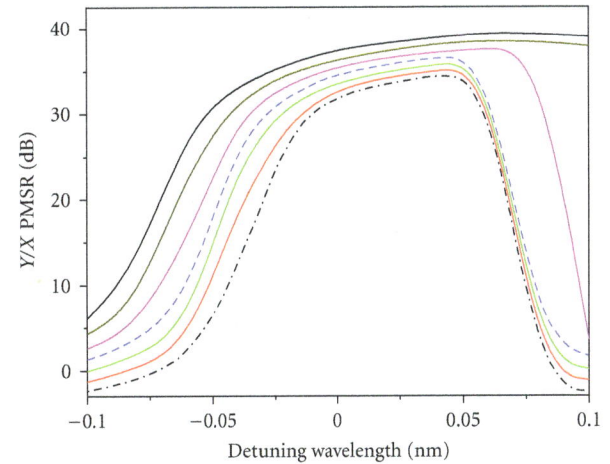

(b)

FIGURE 7: Observed (a) PS dynamics of the single-mode 1.5 μm wavelength VCSEL as a function of the optical injection powers with a driving current slightly above threshold and (b) PMSR as a function of the injection wavelength detuning for various injection powers.

spectral linewidth, and time-averaged pulse width and pulse amplitude of the gain-switched VCSEL output under the external tunable laser beam injection while the injection, beam wavelength was tuned from 1,553.2 nm to 1,554.7 nm. The laser power and 3 dB linewidth of the injection beam were −23 dBm and below the OSA resolution limit of 0.06 nm, respectively. Fine-tuning of the injection laser beam wavelength allowed an optimized condition of the pulse width shortening, pulse amplitude enhancement, spectral linewidth narrowing, and pulse amplitude increasing near the free-running main-mode peak wavelength unlikely to the previously reported results of trade-off relationship between the jitter reduction and pulse width shortening.

Figure 3 shows the measured oscilloscope traces and optical spectra of the gain-switched VCSEL pulses under laser beam injections at various detuned wavelengths. The optimum injection wavelength for the minimum jitter and narrow linewidth matches closely with the main-mode wavelength of 1,553.86 nm which corresponds to the resonant transition between the electron spin sublevels ($J_z = \pm 1/2$) of the conduction band and the heavy hole sublevels ($J_z = \pm 3/2$) of the valence band as illustrated in Figure 4. This low time-jittered gain-switched VCSEL pulse sources with an optimized injection scheme will be useful for power-efficient high-speed optical TDM communications, high-speed all-optical sampling, high-speed all-optical signal processing, and low-error quantum key distribution applications.

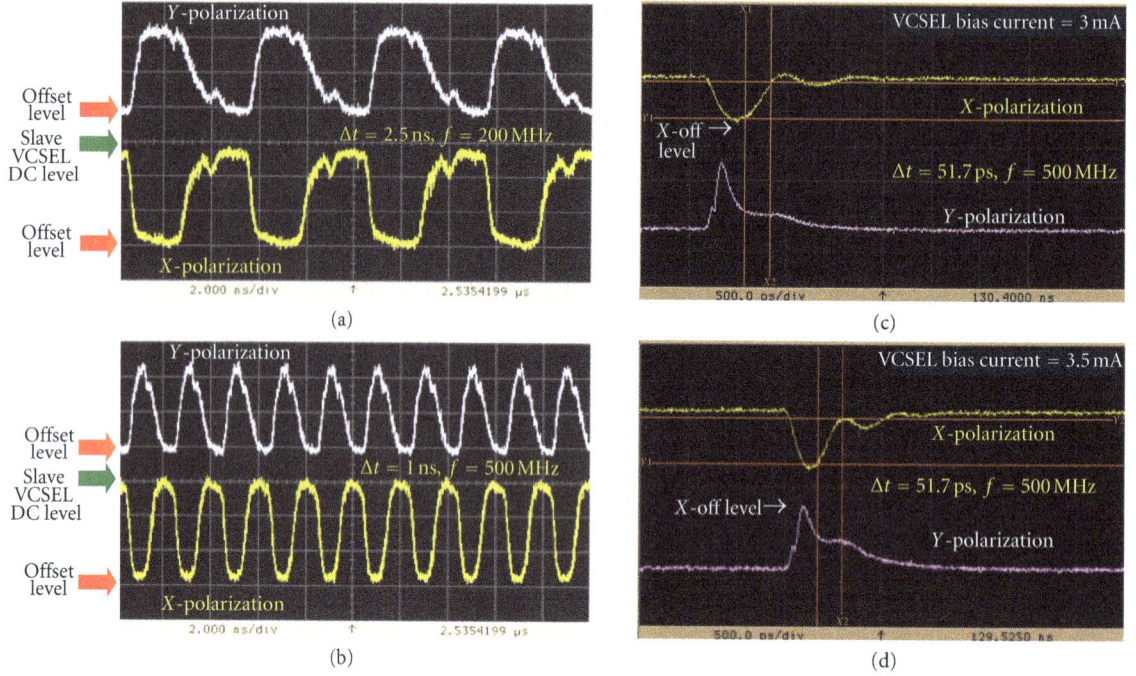

FIGURE 8: Temporal response of the PS dynamics of an SM 1.5 μm wavelength VCSEL with an operating DC bias current of I_b under external beam injection of pulse width (Δt) and repetition rate (f). (a) $\Delta t = 2.5$ ns and $f = 200$ MHz, $I_b = 3.0$ mA, (b) $\Delta t = 1.0$ ns and $f = 500$ MHz, $I_b = 3.0$ mA, (c) $\Delta t = 51.7$ ps and $f = 500$ MHz, $I_b = 3.0$ mA, and (d) $\Delta t = 51.7$ ps and $f = 500$ MHz, $I_b = 3.5$ mA.

3. Polarization Switching Dynamics

Polarization switching (PS) dynamics of the VCSELs with an external optical beam injection have been the focus of significant research effort because of the potential applications to optical flip-flop operation, all-optical signal processing, optical communications, and photonic switching [4–13, 20]. Early stage of the PS research in VCSELs mainly focused on multimode (MM) VCSELs at 850 nm wavelength, which was then followed by the PS dynamics study in 980 nm and 1.3 μm wavelength VCSELs, a special square-shaped SM 1.5 μm wavelength VCSEL, or an MM 1.5 μm VCSEL. The first experimental demonstration of the optical injection-induced PS property of conventional circularly cylindrical-shaped single-transverse-mode 1.5 μm wavelength VCSELs based on InAlGaAs quantum wells on InP wafers was demonstrated in [4]. Even though the circularly cylindrical-shaped SM VCSEL is supposed to have a circular symmetry, it possesses a dominant polarization-mode output at a driving current above threshold as shown in Figure 5. Depending on VCSEL chips, there is a side mode corresponding to the orthogonal polarization of the dominant main mode in a separated wavelength of subnanometer. The side-mode suppression ratio (SMSR) is usually larger than 30 dB.

Figure 6 shows the experimental setup of all-fiber-type configuration used for measurement of polarization switching of an SM 1.5 μm wavelength VCSEL with a CW tunable laser beam injection. Experimentally observed PS dynamics of the VCSEL with a driving current of 2.2 mA slightly above threshold (I_{th} = 1.6 mA) as a function of the optical injection power are shown in Figure 7(a), and

the measured polarization mode suppression ratio (PMSR) between Y- and X-polarizations as a function of the injection wavelength detuning for various injection powers is also shown in Figure 7(b). The tunable LD beam was injected at the polarization direction orthogonal to that of the main-polarization mode of the free-running VCSEL output.

When a gain-switched distributed feedback (DBF) LD was used for optical pulse injection in Figure 6 instead of the CW tunable LD, delayed response of polarization switching and polarization recovery was also observed [4, 7]. For long injection pulses at relatively low repetition rate, no significant time delay of the polarization switching and recovery was observed as shown in Figures 8(a) and 8(b) [21]. However, as the injection pulse width becomes shorter than or close to 50 ps, a significant delay in the polarization switching and recovery was observed as shown in Figures 8(c) and 8(d).

The temporal behavior of polarization dynamics was numerically analyzed using the spin-flip model (SFM) [7, 19]. The SFM describes the electrical fields of the X- and Y-polarizations, E_x and E_y, related to the excited population N and the population difference n between spin-up and spin-down radiation channels as

$$\frac{dE_x}{dt} = k(1 + j\alpha)\left(NE_x - E_x + jnE_y\right)$$

$$- \left(\gamma_a + j\gamma_p + j\Delta\omega\right)E_x + \sqrt{\beta_{sp}}\xi_x,$$

$$\frac{dE_y}{dt} = k(1 + j\alpha)\left(NE_y - E_y - jnE_x\right)$$

$$+ \left(\gamma_a + j\gamma_p - j\Delta\omega\right)E_y + \sqrt{\beta_{sp}}\xi_y + k_{inj}E_{inj},$$

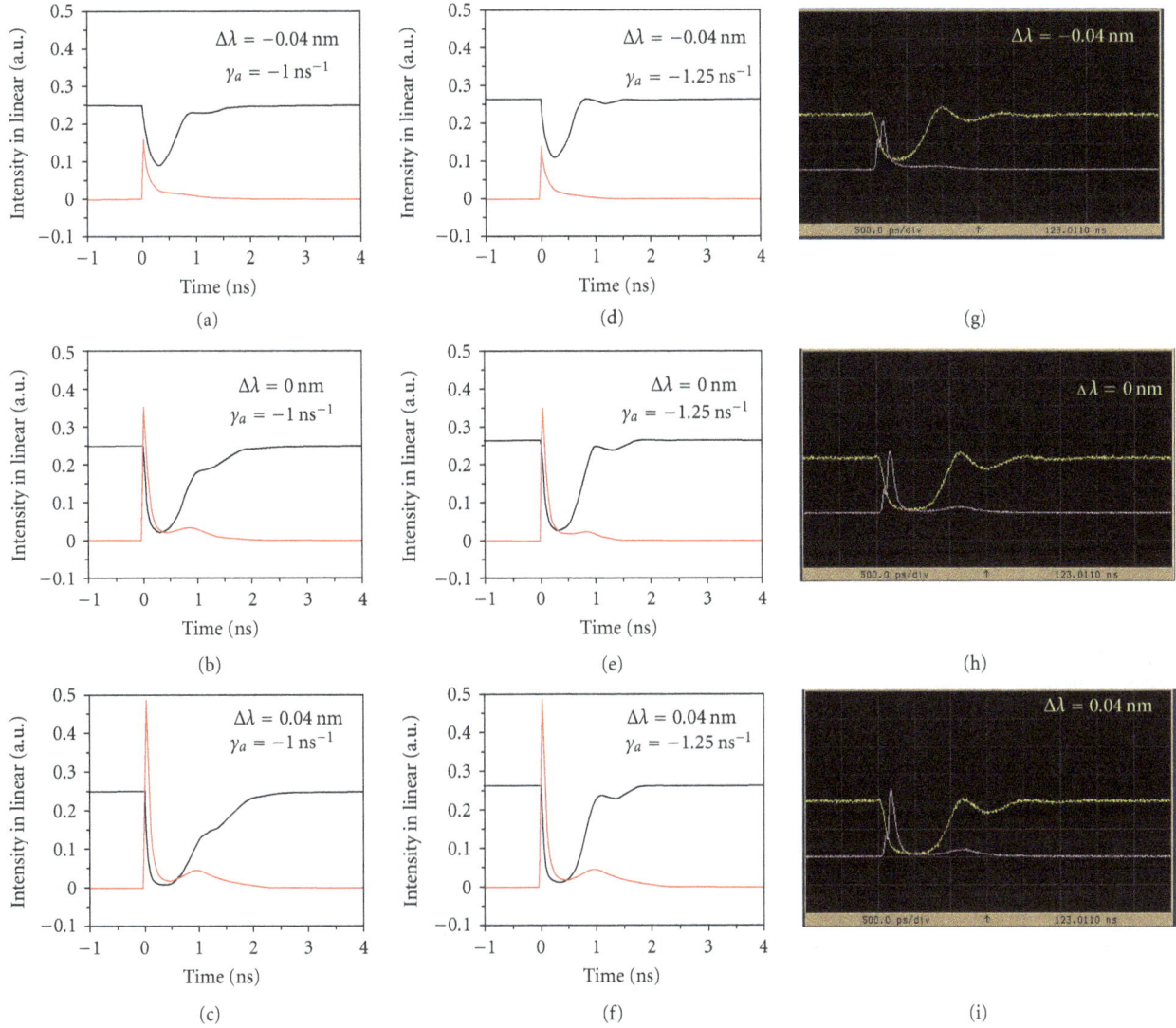

FIGURE 9: Experimentally observed oscilloscope traces compared to numerically calculated plots of the polarization dynamics of an SM VCSEL under injection of short optical pulses of 51.7 ps pulse width for various detuned wavelengths and two linear dichroism parameter values of −1.0 ns^{-1} and −1.25 ns^{-1}.

$$\frac{dN}{dt} = -\gamma_e N(1+P) + \gamma_e \mu - j\gamma_e n\left(E_y E_x^* - E_x E_y^*\right),$$

$$\frac{dn}{dt} = -\gamma_s n - \gamma_e n P - j\gamma_e n\left(E_y E_x^* - E_x E_y^*\right).$$

$$(1)$$

In the above equations, k is the decay rate of the electric field in the VCSEL cavity (≈ 25 ns^{-1}), α is the linewidth enhancement factor (=3), β_{sp} is the strength of the spontaneous emission (=10^{-5}), and μ is the normalized injection current ($\mu = 1$ for threshold). $P = |E_x|^2 + |E_y|^2$ is the normalized output power; ξ_x and ξ_y are independant Gaussian white noise source with a zero mean and a unit variance in the X- and Y-polarization directions, respectively. E_{inj} represents the electric field amplitude of the injection beam with an assumption of a Gaussian-shaped time-dependant pulse, and $\Delta\omega$ is the frequency detuning

between the injection laser wavelength and the suppressed polarization-mode wavelength of the VCSEL. k_{inj} is the coupling coefficient of the injection beam into the VCSEL, γ_a is the linear dichroism, γ_p is the linear birefringence, γ_e is the decay rate of the total population inversion N, and γ_s is the spin-flip rate (=50 ns^{-1}). From the measured two polarization-mode wavelengths in the spectrum shown in Figure 5, the linear birefringence γ_p is taken as a half of the frequency difference value between them, which is −26.09 ns^{-1} [22]. Since the dominant main polarization-mode wavelength is shorter than the other polarization-mode wavelength, the linear dichroism γ_a was taken as a negative value of −1.25 ns^{-1} from the best numerical simulation fit condition to the measured data. Figure 9 shows comparative plots of the measured oscilloscope traces and numerically simulated results of the temporal polarization-switching dynamics of an SM VCSEL under injection of

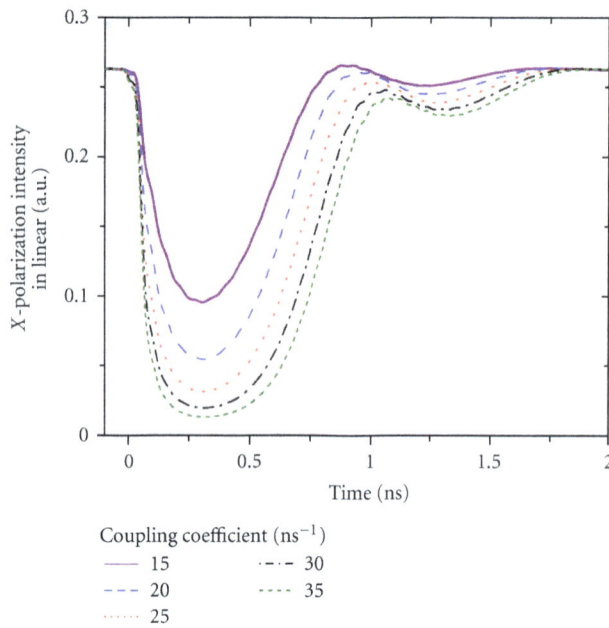

FIGURE 10: Temporal variation of the X-polarized intensity for various values of k_{inj} parameters in ns^{-1} at $k = 25$ ns^{-1}.

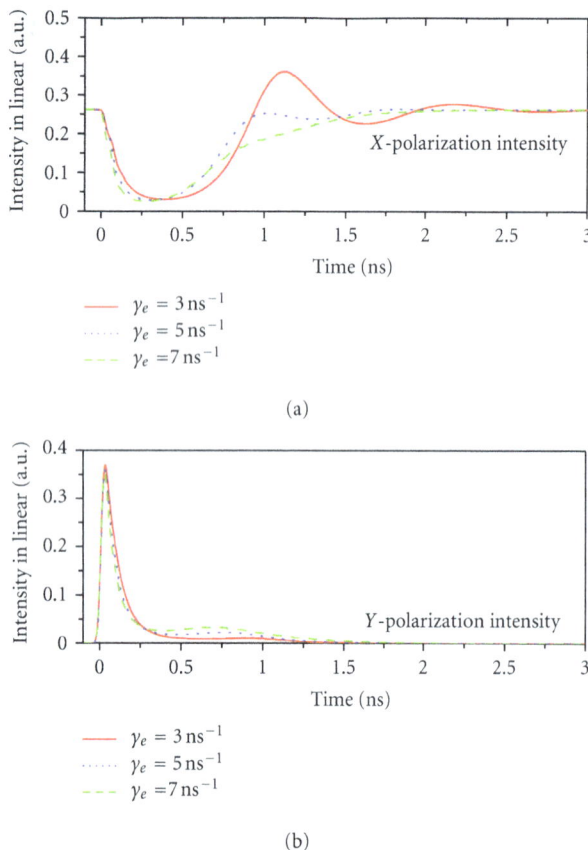

FIGURE 11: Temporal variation of the X- and Y-polarized beam intensities for various values of γ_e parameters in ns^{-1} at $\gamma_a = -1.25$ ns^{-1}.

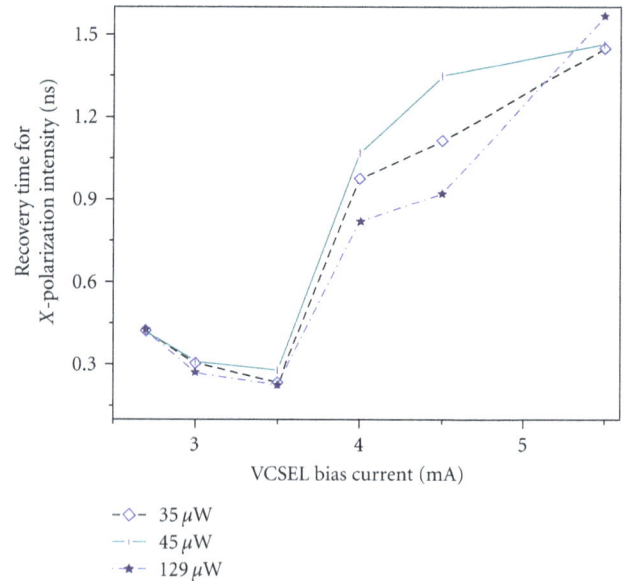

FIGURE 12: Experimentally measured original polarization recovery time as a function of the VCSEL bias current for various peak pulse power of the injection beam.

short optical pulses of 51.7 ps pulse width and 129 μW peak pulse power for three detuned wavelengths and for two linear dichroism parameter values of $\gamma_a = -1.0$ ns^{-1} and -1.25 ns^{-1}. The simulated results with $\gamma_a = -1.25$ ns^{-1} are better fit to the experimental results.

Numerical simulation of the X-polarization recovery dynamics after injection of a Y-polarization pulse of 50 ps width shows that the optimum value of the coupling coefficient k_{inj} is found to be equal to the field decay rate k (=25 ns^{-1}) as illustrated in Figure 10. In addition, the decay rate of the total excited population γ_e is affected by the linear dichroism parameter γ_a because γ_a is related to the power difference between the X- and Y-polarization-mode outputs. Easy polarization switching takes place for a large γ_e when γ_a is small, while a large decay rate γ_e (i.e., a short decay time) is needed for a large γ_a value to ensure the polarization switching and a short polarization recovery time. Figure 11 shows that the simulated temporal response of the polarization recovery is best fit to the measured one when $\gamma_e = 5$ ns^{-1} for $\gamma_a = -1.25$ ns^{-1}.

The delayed polarization recovery after polarization switching on the optical pulse injection could be minimized with an optimum VCSEL bias current. Figure 12 shows the polarization recovery times after the PS due to the external pulse beam injection as a function of the VCSEL bias current for various peak pulse powers of the orthogonal polarization-mode. The results indicates that the optimum VCSEL bias current for the shortest polarization recovery time is 3.5 mA, which is about 1.5 time of threshold current, and is independent of the peak pulse power of the injection beam in the range up to 200 μW. Even at this optimum bias condition, the recovery delay amounts longer than 200 ps. This may be explained with a relatively long photon build-up time in the VCSEL cavity, which is known to be

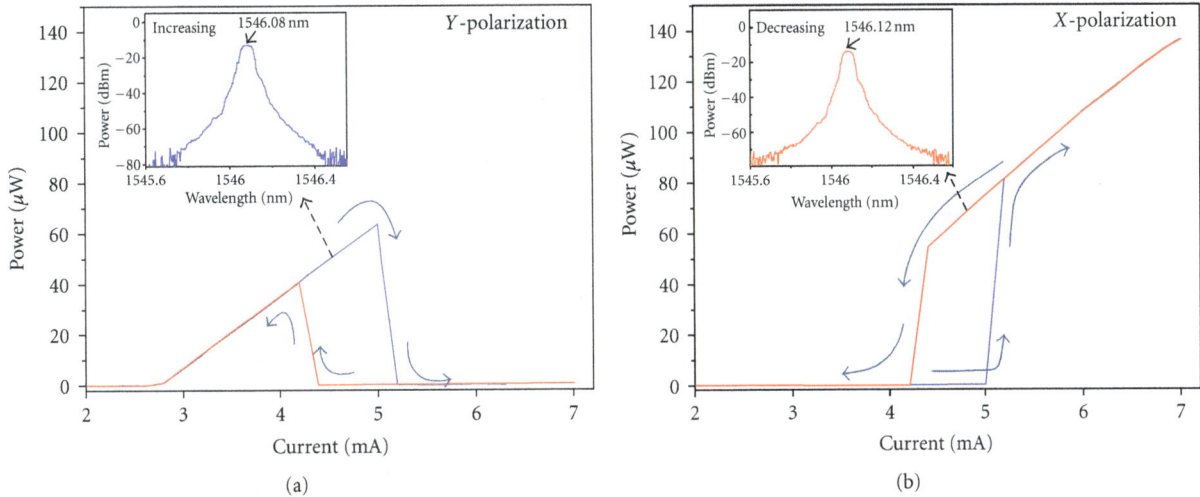

FIGURE 13: Experimentally measured optical laser output bistability of an SM VCSEL of standard circularly cylindrical shape during increasing and decreasing its bias current.

longer than the photon lifetime [23]. From the consideration of reflectivities of the bottom and top distributed Bragg reflectors (DBRs) of 0.999 and 0.9983 and approximated cavity length of about 10 μm, the estimated photon build-up time based on the photon lifetime calculation inside the VCSEL cavity structure is expected to be longer than 90 ps which is also longer than the pulse width of the injection beam.

The delayed rise time and some relaxation oscillation of the polarization-mode recovery processes after the polarization switching with the optical pulse injection were also observed in [24]. The detailed response may vary from chip to chip even though all the chips are taken from the same wafer. Furthermore, it is reported a fast spin-induced polarization oscillation in VCSELs with injection of a circularly polarized beam, whose process can also be described with the spin-flip model and can be applied potentially for high-speed polarization modulation [8, 9].

4. All-Optical Flip-Flop

All-optical logic gate operation based on laser output bistability properties has been investigated by a few research groups with multi-mode, square-shaped, or standard single-mode-type VCSELs [10–13, 24]. The multi-mode and square-shaped VCSELs require relatively high driving currents compared to the SM VCSELs. The standard SM-type VCSELs of an ideal structure of circularly cylindrical-shape are expected to have no bistability because there is no preferred dominant polarization-mode lasing. However, in practical VCSEL chips fabricated, they are more likely to have a small-gain anisotropy existing in their laser cavity. Thus, this causes some optical bistability in their laser output depending on chip-to-chip basis even though they all are taken from a same wafer [12]. The standard SM-type VCSELs with optical bistability laser outputs can be used for the all-optical flip-flop operation.

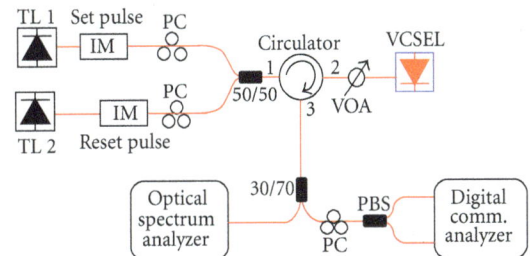

FIGURE 14: Experimental setup used for the all-optical flip-flop operation on an SM VCSEL of standard circularly cylindrical shape with output bistability. TL: tunable laser; IM: intensity modulator; VOA: variable optical attenuator.

Figure 13 shows the measured optical laser output bistability of a selected InAlGaAs/InP SM VCSEL of standard circularly cylindrical-shape during increasing and decreasing its bias current. During the fine adjustment of the driving bias current to the VCSEL near 4.2 to 5.2 mA, the bistable laser output condition between two polarization modes was observed. Since the laser wavelength of each polarization-mode was separated a little bit from the other, the laser output could be switched between two wavelengths (=1,546.08 nm and 1,546.12 nm) corresponding to two orthogonal polarization modes at the bistability region.

The standard SM VCSEL with the optical bistability property was used for the flip-flop operation by using the experimental setup shown in Figure 14. Two tunable lasers with external modulators were used as set and reset pulse generators, both in orthogonal polarization states to each other and each wavelength corresponding to one of the two polarization modes of the VCSEL at the bistable region, and their modulated beams were injected into the VCSEL through an optical circulator after combined with a 3-dB fiber coupler. The two polarization states of the VCSEL output were separated with a polarization beam splitter

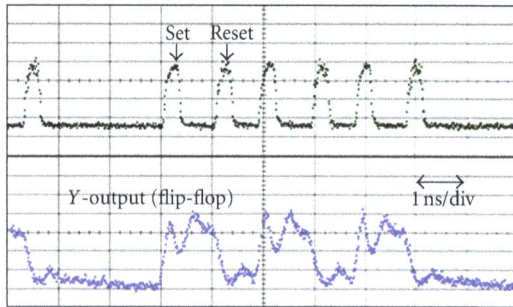

FIGURE 15: Measured oscilloscope traces of the all-optical flip-flop operated X- and Y-polarization mode outputs of a SM VCSEL with 1 GHz switching frequency of the set and reset pulses.

(PBS), and their temporal dynamics were measured with a digital communication analyzer (DCA).

The flip-flop operation of the VCSEL was performed with set and reset signals, each of which was obtained from the modulated tunable laser pulses of 280 ps pulse width and 112 ps rise time. Figure 15 shows the measured oscilloscope traces of the set and reset pulse trains and of the flip-flop-operated Y-polarization modes of the VCSEL at 1 GHz speed. The switch-on rise and switch-off fall times of the flip-flop-operated signals were 166.9 ps and 215.5 ps, respectively. The pulse energies of the set and reset pulses injected into the VCSEL for the flip-flop operation were only 4.5 fJ and 3.5 fJ, respectively. Further improvement of the flip-flop operation speed can be obtained with a high-speed shorter set and reset pulses.

5. Conclusions

Influence of an external laser beam injection to the SM VCSELs on their laser output characteristics has been studied experimentally, and some of them were explained by numerical simulation. Continuous-wave laser beam injection to a gain-switched SM VCSEL near the resonance wavelength corresponding to the main polarization mode lasing reduced the pulse timing jitter, narrowed the linewidth, enhanced the pulse amplitude, and shortened the pulse width. An external laser pulse injection of pulse width shorter than the cavity photon lifetime into the SM VCSEL in an orthogonal polarization direction to its main polarization-mode caused a temporal delay of the polarization recovery after PS due to the photon build-up time, and its delay was the minimum at an optimized bias current, which was about 1.5 time above the threshold current. These polarization dynamics were explained by numerical simulation based on the SFM. Polarization-mode bistability existed even in an SM VCSEL of standard circularly cylindrical shape and was used for all-optical flip-flop operation with set and reset injection pulses of order of 3.5~4.5 fJ pulse energy. These low power consumed high-speed all-optical logic operation can be extended to two-dimensional VCSEL arrays, which may be useful for potential high-capacity all-optical signal processing.

Acknowledgments

This work was supported by the Basic Science Research Programs through the National Research Foundation of Korea (NRF) funded by the Korean Ministry of Education, Science and Technology under Grant no. 2009-0084514. The authors gratefully thank Drs. Byeung-Soo Yoo and Jay Roh of Raycan Co., Ltd. for providing with the VCSELs.

References

[1] E. Kapon and A. Sirbu, "Long-wavelength VCSELs: power-efficient answer," *Nature Photonics*, vol. 3, no. 1, pp. 27–29, 2009.

[2] J. M. Noriega, A. Valle, and L. Pesquera, "Timing jitter reduction in gain-switched VCSELs induced by external optical injection," *Optical and Quantum Electronics*, vol. 40, no. 2–4, pp. 119–129, 2008.

[3] S. H. Lee, K. H. Kim, V. M. Deshmukh, D. W. Kim, and M. H. Lee, "Injection laser wavelength-dependent timing jitter reduction of gain-switched single-mode VCSELs," *IEEE Journal of Quantum Electronics*, vol. 46, no. 9, pp. 1327–1331, 2010.

[4] K. H. Jeong, K. H. Kim, S. H. Lee, M. H. Lee, B. S. Yoo, and K. A. Shore, "Optical injection-induced polarization switching dynamics in 1.5-μm wavelength single-mode vertical-cavity surface-emitting lasers," *IEEE Photonics Technology Letters*, vol. 20, no. 10, pp. 779–781, 2008.

[5] A. Hurtado, A. Quirce, A. Valle, L. Pesquera, and M. J. Adams, "Power and wavelength polarization bistability with very wide hysteresis cycles in a 1550 nm VCSEL subject to orthogonal optical injection," *Optics Express*, vol. 17, no. 26, pp. 23637–23642, 2009.

[6] A. Quirce, A. Valle, and L. Pesquera, "Very wide hysteresis cycles in 1550-nm VCSELs subject to orthogonal optical injection," *IEEE Photonics Technology Letters*, vol. 21, no. 17, pp. 1193–1195, 2009.

[7] V. M. Deshmukh, S. H. Lee, D. W. Kim, K. H. Kim, and M. H. Lee, "Experimental and numerical analysis on temporal dynamics of polarization switching in an injection-locked 1.55-μm wavelength VCSEL," *Optics Express*, vol. 19, no. 18, pp. 16934–16949, 2011.

[8] N. C. Gerhardt, M. Y. Li, H. Jähme, H. H. Höpfner, T. Ackemann, and M. R. Hofmann, "Ultrafast spin-induced polarization oscillations with tunable lifetime in vertical-cavity surface-emitting lasers," *Applied Physics Letters*, vol. 99, no. 15, Article ID 151107, 3 pages, 2011.

[9] N. C. Gerhardt and M. R. Hofmann, "Spin-controlled vertical-cavity surface-emitting lasers," *Advances in Optical Technologies*, vol. 2012, Article ID 268949, 15 pages, 2012.

[10] T. Mori, Y. Yamayoshi, and H. Kawaguchi, "Low-switching-energy and high-repetition-frequency all-optical flip-flop operations of a polarization bistable vertical-cavity surface-emitting laser," *Applied Physics Letters*, vol. 88, no. 10, Article ID 101102, 3 pages, 2006.

[11] T. Katayama, T. Ooi, and H. Kawaguchi, "Experimental demonstration of multi-bit optical buffer memory using 1.55-μm polarization bistable vertical-cavity surface-emitting lasers," *IEEE Journal of Quantum Electronics*, vol. 45, no. 11, pp. 1495–1504, 2009.

[12] S. H. Lee, H. W. Jung, K. H. Kim, and M. H. Lee, "All-optical flip-flop operation based on polarization bistability of conventional-type 1.55-μm wavelength single-mode VCSELs,"

Journal of the Optical Society of Korea, vol. 14, no. 2, pp. 137–141, 2010.

[13] S. H. Lee, H. W. Jung, K. H. Kim et al., "1-GHz All-Optical flip-flop operation of conventional cylindrical-shaped single-mode VCSELs under low-power optical injection," *IEEE Photonics Technology Letters*, vol. 22, no. 23, pp. 1759–1761, 2010.

[14] B. Zhang, X. Zhao, L. Christen et al., "Adjustable chirp injection-locked 1.55-μm VCSELs for enhanced chromatic dispersion compensation at 10-Gbit/s," in *Proceedings on Optical Fiber Communication Conference*, San Diego, Calif, USA, February 2008, paper OWT7.

[15] P. Boffi, A. Boletti, A. Gatto, and M. Martinelli, "VCSEL to VCSEL injection locking for uncompensated 40-km transmission at 10 Gb/s," in *Proceedings on Optical Fiber Communication Conference*, San Diego, Calif, USA, March 2009, paper JThA32.

[16] T. B. Gibbon, K. Prince, T. T. Pham et al., "VCSEL transmission at 10 Gb/s for 20 km single mode fiber WDM-PON without dispersion compensation or injection locking," *Optical Fiber Technology*, vol. 17, no. 1, pp. 41–45, 2011.

[17] D. S. Seo, D. Y. Kim, and H. F. Liu, "Timing jitter reduction of gain-switched DFB laser by external injection-seeding," *Electronics Letters*, vol. 32, no. 1, pp. 44–45, 1996.

[18] K. T. Vu, A. Malinowski, M. A. F. Roelens, and D. J. Richardson, "Detailed comparison of injection-seeded and self-seeded performance of a 1060-nm gain-switched Fabry-Pérot laser diode," *IEEE Journal of Quantum Electronics*, vol. 44, no. 7, pp. 645–651, 2008.

[19] M. San Miguel, Q. Feng, and J. V. Moloney, "Light-polarization dynamics in surface-emitting semiconductor lasers," *Physical Review A*, vol. 52, no. 2, pp. 1728–1739, 1995.

[20] I. Gatare, M. Sciamanna, J. Buesa, H. Thienpont, and K. Panajotov, "Nonlinear dynamics accompanying polarization switching in vertical-cavity surface-emitting lasers with orthogonal optical injection," *Applied Physics Letters*, vol. 88, no. 10, Article ID 101106, 2006.

[21] K. H. Jeong, K. H. Kim, S. H. Lee, M. H. Lee, B. S. Yoo, and K. A. Shore, "Polarization switching in a 1.5 μm wavelength single-mode vertical cavity surface emitting laser under modulated optical beam injection control," in *Proceedings of the Photonics in Switching*, San Francisco, Calif, USA, August 2007, paper TuP1.

[22] J. Martin-Regalado, F. Prati, M. San Miguel, and N. B. Abraham, "Polarization properties of vertical-cavity surface-emitting lasers," *IEEE Journal of Quantum Electronics*, vol. 33, no. 5, pp. 765–783, 1997.

[23] A. E. Siegman, *Lasers*, chapters 25 and 13, University Science Books, 1986.

[24] A. Quirce, J. R. Cuesta, A. Hurtado et al., "Dynamic characteristics of an all-optical inverter based on polarization Switching in long-Wavelength VCSELs," *IEEE Journal of Quantum Electronics*, vol. 48, no. 5, pp. 588–595, 2012.

5

Semiconductor Disk Lasers: Recent Advances in Generation of Yellow-Orange and Mid-IR Radiation

Mircea Guina, Antti Härkönen, Ville-Markus Korpijärvi, Tomi Leinonen, and Soile Suomalainen

Optoelectronics Research Centre, Tampere University of Technology, P.O. Box 692, 33101 Tampere, Finland

Correspondence should be addressed to Mircea Guina, mircea.guina@tut.fi

Academic Editor: Rainer Michalzik

We review the recent advances in the development of semiconductor disk lasers (SDLs) producing yellow-orange and mid-IR radiation. In particular, we focus on presenting the fabrication challenges and characteristics of high-power GaInNAs- and GaSb-based gain mirrors. These two material systems have recently sparked a new wave of interest in developing SDLs for high-impact applications in medicine, spectroscopy, or astronomy. The dilute nitride (GaInNAs) gain mirrors enable emission of more than 11 W of output power at a wavelength range of 1180–1200 nm and subsequent intracavity frequency doubling to generate yellow-orange radiation with power exceeding 7 W. The GaSb gain mirrors have been used to leverage the advantages offered by SDLs to the 2–3 μm wavelength range. Most recently, GaSb-based SDLs incorporating semiconductor saturable absorber mirrors were used to generate optical pulses as short as 384 fs at 2 μm, the shortest pulses obtained from a semiconductor laser at this wavelength range.

1. Introduction

Conceptually, the idea of an optically pumped semiconductor disk laser (OP-SDLs) was suggested already in 1966 by Basov et al. in a paper describing lasers with radiating mirrors [1]. However, it was not until the 1990s that the concept was acknowledged and the first working devices were reported [2–6]. In its essence, the concept of an OP-SDL is based on using an optically pumped semiconductor gain structure (i.e., gain mirror) with vertical emission. We note here that in addition to OP-SDL, also acronyms like OP-VECSEL (optically pumped vertical external-cavity surface-emitting laser) and OPSL (optically pumped semiconductor laser) are commonly used in literature to describe the same type of laser. The laser resonator is typically formed between the gain mirror and one or more external-cavity mirrors. In many ways, this laser architecture is similar to that of traditional solid state disk lasers. An essential difference is that in traditional solid state lasers the emission wavelength is dependent on certain fixed atomic transitions in a host material, whereas in an SDL the wavelength can be specifically tailored in a wide range by engineering the composition of the semiconductor

material. This added wavelength versatility is one of the key factors that have made SDLs successful also commercially.

Technically speaking, the OP-SDL can be considered as a brightness and wavelength converter; it converts low brightness light from multimode diode pump lasers into a high brightness single mode beam at a wavelength that is longer than the pump wavelength. Compared to edge emitting diode lasers and vertical-cavity surface-emitting lasers (VCSELs), the external cavity and optical pumping make the SDLs more complicated but they also bring several benefits. First of all they enable upscaling of the mode area on the gain while still maintaining single transversal mode operation; consequently the output power can be increased to multiwatt levels without risk of catastrophic optical damage due to excessively high optical intensities. In addition, the external cavity allows for cascading multiple gain mirrors thus increasing even more the power scaling capability. The SDL cavity has a high Q-factor and therefore it stores optical energy allowing efficient nonlinear intracavity frequency conversion to visible wavelengths. Another benefit of the external cavity is that it enables incorporation of nonlinear components to initiate ultrashort pulse operation. We should also note that lately, the cost of broad stripe edge emitting pump diodes at 790–980 nm wavelengths has decreased significantly, while at the same time the available power from both single emitters and diode bars has increased markedly. More recently, also high-power pump diodes at other important wavelengths, including 635–690 nm, 1480–1550 nm, and 2000 nm, have been commercialized more actively. These advances in the availability, cost, and performance of pump diodes have made the optical pumping concept even more attractive.

1.1. Cavity Designs. Structurally the SDL gain mirror resembles a half-VCSEL design that comprises a high reflectivity mirror and a semiconductor gain region. The gain region usually includes several quantum-well (QW) or quantum-dot (QD) layers separated by spacer/barrier layers. A typical mirror structure consists of a stack of quarter-wavelength semiconductor layers, forming a distributed Bragg reflector (DBR), although metallic, dielectric, or hybrid [7] mirror structures can be used in some cases as well. While in VCSELs the single transverse mode operation is achieved by confining the laser mode to a very small gain area, in SDLs the same functionality is achieved by controlling the fundamental mode size via cavity design to have it match with the pumped area on the gain. Figure 1 shows various cavity configurations of SDLs.

The simplest conventional SDL cavity has an I-shape that is formed between the gain mirror and a single external output coupler (OC) mirror. However, in practice it is often easier to use a V-shaped cavity formed between the gain mirror, one curved folding mirror, and a planar output coupler. The advantage of the V-cavity is that planar output couplers with various coupling ratios are often cheaper and more widely available on stock than equivalent curved couplers. Another practical advantage of the V-shaped cavity is related to the alignment of the laser; if the final alignment is done by monitoring the output of a photodiode placed behind the output coupler, the folding mirror in a V-shaped cavity collects the light efficiently to the photodiode enhancing the available signal while in I-shaped cavity the spontaneous emission from the gain is rapidly dispersed to all direction. More complex cavity configurations are often used for frequency conversion and mode-locking. For efficient frequency conversion the nonlinear crystal is often placed at a location near or at the mode waist. This is usually easier to do in a V-shaped or Z-shaped cavity than in an I-shaped cavity. More complicated Z-shaped cavities are typically used in mode-locking SDLs to accommodate also a semiconductor saturable absorber mirrors (SESAMs); the challenge here is to produce sufficiently small mode diameter on the absorber mirror, while at the same time maintaining reasonably large mode diameter on the gain [8]. One should notice though that the overall cavity length increases for more complex designs and the mode-locked pulse repetition rate is reduced. This in turn would reduce the efficiency; if the interval of consequent pulses is longer than the carrier-lifetime, which is typically in the ns-range or slightly below, there will be loss of pump energy in time. In other words, the gain element can store energy only for a limited time, and if it is not exploited

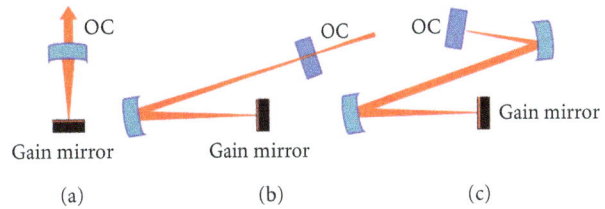

FIGURE 1: Typical SDL cavities. (a) I-shaped cavity; (b) V-shaped cavity; (c) Z-shaped cavity. OC: output coupler.

in that time window by an incoming pulse, a portion of that energy will be lost to spontaneous emission between the consequent pulses. This feature sets a practical upper limit for the cavity length in mode-locked laser with continuous wave pumping.

To scale up the power of an SDL, it is possible to deploy multiple gain elements in single cavity [9–11]. SDLs can also employ ring cavities [12], but probably due to the added complexity such lasers have not gained much popularity. In addition to different external cavity configurations, one can also produce an SDL with a semimonolithic cavity that may include a plane-plane design stabilized by a thermal lens [13, 14]. Such laser may be more limited in power and brightness but does possess an extremely rugged design. Furthermore, the semimonolithic cavity can be processed to have curved surfaces with mirror structures, thus avoiding the need for cavity stabilization by a thermal lens [15].

1.2. Thermal Management of SDLs. Efficient thermal management is a very important aspect required for high-power operation of SDLs. Although heroic in many ways, the early SDL experiments required the use of very low temperatures for high-power operation making the devices unpractical for use outside the laboratory. To large extent this was caused by a lack of adequate heat dissipation techniques. Excess heating reduces the emission efficiency via increased nonradiative recombination and carrier leakage and red-shifts the emission wavelength, which in a resonant periodic gain structure [16] leads also to a mismatch between the emission wavelength and the resonant wavelength further reducing the gain. Effectively, such heat-induced processes create a positive feedback loop with very negative impact on the laser performance. Consequently, the output power of the laser exhibits a roll-over characteristic when the pump power is increased beyond a critical point. For high-power operation, one should implement adequate ways of thermal management. Heating of the gain mirror originates from pump energy, which is converted to useful photons only partially while another part of the pump energy is transferred to phonons due to nonradiative recombination and the quantum defect (i.e., the photon energy difference between the pump photon and the laser photon). As a general strategy, one should try to minimize the heat generation and at the same time maximize the heat transfer from the gain.

Typically the pump photon has markedly higher photon energy than the emitted laser photon; for example, for a 1060 nm laser pumped with 808 nm radiation, the quantum

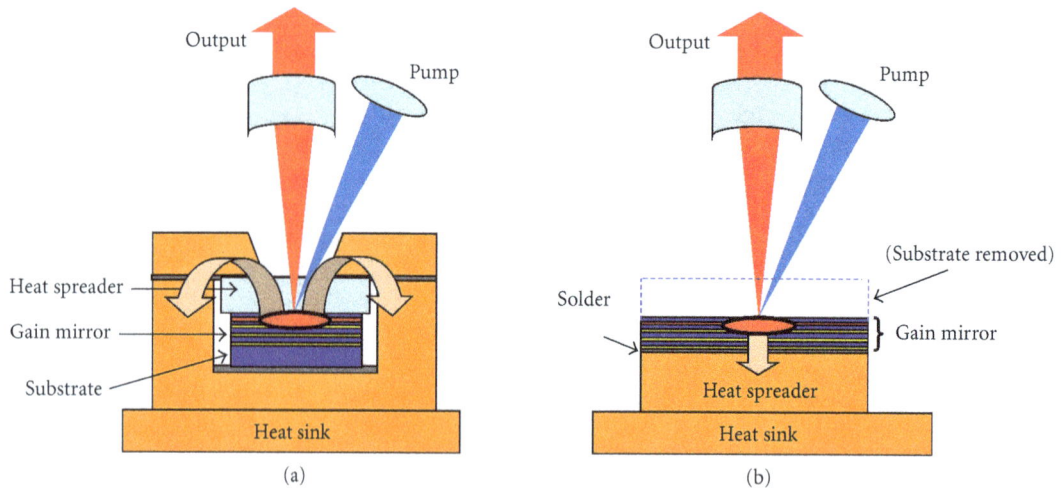

FIGURE 2: Description of two heat extraction strategies. (a) Laser equipped with intracavity heat spreader. (b) Flip-chip mounted thin-device from which the semiconductor substrate is removed.

defect is about 24% of the pump photon energy. In other words, the optical-to-optical conversion efficiency of such laser cannot exceed 76% even under theoretically perfect conditions. Usually, the laser is designed to absorb the pump radiation in the spacer/barrier layers separating the QWs because in this way the interaction length of light in matter is long enough to absorb sufficient amount of pump energy in a single pass. On the other hand, there must be a notable bandgap contrast between the spacer layers and the QWs in order to ensure a good carrier confinement and hence efficient operation at elevated temperatures. In most spacer pumped SDLs, the quantum defect is between 15% and 50% of the pump photon energy. An alternative for spacer pumping is direct "in-well" pumping [17–19] where the spacers are transparent to the pump radiation and the pump wavelength closely matches the QW emission wavelength. This approach minimizes the quantum defect but another technical difficulty arises from a short light-matter interaction length; the thickness of one QW is typically some nanometers and the total absorptive path length is rather small, as a gain mirror would typically include 5–15 QWs. To some extent, the pump absorption can be improved by adding more QWs to the structure but usually either a resonant pumping scheme or external pump recirculation optics is required for efficient pump absorption. We should note that an in-well pumped gain mirror provides by default a high reflection for the unabsorbed pump light, thus avoiding pump absorption in the DBR, and providing double-pass of pump radiation through the gain region. The selection of the pump laser for in-well pumping is more critical (and possibly more expensive) than in a spacer pumped laser where low-cost 808 nm diodes can be used for pumping 920 nm SDLs as well as 2000 nm SDLs. This is particularly true in the case of resonant in-well pumping. Nevertheless, in-well pumping offers an interesting option for reducing the quantum defect and the heating related to it. One should notice that quantum defect optimization makes sense only if the quantum efficiency of the laser is already high. If a significant majority of

pump photons are anyway lost to nonradiative processes, the benefits of quantum defect optimization become marginal to the overall performance of the system. Therefore, high-quality gain materials and proper structural designs are prerequisites for efficient operation of SDLs.

It is also very important to conduct the heat away from the gain region with minimal thermal resistance between the heat sink and the active region. Generally speaking, thermal resistance is dependent on the thermal conductance of the materials used and on the distance that heat needs to be transferred. In short, one should aim to minimize the distance between the heat sink and the gain and at the same time use materials that have high thermal conductance. Using a planar gain mirror geometry, the pumping is concentrated on an area that has typically a diameter of some tens or hundreds of micrometers, whereas the overall thickness of the semiconductor layer structure is only a few microns (e.g., 5–6 μm). In other words, the heated area is very large compared to the thickness of the layers. Thermal simulations show that in such a structure the heat flow is essentially one dimensional and is directed normal to the sample surface [21]. We should point out that the epitaxial layers are grown on a semiconductor substrate that is typically some 200–600 μm thick and presents a major obstacle for the heat flow. Two assembling techniques of the gain mirror to the heat sink are typically employed to overcome this issue. The so-called "intracavity heat spreader" method, is conceptually simple and involves contacting a transparent heat spreader element onto the gain mirror [22] (see Figure 2 for general description). This method does not require substrate removal and the heat spreader is located right next to the gain region. The practical limitations arise from the fact that the heat spreader is located inside the laser cavity and that the number of transparent materials with high thermal conductance is limited; their cost may also be a limiting factor. By far the best material for this purpose is diamond due to its extremely high thermal conductance (up to $\sim$2000 W/m·K) and wide transmission window. Other suitable materials include, for example, silicon carbide (SiC) [23] and sapphire (crystalline Al$_2$O$_3$) [22]. A common technique for contacting the heat spreader and the semiconductor sample is based on capillary bonding [24] with deionized water or other suitable liquid. In this technique, two smooth and flat surfaces (here the gain mirror and heat spreader) are pulled together by surface tension of a liquid, and as the liquid evaporates, the two surfaces are brought to close optical contact and held together by surface forces. Simple mechanical clamping can be also used for optical contacting as long as the surfaces are sufficiently smooth, flat, and free from particles or other contaminants. However, capillary bonding is a good way to make sure that the surfaces meet these requirements and can be brought to close optical and thermal contact. The success of the bonding process can be simply monitored by observing the disappearance of the Newton's interference rings as the surfaces are brought together.

Another option for efficient heat dissipation was presented already in Kuznetsov's paper [25] and it involves growing the mirror and gain structures in reversed order (gain first, then the mirror) and bonding the component "upside-down" on a heat sink after which the substrate is removed by etching. Effectively this method transfers the epitaxial layers from a semiconductor substrate onto a substrate with higher thermal conductance. The process leaves only the Bragg reflector layers between the heated active region and the heat sink, which greatly reduces the thermal resistance in comparison to the situation where the semiconductor substrate would be located between the gain mirror and the heat sink. This process is often referred to as the "flip-chip" process or the "thin-device" process. Sometimes these components are also called bottom emitters, a term that is commonly used in VCSEL processing. From processing point of view, the flip-chip process requires longer overall time but can be done in batches of many devices. The major challenge of the flip-chip approach relates to the fact that without the support of the original substrate the epitaxial layers are mechanically very fragile. The bonding process requires usually the use of temperatures exceeding 150°C. Therefore, any differences in the coefficients of thermal expansion between the epitaxial layers and the heat sink may translate to mechanical stress as the sample cools down and the solder hardens. This is particularly critical issue with large samples bonded with hard solders such as AuSn that have high melting point. To alleviate the mechanical stress, one can resort to soft solders such as indium. However, it is a well-known fact that in high-power diode lasers indium solder tends to fail due to thermal diffusion and other effects. Such effects should be considered in connection with lifetime of high-power SDLs as the gain region is operated at relatively high temperatures. There are also many alternative bonding methods, such as InAu bonding [7, 26], that can be used instead. An important aspect related to soldering concerns the presence of voids within the solder; any voids in the solder will likely result in physical damage to the gain mirror under pumping. The voids can be monitored prior to substrate removal using a scanning acoustic microscope, for example. To further improve the heat dissipation, one can use a heat spreader, such as diamond, between the

sample and the metallic heat sink; the thermal energy is thus rapidly spread from a point source to a larger area over which it is conducted to the actual heat sink. The cost of the heat spreader is also markedly reduced when there is no need for optical quality surface polishing. In addition to soldering the substrate removal is an important step in the flip-chip process. To some extent the substrate can be thinned by lapping, prior to bonding, but in any case tens or hundreds of microns of semiconductor substrate must be removed by etching. This is done usually by wet etching employing an etch stop layer. It is important for the success of the process that the selectiveness of the etching is sufficiently high and that the process can be carried out in reasonable time. For GaAs removal one can use, for example, InGaP or Al(Ga)As etch stop layers and NH$_4$OH : H$_2$O$_2$-based etchants. InP etchants are often based on HCl [27], which may limit or hinder the use of indium as a solder for sample bonding. For GaSb-based compounds, good etchant-etch stop combinations are less developed; successful flip-chip SDLs based on this material system were just recently reported [26].

When compared to the flip-chip design, the intracavity heat spreader approach has proved to be very quick and simple to do in laboratory conditions. The heat spreaders can be also recycled almost endlessly, which overcomes their high initial cost at least for research use. The intrinsic disadvantage of the intracavity heat spreader is that it introduces a loss element in the cavity that can also act as an etalon. The etalon effect modulates the optical spectrum affecting the mode-locking mechanisms and making continuous wavelength tuning difficult. The etalon effect can be suppressed by using a wedged heat spreader with an antireflective coating [28]. The wedge angle usually increases the reflection losses despite the AR layer and hence decreases the output power. One should notice that inside the laser resonator, etalon effects may arise also from unexpected sources such as double-side polished semiconductor wafers onto which the gain mirror or the saturable absorber mirror may have been grown. Although the reflectivity of the gain mirror DBR is usually over 99.5%, the small transmitted portion of light may penetrate to the substrate (if transparent) and can be reflected back from its second surface, in which case an etalon is established in the system and the spectrum of the laser is affected. Both heat management strategies have been successfully used to achieve over 10 W output from standard InGaAs/GaAs gain structures, though the highest output powers have been achieved with the flip-chip components [29, 30]. However, if the thickness of the DBR stack increases (due to longer operation wavelength or poor index contrast of the materials), or if the DBR layers have very poor thermal conductance, it may be more advantageous to use the intracavity heat spreader technique. This is particularly true for InP- and GaSb-based SDLs. The thermal issues of SDLs have been discussed in detail in a number of papers reporting simulations and experimental results on the subject [21, 31–34].

To summarize, efficient heat removal is highly important for high-power operation of SDLs. Use of high thermal conductance heat spreader materials, such as diamond,

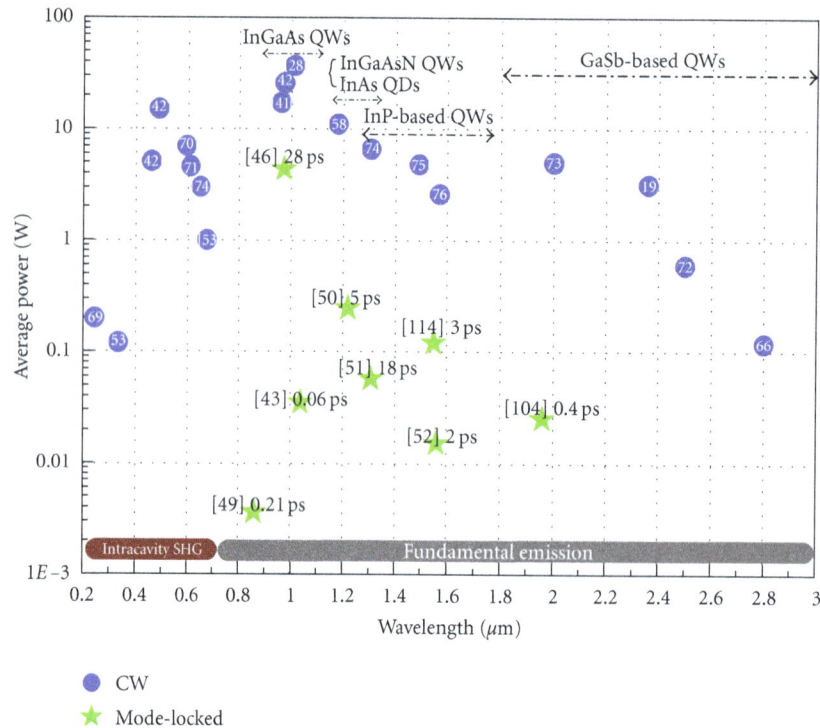

FIGURE 3: Selection of SDL results showing the maximum average power reached at different wavelengths. InGaAs(N)-based gain materials dominate the results up to 1.3 μm above which InP-based and GaSb-based QWs are used. Only SDLs incorporating single gain chips are included. Pulse durations related to mode-locked results are given in picoseconds.

greatly improves the heat extraction from the point source. The distance from the gain region to the heat spreader can be minimized by optically contacting the heat spreader onto the sample or by flip-chip bonding the component on a heat spreader/heat sink. The application and the type of gain material determine which process is more suitable. Flip-chip processing suits well for mode-locking, continuous spectral tuning, and single-frequency operation since the laser spectrum is not affected by the intracavity heat spreader element. The intracavity heat spreader approach suits particularly well for long wavelength (GaSb and InP) lasers and applications that are not spectrally sensitive.

1.3. Wavelength Coverage. During the last decade, the SDL research has been largely channelized along three major directions, namely, (i) power scaling, (ii) extending the wavelength coverage, and (iii) generation of ultrashort pulses. Along the way, many demonstrations concerned widely tunable [35–38] and narrow band lasers [39–41]. In terms of available output power, the 10 W level has been reached and exceeded using both single and multiple gain elements [29, 30, 42, 43]. Excellent results have been obtained lately in ultrashort pulse generation [44, 45], as well as in generation of pulses with high average output power [46, 47] and high repetition rate [48]. The spectral coverage of mode-locked [49] SDLs has also extended [50–53] outside the common InGaAs wavelengths near 1 μm. Interestingly, the spectral coverage of continuous wave SDLs (fundamental and frequency converted emission) spans today from 244 nm to 5000 nm [18, 20, 22, 23, 37, 54–76], although not without

gaps. Figure 3 gives an overview of maximum output powers achieved as a function of wavelength for both continuous-wave (CW) and mode-locking operation regimes. It also provides a correspondence to main classes of material systems used to reach a certain wavelength region.

In terms of more recent efforts and development directions, the 580–630 nm wavelength range is particularly interesting as it cannot be reached via direct emission from semiconductors. Nevertheless, it can be covered conveniently by frequency doubled 1160–1260 nm infrared lasers. We should note that it is also difficult to find suitable solid state materials for these visible and IR ranges. Because of these reasons, there has been a lot of scientific and commercial interest in extending the SDL technology to this particular wavelength range. In the following we will review different options for reaching emission at ~1150–1300 nm with semiconductors. First, it is important to understand the main features of the semiconductor structures we are considering for fabricating SDL mirrors. The gain mirror is essentially a stack of epitaxially grown semiconductor thin-films, fabricated on a GaAs, InP, GaSb, or other suitable semiconductor substrate by epitaxial growth. It is quite essential that one is able to grow high-quality gain material (QW, QD, or bulk) with desired bandgap energy, while keeping the material strain within reasonable limits. Secondly, the DBR should provide sufficient reflectance with a reasonable stack thickness and level of strain. Excessive material strain, arising from the difference between the lattice constants of the semiconductor layers, can lead to formation of crystalline defects and ultimately to relaxation of the layered structure. The

TABLE 1: Different technologies used for fabricating SDLs with emission at 1150–1300 nm.

Strategies for wavelength extension to 1150–1300 nm	Challenges
GaAsSb/GaAs QW gain material [81]	Low confinement of carriers in the QWs. Poor temperature behavior
InP-based gain with InP-based Bragg reflector [64]	Compromised reflectivity, Increased stack thickness, low thermal conductance of the DBR
Hybrid mirrors with InP-based gain [7]	Compromised thermal conductance
Wafer fusion of different gain and active regions [82]	More expensive processing. Two growths required for one component
InAs/GaAs QDs [83, 84]	Reduced design flexibility and low modal gain
Strain compensated high indium content InGaAs QWs [85]	Strain-related lifetime issues
Dilute nitride GaInNAs/GaAs QWs [20, 62]	Formation of nitrogen-related defects

1150–1300 nm wavelength range has previously been very challenging for the growth of SDLs for two main reasons. First, for conventional InGaAs/GaAs QW material a relatively large content of indium must be used to reduce the bandgap energy to the desired value and the high indium content increases the lattice constant of the material causing buildup of strain in the layer structure. Alternatively, one could also resort to the use of InP-based QWs which work at $1.2\,\mu m$–$1.6\,\mu m$, but unlike with GaAs, the DBR materials lattice matched to InP have very low index contrast [77–80]. Therefore, the thickness of the Bragg reflector must be increased significantly in order to achieve high reflectance. A number of techniques have been proposed to extend the emission wavelength of GaAs-based structures beyond the typical InGaAs spectral window near $1\,\mu m$ or to enable the use of InP-based gain regions in surface normal lasers; the main techniques have been listed in Table 1 with related challenges.

In the next sections, we will review the basic technological aspects regarding the development of dilute nitrides gain mirrors and the recent achievements concerning dilute-nitride SDLs with yellow-orange emission. GaSb-based SDL emitting at around $2\,\mu m$ will be discussed in Section 3.

2. High-Power Yellow-Orange SDLs Based on Dilute Nitride Gain Mirrors

2.1. Dilute Nitrides: Band-Gap Engineering and Gain Mirror Technology. To produce yellow emission by second harmonic emission, the indium content of the conventional InGaAs/GaAs QWs needs to be relatively high ($x > 35\%$). The high indium content increases the compressive lattice strain close to the point where misfit dislocations start to appear. The high lattice strain, together with high operation temperatures, can strongly deteriorate the lifetime of a device based on such QWs [86]. By adding a small amount of N (typically less than 3%) to InGaAs, one can reduce at the same time the lattice constant and the band-gap of the material. This opens up great opportunities for GaAs-based technology. For example, the compressive strain of InGaAs/GaAs material system can be compensated by N incorporation enabling emission at wavelengths up to $1.55\,\mu m$ [87]. These dilute nitride compounds can be in fact lattice matched to GaAs;

$Ga_{1-x}In_xN_yAs_{1-y}$ with $x \approx 2.8\,y$ is lattice matched to GaAs whereas compositions with $x > 2.8\,y$ and $x < 2.8\,y$ lead to compressively and tensile strained compounds, respectively. Furthermore, GaNAs layers exhibit a tensile strain that can be used for balancing the compressive strain of GaInAs layers.

The dramatic effect of nitrogen on the band gap is generally explained as being caused by the small size and large electronegativity of N atoms (radius $\sim 0.068\,\text{Å}$, electronegativity ~ 3.04 in units of Pauling scale) as compared to As atoms (radius $\sim 0.121\,\text{Å}$, electronegativity ~ 2.18) of the host crystal. Such impurity atoms create localized energy levels close to the conduction band edge and, as a result, modify the conduction band structure of the alloy. The interaction between the localized states and the conduction band is usually modeled using a so-called band anticrossing model (BAC). BAC has been very successful in explaining anomalous properties of the dilute nitrides, especially the conduction band structure and the related electron effective mass [88–91]. The theoretical dependence of the GaInNAs band-gap as a function of N and In composition is shown in Figure 4 (the material parameters used for calculation are taken from [91]). The band gap decreases strongly by incorporating only a few percent of nitrogen and the 1200 nm wavelength range is readily achievable by using GaInNAs with relatively low N content. We should also note that nitrogen incorporation is associated with an increase of the nonradiative recombination centers [92]. Incorporation of higher amounts of nitrogen can cause clustering and phase separation [93] having a detrimental effect on the optical quality of the material. To some extent, this effect can be alleviated by rapid thermal annealing (RTA) which, however, leads to a considerable blue shift of the PL wavelength [94], an effect that should be taken into account in order to achieve the desired laser performance.

In general, the control and understanding of epitaxial processes used to fabricate dilute nitrides is rather challenging. For example, the range of suitable growth temperatures for fabricating high-quality dilute nitrides is narrower than that for growing GaInAs. The typical growth temperature for GaInNAs is in the range of $\sim 460°C$, while GaInAs QWs are grown typically at $\sim 520°C$. The highest performance InGaAsN-based heterostructures are routinely fabricated by molecular beam epitaxy [95]. The standard

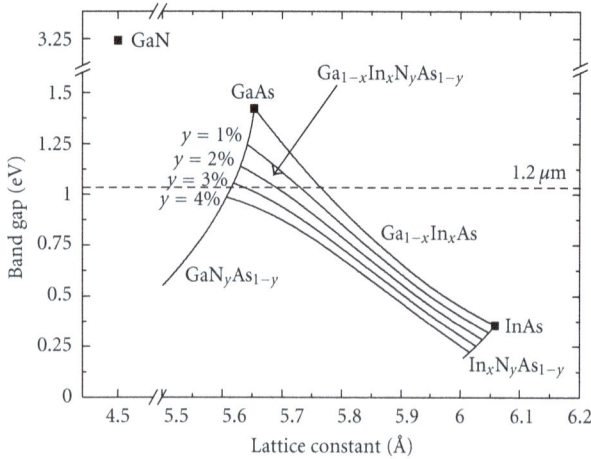

FIGURE 4: Band gap and lattice constant of dilute nitride GaInNAs.

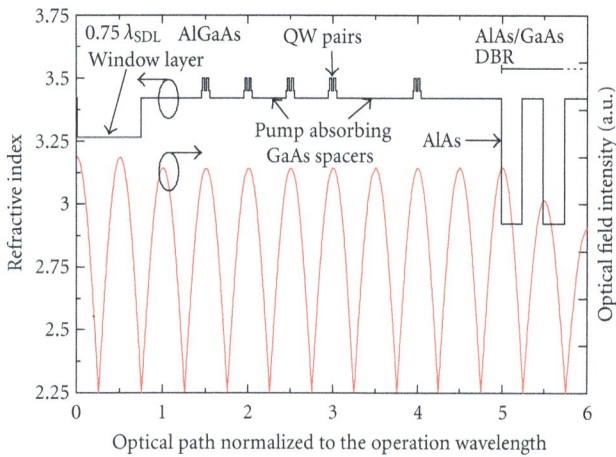

FIGURE 6: Reflectance and photoluminescence spectra of the 1180 nm GaInNAs gain mirror measured at different temperatures.

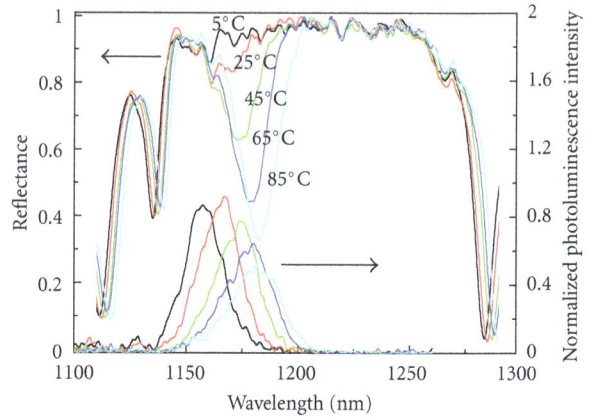

FIGURE 5: Typical structure of the quantum well region of a GaInNAs gain mirror [20].

technique used to incorporate N is dissociation of atomic nitrogen from molecular nitrogen using a radio-frequency (RF) plasma source attached to the MBE growth chamber [96]. Optimization of the plasma operation is one of the key issues that need to be addressed in order to fabricate high-quality dilute nitride heterostructures. The state of the nitrogen plasma depends on the RF power, the flow of N_2, and pressure. The main constituents of the plasma are the molecular nitrogen, atomic nitrogen, and nitrogen ions, each of them having a specific spectral signature that can be used for optimizing the plasma operation [97]. Although the energy of the ions is small, they can cause significant degradation of the optical quality as they impinge on the semiconductor structure during the formation of the QWs [94]. Another important growth parameter affecting the quality of dilute nitrides is the As pressure [98].

Figure 5 displays the structure of a typical dilute nitride gain mirror comprising 10 $Ga_{0.33}In_{0.67}N_{0.006}As_{0.994}$ QWs placed in five pairs. The $GaN_{0.006}As_{0.994}$ layers surrounding the QWs shift their ground state to lower energy and compensate for the compressive strain. For achieving lasing at

around 1180 nm, the room temperature emission wavelength of the QWs was designed to be ~1145 nm. The first four QW pairs were equally spaced at one half wavelength distance apart from each other. The last QW pair was located a full wavelength distance apart from the fourth pair in order to compensate for the pump intensity drop along the gain structure. A 0.75-λ $Al_{0.25}Ga_{0.75}As$ window layer was grown on top of the active region. The active region was grown on top of a 25.5-pair AlAs/GaAs DBR. The growth rate was 0.95 μm/hour and the As/III beam equivalent ratio was 25. After the growth, the sample was kept in the growth chamber under As pressure for a 7 min *in situ* anneal at 680°C to improve the luminescence properties.

The reflectance and photoluminescence (PL) spectra measured for different temperatures of the gain mirror are displayed in Figure 6. The PL graph reveals a temperature dependent red-shift of about 0.3 nm/K. The reason for the decrease in the PL intensity is the increase in the nonradiative recombination rate with increasing temperature resulting in a quantum efficiency drop. The DBR exhibits a temperature red-shift of about 0.06 nm/K. The reflectance spectra were recorded from an as-grown sample, and the photoluminescence spectra were recorded from a sample with diamond heat spreader having an anti-reflective coating on it.

2.2. Operation at Fundamental Wavelength. The gain mirror wafer was cut into 2.5 × 2.5 mm^2 chips, which were then capillary-bonded to synthetic diamond heat spreaders with a wedge angle of about 2° to alleviate the spectral modulation caused by the etalon effect. In addition, we applied a 2-layer TiO_2/SiO_2 antireflective coating on top of the diamond. The laser chip was clamped onto a copper heat sink having small water cooling channels. Despite the flow of cooling water within the heat sink, the heat load generated by pumping the gain mirror led to a slight increase in the mount temperature (T_{mount}). The dependence of T_{mount} on the pump power is shown in Figure 7 for three different pump spot diameters (ϕ_{pump}) and two different temperatures of the cooling water (T_{water}). For laser characterization, the gain chips were tested in a V-shaped SDL cavity shown in Figure 8. The distance

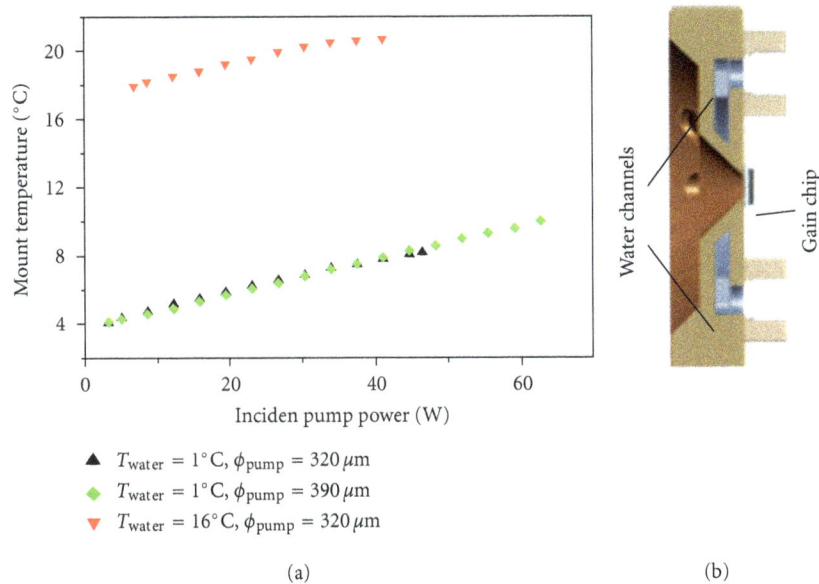

FIGURE 7: (a) The dependence of the mount temperature on the pump power for three diameters of the pump spot and two temperatures of the cooling water. (b) Drawing of the water-cooled mount.

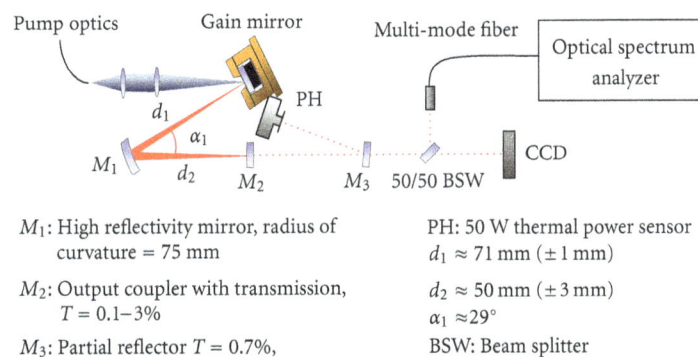

M_1: High reflectivity mirror, radius of curvature = 75 mm

M_2: Output coupler with transmission, $T = 0.1$–3%

M_3: Partial reflector $T = 0.7\%$,

PH: 50 W thermal power sensor

$d_1 \approx 71$ mm (± 1 mm)

$d_2 \approx 50$ mm (± 3 mm)

$\alpha_1 \approx 29°$

BSW: Beam splitter

FIGURE 8: Description of the setup used for the spectrum and beam shape measurements.

between M_1 and M_2 was adjusted to match the size of the TEM_{00} mode to the pump spot on the gain chip, while monitoring the intensity of the output beam to resemble as close as possible a circular Gaussian geometry. In the experiments presented here the gain mirror was pumped by an 808 nm diode bar coupled to a 200 μm multimode fiber. The incidence angle of the pump beam was about 27°.

The SDL output characteristics for different output couplers are shown in Figure 9. Here the water temperature was set to 16°C and the diameter of the pump spot to 320 μm. The maximum output power before thermal roll-over was achieved with 1.5% transmissive output coupler. The highest slope efficiency, of 27%, corresponded to a coupling ratio of 3%. The threshold pump power varied in the range of 3–7 W when the output coupling ratio was varied from 0.1 to 3%.

Next, in order to optimize the pump spot for reaching highest possible power, T_{water} was set to 1°C. The results shown in Figure 10 reveal that the maximum output power increased when ϕ_{pump} was increased from 320 μm to 390 μm.

Also the pump power at which the thermal roll-over was observed was increased from 45 W to 63 W. An output power of slightly more than 11 W was reached with a pump spot of $\phi_{pump} = 390$ μm; the effective mount temperature was 10°C. When ϕ_{pump} was increased from 390 μm to 460 μm, the thermal roll-over point increased slightly to 70 W. However, the slope efficiency dropped and the output power stayed below 11 W, allegedly because of nonideal heat extraction from the gain mirror [99] or overlapping of the larger pump beam with defects on the gain mirror.

2.3. Frequency Doubling. To generate yellow-orange radiation via frequency doubling, we have used V-shape cavity as shown in Figure 11. The nonlinear conversion experiments were performed in free-running mode, that is, without any wavelength control. Compared to the cavity used for fundamental wavelength, the output coupler has been replaced by a mirror that was highly reflective for both IR and visible, whereas the folding mirror reflects infrared but transmits

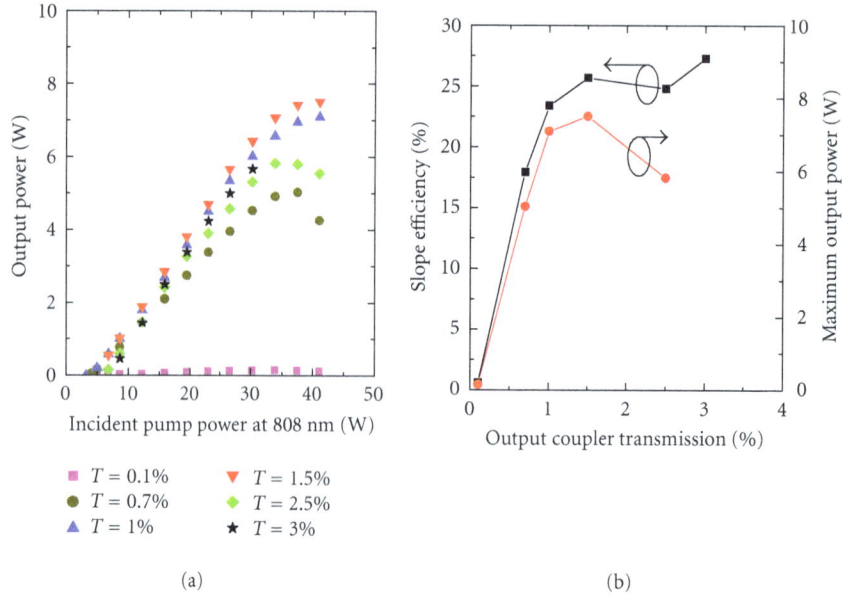

(a)

(b)

FIGURE 9: Output characteristics of the 1.18 μm SDL for different output couplers. The temperature of the cooling water was set to 16°C and the diameter of the pump spot was ~320 μm [20].

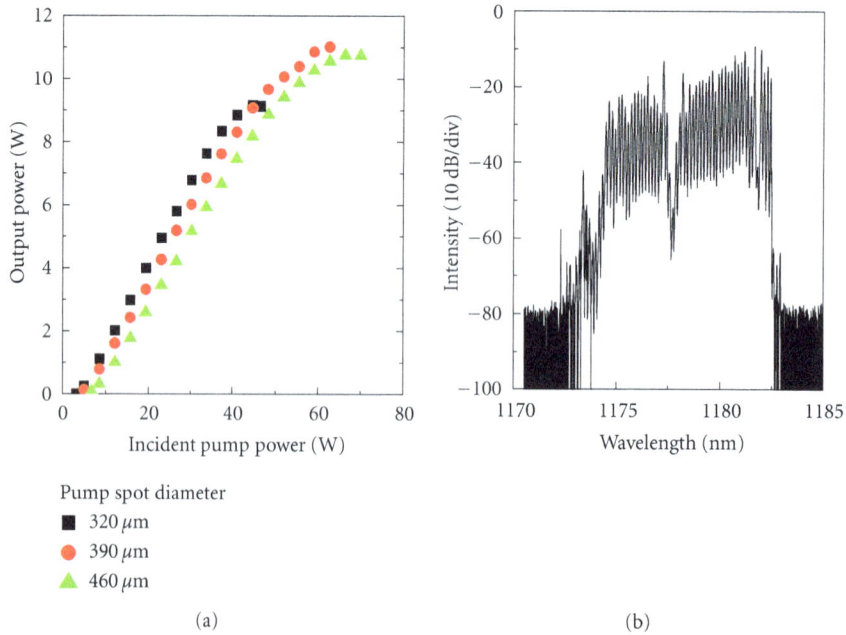

(a)

(b)

FIGURE 10: Output characteristic (a) and typical spectrum for an output power of 5 W (b). The temperature of the cooling water was set to 1°C and the transmission of output coupler was 1.5% [20].

visible light. The frequency conversion was achieved using a 4 mm long type-I critically phase-matched BBO crystal. Figure 12 shows a power transfer graph comparison between the SDL emitting at fundamental infrared wavelength of ~1180 nm and frequency-doubled light at 590 nm. For a pump power of 41.5 W, we demonstrated a maximum

conversion efficiency (absorbed pump light to frequency-converted light) of 17%, which to our knowledge is the highest efficiency reported for a yellow SDL. The ratio of absorbed light to incident pump power was estimated to be ~0.94. At this pump power level, the output power of frequency-doubled light was about 77% of that obtained at

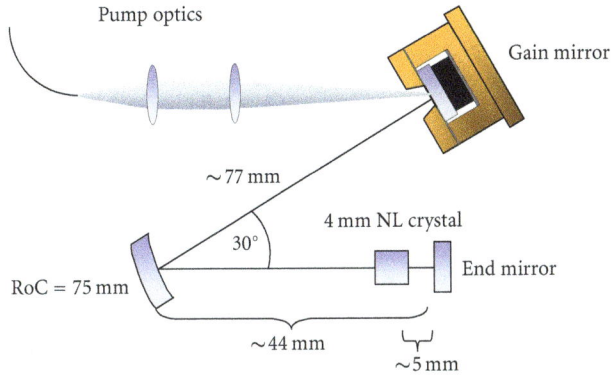

FIGURE 11: SDL setup used for yellow second-harmonic generation in free-running mode.

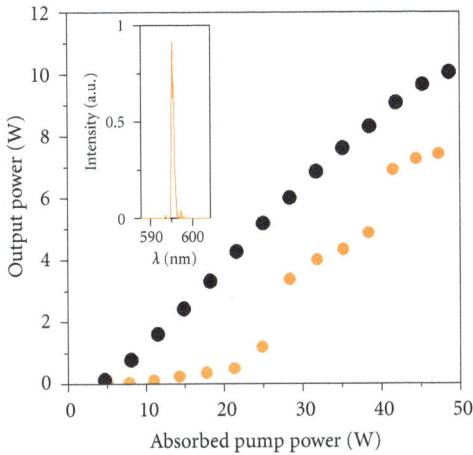

(a)

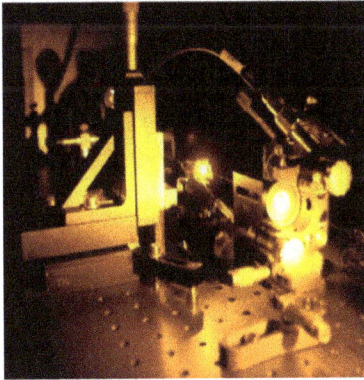

(b)

FIGURE 12: Power transfer graphs of an SDL emitting fundamental 1180 nm light (black) and of a frequency doubled SDL (orange). The inset shows the output spectrum of the frequency-doubled SDL. The photograph at right shows an SDL in operation.

the fundamental wavelength with similar lasing conditions. The inset of Figure 12 shows the emission spectrum of the frequency-doubled radiation at 7 W of output.

Based on the result discussed previously we can conclude that GaInNAs gain mirror technology is the leading

candidates for realizing high-power SDLs with emission at yellow-orange wavelengths.

3. GaSb-Based SDLs for 2–3 μm Wavelength Range

Another commercially and scientifically very interesting spectral domain is that located between 2 and 3 μm. This spectral range can be accessed using GaSb material system. GaSb-based SDLs with high-power (>1 W) and widely tunable (up to ~160 nm) operations have been reported by several groups [19, 37, 65, 100, 101]. In addition to continuous wave lasers also ultrashort pulse SDLs in this wavelength range are of interest, as they could be used as seed sources for mid-IR supercontinuum sources [102] or for pumping of mid-IR optical parametric oscillators. However, because of limited availability of some essential components, such as like semiconductor saturable absorber mirrors (SESAMs), the first passively mode-locked 2 μm GaSb-based SDLs were reported only very recently [103, 104].

The development of GaSb-based (AlGaIn)(AsSb) heterostructures designed for 2–3 μm wavelength range has struggled with many obstacles, such as increased Auger recombination, typical in narrow bandgap semiconductors, and reduced carrier confinement leading to type-II band alignment in QWs instead of preferred type-I. Regardless, electrically pumped, edge-emitting lasers based on GaSb epitaxy have demonstrated CW operation even beyond 3 μm [105, 106] with careful band-gap engineering and utilization of quinternary AlGaInAsSb waveguides. For optical pumping, the decreased thermal properties of GaSb compared to conventional GaAs make thermal management more demanding for mid-IR SDLs. Typically, the optical pumping is based on commercially available diode pumps at 780–980 nm wavelength range, causing excessive heating due to a large quantum defect. The power scalability of GaSb-based SDLs is therefore limited by the effectiveness of thermal management [21]. To reduce the thermal load, different methods have been investigated, such as in-well pumping [19], a flip-chip process with GaSb substrate removal [26], and the use of high thermal conductivity substrate, such as Si or GaAs, in combination with metamorphic growth [107].

The benefits brought by GaSb-based material system to SDLs are the high index contrast ($\Delta n \sim 0.6$) of lattice-matched $AlAs_{0.08}Sb_{0.92}$/GaSb DBR layers, which enables to achieve high reflectivity in exceptionally broad band (~300 nm [100]) with a relatively small number of layer pairs. This makes GaSb SDLs very attractive for spectroscopic application where broad tunability of the laser is needed. High-quality AlAsSb/GaSb DBR can also be used for SESAMs [108]. GaAs-based 1-μm SESAMs have been exploited extensively and their properties can be nowadays tailored to produce ultrashort pulses in various laser types. However, investigation of GaSb-based SESAMs has received far less attention [109, 110]. SESAMs operating at wavelengths around 2 μm and above would have a significant impact on the development of practical ultrafast lasers required in medical applications and time-resolved molecular spectroscopy, or as seeders for optical amplifiers and mid-IR supercontinuum lasers.

To our knowledge the first diode-pumped GaSb-based SDL was demonstrated by Cerutti et al. [111] in 2004; the 2.3 μm laser reached lasing at temperatures up to 350 K with quite moderate output powers. Currently, the emission wavelengths of GaSb SDLs cover the 1.96–2.8 μm spectral range [66, 100]. At 2–2.35 μm the CW power levels have reached multiple watts [37, 112] for near room temperature operation. The achieved output powers of 0.6 W at 2.5 μm and 0.1 W at 2.8 μm [66, 72] have not yet reclaimed the position as such SDLs as high-power lasers. Here, our work had two primary targets: (1) obtain as high CW power as possible at 2 μm and (2) produce ultrashort pulses by passive mode locking at 2 μm. We have developed a gain mirror structure grown on GaSb substrate by solid source MBE. The design included an 18.5-pair DBR made of lattice matched AlAsSb/GaSb layers, and a gain region with 15 InGaSb QWs. For continuous wave experiments, the SDL mirrors were bonded to a planar intracavity diamond heat spreader in a similar manner as the GaInNAs samples described previously. Details of the fabrication process are provided in [37].

3.1. Continuous Wave GaSb Disk Laser.

Our 2 μm range SDLs employed a V-shaped laser cavity. The output characteristics obtained with 99–97% reflective couplers are shown in Figure 13. Here the cooling water temperature was set to 3.5°C and the pump spot diameter was about 350 μm. The emission wavelength was about 1990 nm (Figure 13), slightly depending on the power and output coupler. We observed a general tendency for a spectrum shift towards longer wavelengths with increased coupler reflectance, which could be caused by different heat loads on the gain.

The output characteristics were also measured as a function of the cooling water temperature (Figure 14) using a 98% reflective coupler. The available maximum power was reduced with increasing temperature but it is worth noting that for a coolant temperature of 45°C the laser still produced nearly 1 W of output power. In order to further increase the output power, the pump spot diameter was increased from 350 μm to 440 μm which enabled a maximum power of 5.75 W to be achieved at a water temperature of 3.5°C.

While the power from a single gain chip was limited to less than 6 W, we studied also the possibility to increase the laser output by cascading 2 gain chips in one laser cavity. A W-shaped laser cavity was set up as shown in Figure 15. The pump spot diameter was further increased to 500 μm and the temperature was reduced to −2.5°C. For this purpose a mixture of water and alcohol had to be used as coolant and a flow of nitrogen was provided to the samples to prevent condensation of water from the surrounding air. Eventually we were able to increase the power to 8.6 W (Figure 15) but could not achieve linear power scaling that should have theoretically resulted in more than 11 W of power with these two particular chips. Reasons for this can be many. One important contributing factor may be the output coupler that was 94.6% reflective; out of all available output couplers it enabled the highest output power but might not have been

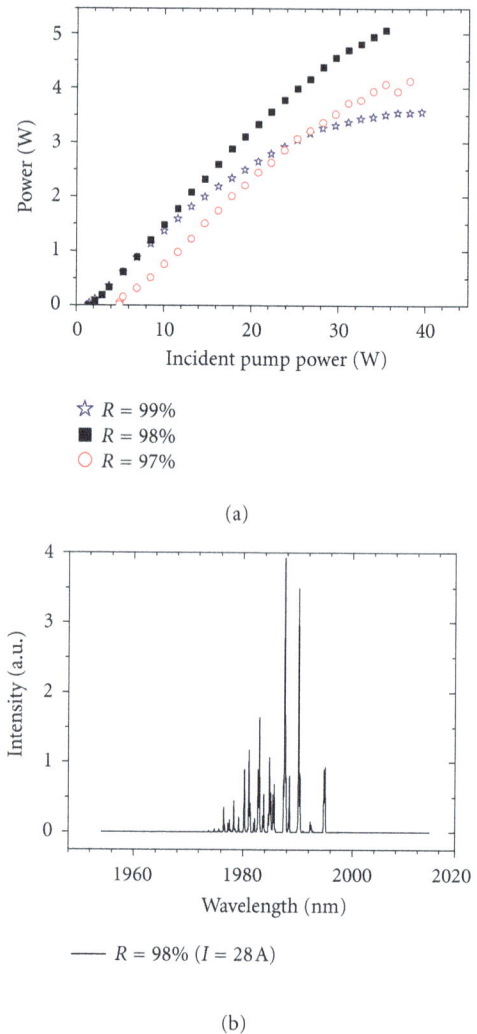

☆ R = 99%
■ R = 98%
○ R = 97%

(a)

—— R = 98% (I = 28 A)

(b)

FIGURE 13: (a) Laser output for different output couplers. $T = 3.5°C$ (cooling water temperature). Pump spot ∅ = 350 μm. R : output coupler reflectivity. (b) Laser spectrum for a pump power of 35.2 W.

optimal for this particular laser configuration. The coolant mixture had also lower specific heat than pure water and therefore the cooling may not be as effective as the coolant temperature could suggest. The 2-gain laser alignment is also somewhat more complicated than a single chip laser which may hinder power scaling. Despite these difficulties, reasonable results were obtained in terms of output power at 2 μm wavelength.

3.2. Femtosecond Pulse Generation.

As discussed earlier, ultrashort optical pulses have been generated in GaAs- and InP-based disk lasers in various configurations using both active and passive mode-locking schemes [2, 4, 6, 44, 45, 113, 114]. On the contrast, the development of ultrafast GaSb disk lasers has been much slower, possibly due to lack of SESAMs and more demanding SESAM characterization. We have shown only very recently that also GaSb-based disk laser can generate sub-picoseconds pulses at 2 μm [104]. The development of low-nonlinearity GaSb-based SESAMs was

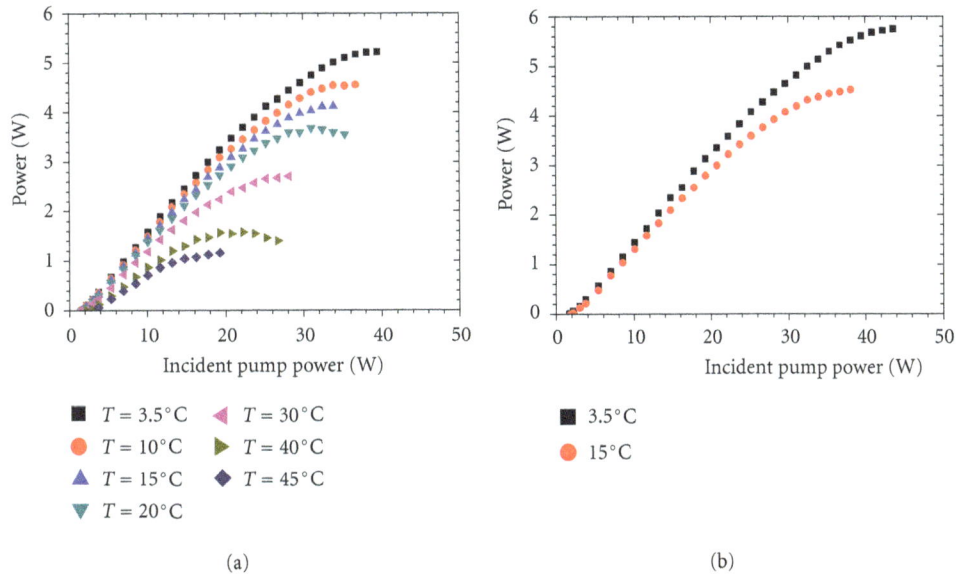

FIGURE 14: (a) Laser output characteristics at different cooling water temperatures. Output coupler R is 98%, and pump spot diameter is 350 μm. (b) Laser output characteristics with 440 μm pump spot diameter. Output coupler R is 98%.

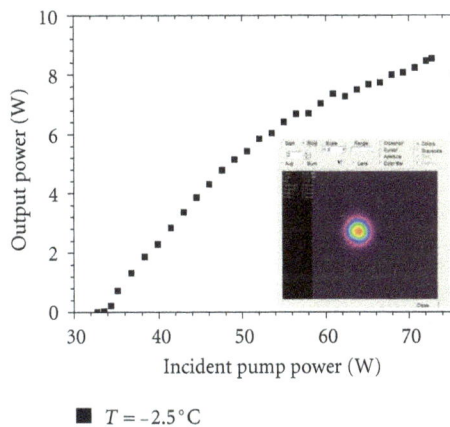

FIGURE 15: (a) Photograph of the W-shaped 2-gain laser cavity. (b) 2-gain laser output characteristics with 500 μm pump spot diameter. Output coupler R is 94.6%. Inset: profile of the output beam.

instrumental for demonstrating ultrafast 2 μm SDLs. The right combination of the dynamic properties of the SESAM (saturation fluence, absorption recovery time, nonlinear reflectivity, and nonsaturable losses) is quite essential for achieving stable mode-locking. An interesting finding was made that as-grown 2 μm GaSb saturable absorber mirrors had much faster recovery time than typical as-grown GaAs-based components operating in the 1 μm regime. Fabrication details of the SESAMs we have used are provided in reference [110]. The GaSb absorber mirrors were studied with pump probe measurements. The growth temperature and amount of strain were used as controlled variables. For the mode-locking experiments, we used a Z-shaped cavity (Figure 16)

that allowed convenient alignment of the SESAM and gain mirror.

The gain chip was cooled in a similar manner as the continuous wave 2 μm laser, but the heat spreader diamond had a 2° wedge and AR coating to suppress the etalon effect arising from it. The gain mirror was pumped with a fiber-coupled 980 nm diode laser and the output coupler had a reflectivity of $R = 99\%$ at the operation wavelength. Simulated mode diameter was ~230 μm on the gain and ~25 μm on the SESAM. The pulse repetition rate, defined by the cavity length, was in the order of 890 MHz. The output of the laser was monitored with an optical spectrum analyzer and a 2.5 GHz photodiode from which the signal

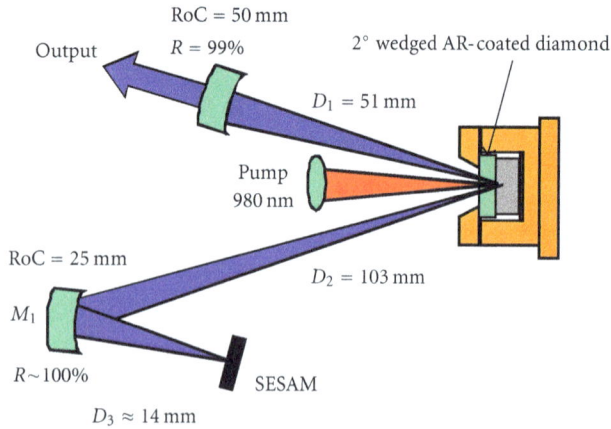

FIGURE 16: A Z-shaped cavity of the mode-locked GaSb laser.

was coupled to an RF-analyzer and oscilloscope. The laser beam profile could be monitored with a pyroelectric camera for proper alignment. The pulses were characterized with an interferometric autocorrelator that was based on two-photon absorption in a silicon detector. Depending on the laser alignment, we were able to measure pulses with duration varying from slightly less than 400 fs to slightly more than 400 fs, with the shortest measured pulse being 384 fs. A typical autocorrelation trace is shown in Figure 17 with the optical spectrum, RF-spectrum, and retrieved pulse shape.

Quite surprisingly the output power level of the mode-locked laser was only some tens of milliwatts despite many watts of pump power. In continuous wave mode over 5 W of power was obtained from other devices having similar gain material. To some extent the differences can be explained by variations between individual chips, nonoptimal output coupling ratio, lossy cavity, and smaller pump spot diameter but clearly the average power in mode-locked operation should have been markedly higher. In order to study the potential of the laser, we replaced the SESAM with a high reflective mirror and then we measured the output power in continuous wave mode. As shown in Figure 18, for an incident pump power of 7.6 W, the power was ~23 mW with the SESAM, and slightly over 40 mW with the HR mirror.

The precise position of the mirror could be determined from the RF-spectrum, which helped to monitor the output power as a function of the mirror position in regard to the original position of the SESAM. When the HR mirror was repositioned about 60 μm closer to the curved mirror than the original SESAM position, the power increased from ~ 40 mW to 130 mW (see Figure 19 for details). The position of the HR/absorber mirror changes the mode diameter also on the gain and therefore it has an impact on the efficiency of the laser. The study revealed that the mirror position that was optimal for mode-locking may not have been optimal for achieving the highest average power. The result suggests that with further optimization of the cavity geometry and adjustments of the pump spot diameter and output coupling ratio, it should be possible to increase the average power to >100 mW also in mode-locked operation.

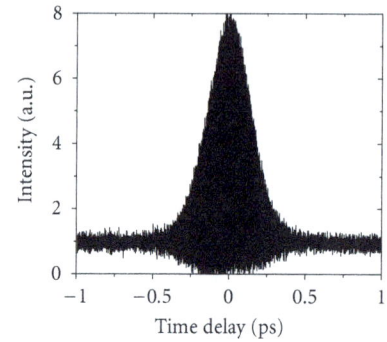

(a)

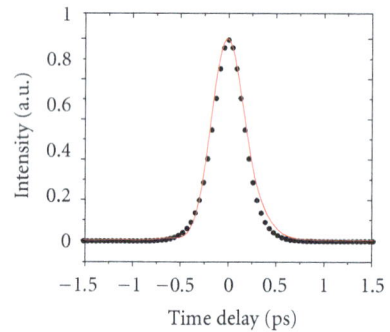

● Fourier limit
— Measured pulse

(b)

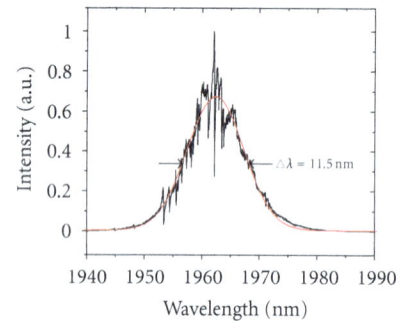

(c)

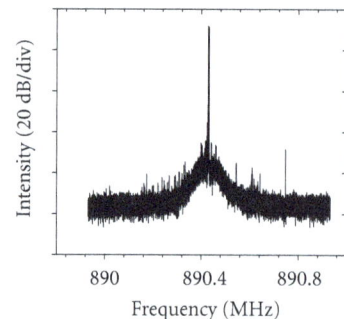

(d)

FIGURE 17: (a) Interferometric autocorrelation trace. (b) Intensity autocorrelation retrieved from the interferometric data. shown together with calculated Fourier limit. (c) Optical spectrum and a fitting. (d) Measured RF-spectrum.

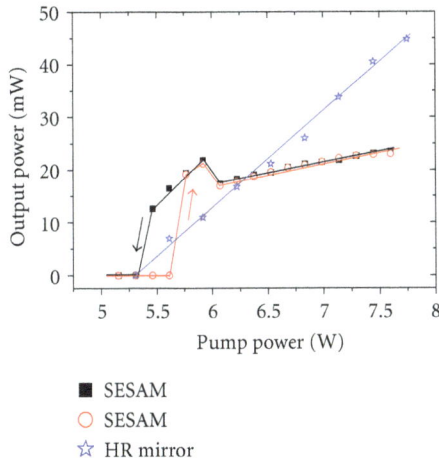

FIGURE 18: Average power of the GaSb SDL with a SESAM, and with a high reflective mirror. With the SESAM we can observe a hysteresis characteristic, typical for mode-locking.

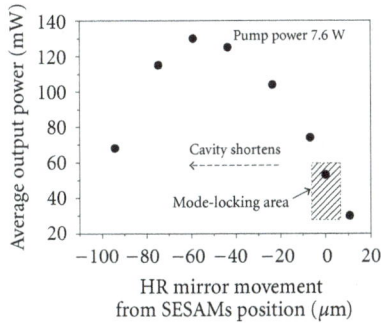

FIGURE 19: Output power of the GaSb disk laser with the high reflective mirror, given as a function of the HR mirror's position in regard to the original SESAM's position (point of optimal mode-locking).

To summarize, we have shown that nearly transform limited femtosecond pulses can be obtained from SESAM mode-locked GaSb disk lasers without use of additional dispersion compensating elements. So far the power levels are modest but there is reason to believe that the average power could be increased beyond 100 mW with further optimization of the laser.

4. Future Outlook

SDLs (or VECSELs, or OPSLs) combine a small footprint, multiwatt output power capability, high beam quality, and the capability to fill spectral gaps that cannot be reached by traditional solid state disk lasers. In terms of semiconductor technology, OP-SDLs are in many aspects simpler than VCSELs; they do not require doping of the mirrors, usually make use of only one semiconductor mirror, and their processing is simpler. Owing to these advantageous features and intense developments efforts in the last decade, SDLs have reached a development stage that makes them very attractive for application deployments. While SDLs with emission at green or blue have been commercialized for several years by Coherent Inc., there are other wavelength regions where SDLs could have a tremendous impact on the development of new applications. Leveraging the advantages of SDLs technology to broader wavelength domains is inherently linked to the development of new semiconductor structures enabling wavelength tailoring and increased functionality.

In this paper, we *reviewed* the main advances in the development of SDLs producing yellow-orange and $2\,\mu m$ radiation, which are required in medicine, astronomy, life science research, sensing, or infrared countermeasures. GaInNAs/GaAs gain mirrors are now a reliable approach for the development of SDLs with fundamental emission of more than 11 W at around 1180–1200 nm. This material system has enabled generation of yellow-orange laser radiation with excellent beam quality and output power exceeding 7 W (the highest power obtained from a semiconductor-based laser at this wavelength range). The GaSb gain mirrors have been used to leverage the advantages offered by SDLs to the 2–$3\,\mu m$ wavelength range. This material system is very robust in terms of reliability and life-time, enabling one to reach output powers in excess of several watts at wavelengths of 2–$2.3\,\mu m$. Most recently, we have demonstrated that GaSb SDLs are suitable for generation of femtosecond pulses at 2-μm, a wavelength range that is particularly attractive for surgery, infrared counter measures, or LIDAR, and where there is a lack of compact high-power ultrashort pulse sources. Despite these achievements, there are certainly several development steps that are required for reaching new functionality and for advancing the technology to levels suitable for application deployment. The main development directions we undertake for the advances of dilute-nitride and GaSb-based SDLs are briefly discussed as follows.

Development of flip-chip technology for dilute nitride SDLs, would enable further improvements of the spectral and power characteristics. The main limitation of using intracavity heat spreader is related to spectral modulation caused by spurious etalon effects, which have a detrimental effect on wavelength tuning and mode-locking. Allegedly, the flip-chip technology would also enable a more predictable power scaling with increasing the area of the pump region on the gain mirror. The main difficulty related to the use of flip-chip technology for dilute-nitrides is apparently related to the high level of residual strain corresponding to the GaInNAs active region; the strain leads to occurrence of structural defects due to mechanical deformation once the substrate is removed. Advanced strain compensation techniques are expected to alleviate this problem.

Flip-chip technology could provide advantages to GaSb-based SDLs for increased functionality and development of a process that is more suitable for volume production. In particular, we should note that wavelength-tuning capability is very important for mid-IR SDLs as many of the applications could be related to spectroscopy. The main difficulty related to the use of flip-chip technology for GaSb is related to the fact that this material system is less developed from processing point of view. Successful steps in GaSb SDL flip-chip processing and development of adequate etch stop layers for substrate removal have been already made [26]. Very recently we have also demonstrated that InPSb can be used effectively as an etch stop layer for the GaSb substrate removal; we have achieved an etch selectivity of GaSb substrate as high as 244 and excellent substrate removal rate of 32.4 μm/min [115]. The flip-chip GaSb gain mirrors would be beneficial for ultrashort pulse operation and could ultimately enable to take full advantage of the broad gain bandwidth of GaSb.

Development of electrically pumped GaSb SDLs. While optically pumped SDLs can produce multiple watts of output power, they require a separate pump source that adds to the cost and complexity of the device. Direct electrical-pumping offers an interesting alternative that simplifies the overall laser scheme. If the power level is not the main target, electrically pumping of SDLs, more often termed as EP-VECSELs, can be realized conveniently [116]. The essential challenges of electrical pumping relate to nonuniform current spreading and optical losses in doped semiconductor material [117–119]. Doping is necessary for achieving low electrical resistance, but at the same time it does increase absorption. On the other hand, the current spreading problems limit the size of usable gain area and therefore hinder power scaling. Despite several technical challenges, EP-VECSELs have been studied actively and there have been also serious attempts to commercialize this type of laser [120, 121]. Using standard GaAs gain mirrors output power levels in excess of 400 mW have been reported [122]. On the contrary GaSb-based EP-VECSELs have been demonstrated only recently [123]. For this preliminary demonstration, we have used a (1/2)-VCSEL gain mirror ($\lambda \sim 2.3 \mu$m) that was fabricated at the Walter Schottky Institute in Germany. An I-shaped cavity was formed between the gain mirror and a curved output coupler. In a first study we tested 7 different components with diameters of 30–90 μm. Lasing was obtained from all components at 15°C mount temperature using pulsed current with 1 μs pulse width and 3% duty cycle. A maximum peak power of 1.5 mW was obtained from the 60 μm component. Thermal issues seemed to be the major factor limiting the power. We should note here that electrical pumping is particularly attractive for mid-IR GaSb VECSELs, as the requirements for deployment in spectroscopic applications are mainly related to compactness, tunability, single-frequency lasers, and less to the power level. An EP-VECSEL would be compact but at the same time would enable to include intracavity elements for wavelength tuning in a broad wavelength range or would enable the use of intracavity spectroscopy in compact and efficient laser architectures.

Acknowledgments

The authors would like to thank several colleagues and external collaborators for continuous contribution to the development of SDLs reviewed here. In particular, Janne Puustinen, Jonna Paajaste, and Riku Koskinen are acknowledged for MBE fabrication of the semiconductor structures, and Lasse Orsila and Jari Nikkinen are acknowledged for their help with the deposition of antireflection coatings. They thank Jukka-Pekka Alanko, Christian Grebing, and Professor Günter Steinmeyer for their contribution to the demonstration of 2-μm mode-locked SDL. They acknowledge Professor Anne Tropper from the University of Southampton for useful discussion and sharing some of the data presented in Figure 3. They thank Professor Amann and his group from the Walter Schottky Institute for fabricating the 2.3 μm EP-VECSEL samples they reported in [123]. They gratefully acknowledge the financial support provided by by Areté Associates, Pirkanmaa TE-center, the Finnish Funding Agency for Development and Innovation (TEKES), the Academy of Finland (no. 128364), the Graduate Schools in Material Science and GETA, and the United States Office of Naval Research Global (ONRG) under the Grant no. N62909-10-1-7030.

References

[1] N. Basov, O. Bogdankevich, and A. Grasyuk, "9B4—semiconductor lasers with radiating mirrors," *Quantum Electronics*, vol. 2, no. 9, pp. 594–597, 1966.

[2] W. B. Jiang, S. R. Friberg, H. Iwamura, and Y. Yamamoto, "High powers and subpicosecond pulses from an external-cavity surface-emitting InGaAs/InP multiple quantum well laser," *Applied Physics Letters*, vol. 58, no. 8, pp. 807–809, 1991.

[3] H. Q. Le, S. Di Cecca, and A. Mooradian, "Scalable high-power optically pumped GaAs laser," *Applied Physics Letters*, vol. 58, no. 18, pp. 1967–1969, 1991.

[4] W. H. Xiang, S. R. Friberg, K. Watanabe et al., "Sub-100 femtosecond pulses from an external-cavity surface-emitting InGaAs/InP multiple quantum well laser with soliton-effect compression," *Applied Physics Letters*, vol. 59, no. 17, pp. 2076–2078, 1991.

[5] D. C. Sun, S. R. Friberg, K. Watanabe, S. MacHida, Y. Horikoshi, and Y. Yamamoto, "High power and high efficiency vertical cavity surface emitting GaAs laser," *Applied Physics Letters*, vol. 61, no. 13, pp. 1502–1503, 1992.

[6] W. B. Jiang, R. Mirin, and J. E. Bowers, "Mode-locked GaAs vertical cavity surface emitting lasers," *Applied Physics Letters*, vol. 60, no. 6, pp. 677–679, 1992.

[7] C. Symonds, J. Dion, I. Sagnes et al., "High performance 1.55 μm vertical external cavity surface emitting laser with broadband integrated dielectric-metal mirror," *Electronics Letters*, vol. 40, no. 12, pp. 734–735, 2004.

[8] D. Lorenser, H. J. Unold, D. J. H. C. Maas et al., "Towards wafer-scale integration of high repetition rate passively mode-locked surface-emitting semiconductor lasers," *Applied Physics B*, vol. 79, no. 8, pp. 927–932, 2004.

[9] M. Kuznetsov, F. Hakimi, R. Sprague, and A. Mooradian, "Design and characteristics of high-power (>0.5-W CW) diode-pumped vertical-external-cavity surface-emitting semiconductor lasers with circular TEM$_{00}$ beams," *IEEE Journal on Selected Topics in Quantum Electronics*, vol. 5, no. 3, pp. 561–573, 1999.

[10] E. J. Saarinen, A. Härkönen, S. Suomalainen, and O. G. Okhotnikov, "Power scalable semiconductor disk laser using multiple gain cavity," in *Proceedings of the Conference on Lasers and Electro-Optics (CLEO '07)*, 2007.

[11] L. Fan, M. Fallahi, J. Hader et al., "Multichip vertical-external-cavity surface-emitting lasers: a coherent power scaling scheme," *Optics Letters*, vol. 31, no. 24, pp. 3612–3614, 2006.

[12] T. J. Ochalski, A. De Burea, G. Huyet et al., "Passively mod-elocked bi-directional vertical external ring cavity surface emitting laser," in *Proceedings of the Conference on Quantum Electronics and Laser Science Conference on Lasers and Electro-Optics (CLEO/QELS '08)*, May 2008.

[13] J. E. Hastie, J. M. Hopkins, C. W. Jeon et al., "Microchip verti-cal external cavity surface emitting lasers," *Electronics Letters*, vol. 39, no. 18, pp. 1324–1326, 2003.

[14] S. A. Smith, J. M. Hopkins, J. E. Hastie et al., "Diamond-microchip GaInNAs vertical external-cavity surface-emitting laser operating CW at 1315 nm," *Electronics Letters*, vol. 40, no. 15, pp. 935–936, 2004.

[15] N. Laurand, C. L. Lee, E. Gu, J. E. Hastie, S. Calvez, and M. D. Dawson, "Microlensed microchip VECSEL," *Optics Express*, vol. 15, no. 15, pp. 9341–9346, 2007.

[16] S. W. Corzine, R. S. Geels, J. W. Scott, R. H. Yan, and L. A. Coldren, "Design of Fabry-Perot surface-emitting lasers with a periodic gain structure," *IEEE Journal of Quantum Electron-ics*, vol. 25, no. 6, pp. 1513–1524, 1989.

[17] M. Schmid, S. Benchabane, F. Torabi-Goudarzi, R. Abram, A. I. Ferguson, and E. Riis, "Optical in-well pumping of a vertical-external-cavity surface-emitting laser," *Applied Let-ters*, vol. 84, no. 24, pp. 4860–4862, 2004.

[18] S. S. Beyertt, U. Brauch, F. Demaria et al., "Efficient gallium-arsenide disk laser," *IEEE Journal of Quantum Electronics*, vol. 43, no. 10, pp. 869–875, 2007.

[19] N. Schulz, M. Rattunde, C. Ritzenthaler et al., "Resonant optical in-well pumping of an (AlGaIn)(AsSb)-based vertic-al-external-cavity surface-emitting laser emitting at 2.35 μm," *Applied Physics Letters*, vol. 91, no. 9, Article ID 091113, 2007.

[20] V. M. Korpijärvi, T. Leinonen, J. Puustinen, A. Härkönen, and M. D. Guina, "11 W single gain-chip dilute nitride disk laser emitting around 1180 nm," *Optics Express*, vol. 18, no. 25, pp. 25633–25641, 2010.

[21] A. J. Kemp, G. J. Valentine, J. M. Hopkins et al., "Thermal management in vertical-external-cavity surface-emitting la-sers: finite-element analysis of a heatspreader approach," *IEEE Journal of Quantum Electronics*, vol. 41, no. 2, pp. 148–155, 2005.

[22] W. J. Alford, T. D. Raymond, and A. A. Allerman, "High power and good beam quality at 980 nm from a vertical external-cavity surface-emitting laser," *Journal of the Optical Society of America B*, vol. 19, no. 4, pp. 663–666, 2002.

[23] J. E. Hastie, J. M. Hopkins, S. Calvez et al., "0.5-W single transverse-mode operation of an 850 nm diode-pumped surface-emitting semiconductor laser," *IEEE Photonics Tech-nology Letters*, vol. 15, no. 7, pp. 894–896, 2003.

[24] Z. L. Liau, "Semiconductor wafer bonding via liquid capillar-ity," *Applied Physics Letters*, vol. 77, no. 5, pp. 651–653, 2000.

[25] M. Kuznetsov, F. Hakimi, R. Sprague, and A. Moora-dian, "High-power (>0.5-W CW) diode-pumped vertical-external-cavity surface-emitting semiconductor lasers with circular TEM$_{00}$ beams," *IEEE Photonics Technology Letters*, vol. 9, no. 8, pp. 1063–1065, 1997.

[26] J. P. Perez, A. Laurain, L. Cerutti, I. Sagnes, and A. Garnache, "Technologies for thermal management of mid-IR Sb-based surface emitting lasers," *Semiconductor Science and Technology*, vol. 25, no. 4, Article ID 045021, 2010.

[27] A. R. Clawson, "Guide to references on III-V semiconductor chemical etching," *Materials Science and Engineering R*, vol. 31, no. 1, pp. 1–438, 2001.

[28] A. J. Maclean, A. J. Kemp, S. Calvez et al., "Continuous tun-ing and efficient intracavity second-harmonic generation in a semiconductor disk laser with an intracavity diamond

[29] T. L. Wang, Y. Kaneda, J. M. Yarborough et al., "High-power optically pumped semiconductor laser at 1040 nm," *IEEE Photonics Technology Letters*, vol. 22, no. 9, Article ID 5422654, pp. 661–663, 2010.

heatspreader," *IEEE Journal of Quantum Electronics*, vol. 44, no. 3, pp. 216–225, 2008.

[30] J. Chilla, Q. Shu, H. Zhou, E. Weiss, M. Reed, and L. Spinelli, "Recent advances in optically pumped semiconductor lasers," in *Solid State Lasers XVI: Technology and Devices*, vol. 6451 of *Proceedings of SPIE*, San Jose, Calif, USA, 2007.

[31] H. Lindberg, M. Strassner, E. Gerster, J. Bengtsson, and A. Larsson, "Thermal management of optically pumped long-wavelength InP-based semiconductor disk lasers," *IEEE Jour-nal on Selected Topics in Quantum Electronics*, vol. 11, no. 5, pp. 1126–1134, 2005.

[32] A. J. Kemp, A. J. MacLean, J. E. Hastie et al., "Thermal lens-ing, thermal management and transverse mode control in microchip VECSELs," *Applied Physics B*, vol. 83, no. 2, pp. 189–194, 2006.

[33] A. J. Maclean, R. B. Birch, P. W. Roth, A. J. Kemp, and D. Burns, "Limits on efficiency and power scaling in semicon-ductor disk lasers with diamond heatspreaders," *Journal of the Optical Society of America B*, vol. 26, no. 12, pp. 2228–2236, 2009.

[34] A. R. Zakharian, J. Hader, J. V. Moloney, S. W. Koch, P. Brick, and S. Lutgen, "Experimental and theoretical analysis of optically pumped semiconductor disk lasers," *Applied Physics Letters*, vol. 83, no. 7, pp. 1313–1315, 2003.

[35] L. Fan, M. Fallahi, J. T. Murray et al., "Tunable high-power high-brightness linearly polarized vertical-external-cavity surface-emitting lasers," *Applied Physics Letters*, vol. 88, no. 2, Article ID 021105, pp. 1–3, 2006.

[36] L. Fan, M. Fallahi, A. R. Zakharian et al., "Extended tunability in a two-chip VECSEL," *IEEE Photonics Technology Letters*, vol. 19, no. 8, pp. 544–546, 2007.

[37] J. Paajaste, S. Suomalainen, R. Koskinen, A. Härkönen, M. Guina, and M. Pessa, "High-power and broadly tunable GaSb-based optically pumped VECSELs emitting near 2 μm," *Journal of Crystal Growth*, vol. 311, no. 7, pp. 1917–1919, 2009.

[38] C. Borgentun, J. Bengtsson, A. Larsson, F. Demaria, A. Hein, and P. Unger, "Optimization of a broadband gain element for a widely tunable high-power semiconductor disk laser," *IEEE Photonics Technology Letters*, vol. 22, no. 13, Article ID 5451056, pp. 978–980, 2010.

[39] M. A. Holm, D. Burns, A. I. Ferguson, and M. D. Dawson, "Actively stabilized single-frequency vertical-external-cavity AlGaAs laser," *IEEE Photonics Technology Letters*, vol. 11, no. 12, pp. 1551–1553, 1999.

[40] H. Lindberg, A. Larsson, and M. Strassner, "Single-frequency operation of a high-power, long-wavelength semiconductor disk laser," *Optics Letters*, vol. 30, no. 17, pp. 2260–2262, 2005.

[41] A. Laurain, M. Myara, G. Beaudoin, I. Sagnes, and A. Gar-nache, "Multiwatt-power highly-coherent compact single-frequency tunable vertical-external-cavity-surface-emitting-semiconductor-laser," *Optics Express*, vol. 18, no. 14, pp. 14627–14636, 2010.

[42] B. Rudin, A. Rutz, M. Hoffmann et al., "Highly efficient opti-cally pumped vertical-emitting semiconductor laser with more than 20 W average output power in a fundamental transverse mode," *Optics Letters*, vol. 33, no. 22, pp. 2719–2721, 2008.

[43] J. Chilla, S. Butterworth, A. Zeitschel et al., "High power optically pumped semiconductor lasers," in *Solid State Lasers*

XIII: Technology and Devices, vol. 5332 of *Proceedings of SPIE*, pp. 143–150, San Jose, Calif, USA, 2004.

[44] A. H. Quarterman, K. G. Wilcox, V. Apostolopoulos et al., "A passively mode-locked external-cavity semiconductor laser emitting 60 fs pulses," *Nature Photonics*, vol. 3, no. 12, pp. 729–731, 2009.

[45] P. Klopp, F. Saas, M. Zorn, M. Weyers, and U. Griebner, "290 fs pulses from a semiconductor disk laser," *Optics Express*, vol. 16, no. 8, pp. 5770–5775, 2008.

[46] A. Aschwanden, D. Lorenser, H. J. Unold, R. Paschotta, E. Gini, and U. Keller, "2.1 W picosecond passively mode-locked external-cavity semiconductor laser," *Optics Letters*, vol. 30, no. 3, pp. 272–274, 2005.

[47] B. Rudin, V. J. Wittwer, D. J. H. C. Maas et al., "High-power MIXSEL: an integrated ultrafast semiconductor laser with 6.4 W average power," *Optics Express*, vol. 18, no. 26, pp. 27582–27588, 2010.

[48] D. Lorenser, D. J. H. C. Maas, H. J. Unold et al., "50 GHz passively mode-locked surface-emitting semiconductor laser with 100 mW average output power," *IEEE Journal of Quantum Electronics*, vol. 42, no. 8, Article ID 01658136, pp. 838–847, 2006.

[49] U. Keller and A. C. Tropper, "Passively modelocked surface-emitting semiconductor lasers," *Physics Reports*, vol. 429, no. 2, pp. 67–120, 2006.

[50] W. Zhang, A. McDonald, T. Ackemann, E. Riis, and G. McConnell, "Femtosecond synchronously in-well pumped vertical-external-cavity surface-emitting laser," *Optics Express*, vol. 18, no. 1, pp. 187–192, 2010.

[51] J. Rautiainen, V. M. Korpijärvi, J. Puustinen, M. Guina, and O. G. Okhotnikov, "Passively mode-locked GaInNAs disk laser operating at 1220 nm," *Optics Express*, vol. 16, no. 20, pp. 15964–15969, 2008.

[52] A. Rutz, V. Liverini, D. J. H. C. Maas et al., "Passively mod-elocked GaInNAs VECSEL at centre wavelength around 1.3 μm," *Electronics Letters*, vol. 42, no. 16, pp. 926–927, 2006.

[53] A. Khadour, S. Bouchoule, G. Aubin, J. C. Harmand, J. Decobert, and J. L. Oudar, "Ultrashort pulse generation from 1.56 μm modelocked VECSEL at room temperature," *Optics Express*, vol. 18, no. 19, pp. 19902–19913, 2010.

[54] J. E. Hastie, L. G. Morton, A. J. Kemp, M. D. Dawson, A. B. Krysa, and J. S. Roberts, "Tunable ultraviolet output from an intracavity frequency-doubled red vertical-external-cavity surface-emitting laser," *Applied Physics Letters*, vol. 89, no. 6, Article ID 061114, 2006.

[55] S. H. Park, J. Kim, H. Jeon et al., "Room-temperature GaN vertical-cavity surface-emitting laser operation in an extended cavity scheme," *Applied Physics Letters*, vol. 83, no. 11, pp. 2121–2123, 2003.

[56] J. Y. Kim, S. Cho, S. J. Lim et al., "Efficient blue lasers based on gain structure optimizing of vertical-external-cavity surface-emitting laser with second harmonic generation," *Journal of Applied Physics*, vol. 101, no. 3, Article ID 033103, 2007.

[57] J. Lee, S. Lee, T. Kim, and Y. Park, "7 W high-efficiency continuous-wave green light generation by intracavity frequency doubling of an end-pumped vertical external-cavity surface emitting semiconductor laser," *Applied Physics Letters*, vol. 89, no. 24, Article ID 241107, 2006.

[58] S. Hilbich, W. Seelert, V. Ostroumov et al., "New wavelengths in the yellow orange range between 545 nm to 580 nm generated by an intracavity frequency-doubled Optically Pumped Semiconductor Laser," in *Solid State Lasers XVI: Technology and Devices*, vol. 6451 of *Proceedings of SPIE*, San Jose, Calif, USA, 2007.

[59] A. Härkönen, J. Rautiainen, M. Guina et al., "High power frequency doubled GaInNAs semiconductor disk laser emitting at 615 nm," *Optics Express*, vol. 15, no. 6, pp. 3224–3229, 2007.

[60] M. I. Müller, N. Linder, C. Karnutsch et al., "Optically pumped semiconductor thin-disk laser with external cavity operating at 660 nm," in *Vertical-Cavity Surface-Emitting Lasers VI*, vol. 4649 of *Proceedings of SPIE*, pp. 265–271, San Jose, Calif, USA, 2002.

[61] K. S. Kim, J. R. Yoo, S. H. Cho et al., "1060 nm vertical-external-cavity surface-emitting lasers with an optical-to-optical efficiency of 44% at room temperature," *Applied Physics Letters*, vol. 88, no. 9, Article ID 091107, 2006.

[62] J. Konttinen, A. Härkönen, P. Tuomisto et al., "High-power (>1 W) dilute nitride semiconductor disk laser emitting at 1240 nm," *New Journal of Physics*, vol. 9, article 140, 2007.

[63] J. M. Hopkins, S. A. Smith, C. W. Jeon et al., "0.6 W CW GaInNAs vertical external-cavity surface emitting laser operating at 1.32 μm," *Electronics Letters*, vol. 40, no. 1, pp. 30–31, 2004.

[64] H. Lindberg, M. Strassner, E. Gerster, and A. Larsson, "0.8 W optically pumped vertical external cavity surface emitting laser operating CW at 1550 nm," *Electronics Letters*, vol. 40, no. 10, pp. 601–602, 2004.

[65] J. Nikkinen, J. Paajaste, R. Koskinen, S. Suomalainen, and O. G. Okhotnikov, "GaSb-based semiconductor disk laser with 130 nm tuning range at 2.5 μm," *IEEE Photonics Technology Letters*, vol. 23, no. 12, Article ID 5723696, pp. 777–779, 2011.

[66] B. Rösener, M. Rattunde, R. Moser et al., "Continuous-wave room-temperature operation of a 2.8 μm GaSb-based semiconductor disk laser," *Optics Letters*, vol. 36, no. 3, pp. 319–321, 2011.

[67] M. Rahim, A. Khiar, F. Felder, M. Fill, and H. Zogg, "4.5 μm wavelength vertical external cavity surface emitting laser operating above room temperature," *Applied Physics Letters*, vol. 94, no. 20, Article ID 201112, 2009.

[68] M. Rahim, F. Felder, M. Fill, and H. Zogg, "Optically pumped 5 μm IV-VI VECSEL with Al-heat spreader," *Optics Letters*, vol. 33, no. 24, pp. 3010–3012, 2008.

[69] Y. Kaneda, J. M. Yarborough, L. Li et al., "Continuous-wave all-solid-state 244 nm deep-ultraviolet laser source by fourth-harmonic generation of an optically pumped semiconductor laser using $CsLiB_6O_{10}$ inanexternalresonator," *Optics Letters*, vol. 33, no. 15, pp. 1705–1707, 2008.

[70] T. Leinonen, J. Puustinen, V. -M. Korpijärvi, A. Härkönen, M. Guina, and R. J. Epstein, "Generation of high power (>7 W) yellow-orange radiation by frequency doubling of GaInNAs-based semiconductor disk laser," in *Proceedings of the Conference on Lasers and Electro-Optics Europe and 12th European Quantum Electronics Conference (CLEO EUROPE/EQEC '11)*, 2011.

[71] J. Rautiainen, A. Härkönen, V.-M. Korpijärvi et al., "Red and UV generation using frequency-converted GaInNAs-based semiconductor disk laser," in *Proceedings of the Conference on Lasers and Electro-Optics and 2009 Conference on Quantum Electronics and Laser Science Conference (CLEO/QELS '09)*, 2009.

[72] J. Paajaste, R. Koskinen, J. Nikkinen, S. Suomalainen, and O. G. Okhotnikov, "Power scalable 2.5 μm (AlGaIn)(AsSb) semiconductor disk laser grown by molecular beam epitaxy," *Journal of Crystal Growth*, vol. 323, pp. 454–456, 2010.

[73] J. M. Hopkins, N. Hempler, B. Rösener et al., "5 W Mid-IR optically-pumped semiconductor disk laser," in *Proceedings of the Conference on Quantum Electronics and Laser Science*

Conference on Lasers and Electro-Optics (CLEO/QELS '08), May 2008.

[74] A. Rantamäki, A. Sirbu, A. Mereuta, E. Kapon, and O. G. Okhotnikov, "3 W of 650 nm red emission by frequency doubling of wafer-fused semiconductor disk laser," *Optics Express,* vol. 18, no. 21, pp. 21645–21650, 2010.

[75] J. Lyytikäinen, J. Rautiainen, A. Sirbu et al., "High-power 1.48 µm wafer-fused optically pumped semiconductor disk laser," vol. 23, no. 13, pp. 917–919, 2011.

[76] J. Rautiainen, J. Lyytikäinen, A. Sirbu et al., "2.6 W optically-pumped semiconductor disk laser operating at 1.57-µm using wafer fusion," *Optics Express,* vol. 16, no. 26, pp. 21881–21886, 2008.

[77] B. Lambert, Y. Toudic, Y. Rouillard et al., "High reflectivity 1.55 µm (Al)GaAsSb/AlAsSb Bragg reflector lattice matched on InP substrates," *Applied Physics Letters,* vol. 66, pp. 442–444, 1995.

[78] I. F. L. Dias, B. Nabet, A. Kohl, J. L. Benchimol, and J. C. Harmand, "Electrical and optical characteristics of n-typed-doped distributed bragg mirrors on InP," *IEEE Photonics Technology Letters,* vol. 10, no. 6, pp. 763–765, 1998.

[79] Y. Imajo, A. Kasukawa, S. Kashiwa, and H. Okamoto, "GaInAsP/InP semiconductor multilayer reflector grwon by metalorganic chemical vapor deposition and its application to surface emitting laser diode," *Japanese Journal of Applied Physics,* vol. 29, no. 7, pp. 1130–1132, 1990.

[80] J. H. Baek, I. H. Choi, B. Lee, W. S. Han, and H. K. Cho, "Precise control of 1.55 µm vertical-cavity surface-emitting laser structure with InAlGaAs/InAlAs Bragg reflectors by in situ growth monitoring," *Applied Physics Letters,* vol. 75, no. 11, pp. 1500–1502, 1999.

[81] E. Gerster, I. Ecker, S. Lorch, C. Hahn, S. Menzel, and P. Unger, "Orange-emitting frequency-doubled GaAsSb/GaAs semiconductor disk laser," *Journal of Applied Physics,* vol. 94, no. 12, pp. 7397–7401, 2003.

[82] J. Lyytikäinen, J. Rautiainen, L. Toikkanen et al., "1.3-µm optically-pumped semiconductor disk laser by wafer fusion," *Optics Express,* vol. 17, no. 11, pp. 9047–9052, 2009.

[83] T. D. Germann, A. Strittmatter, U. W. Pohl et al., "Quantum-dot semiconductor disk lasers," *Journal of Crystal Growth,* vol. 310, no. 23, pp. 5182–5186, 2008.

[84] J. Rautiainen, I. Krestnikov, M. Butkus, E. U. Rafailov, and O. G. Okhotnikov, "Optically pumped semiconductor quantum dot disk laser operating at 1180 nm," *Optics Letters,* vol. 35, no. 5, pp. 694–696, 2010.

[85] L. Fan, C. Hessenius, M. Fallahi et al., "Highly strained InGaAsGaAs multiwatt vertical-external-cavity surface-emitting laser emitting around 1170 nm," *Applied Physics Letters,* vol. 91, no. 13, Article ID 131114, 2007.

[86] K. J. Beernink, P. K. York, J. J. Coleman, R. G. Waters, J. Kim, and C. M. Wayman, "Characterization of InGaAs-GaAs strained-layer lasers with quantum wells near the critical thickness," *Applied Physics Letters,* vol. 55, no. 21, pp. 2167–2169, 1989.

[87] G. Jaschke, R. Averbeck, L. Geelhaar, and H. Riechert, "Low threshold InGaAsN/GaAs lasers beyond 1500 nm," *Journal of Crystal Growth,* vol. 278, no. 1–4, pp. 224–228, 2005.

[88] W. Walukiewicz, W. Shan, J. Wu, K. M. Yu, and J. W. Ager, "Band anticrossing and related electronic structure in III-N-V alloys," in *Dilute Nitride Semiconductors,* M. Henin, Ed., pp. 325–359, Elsevier, 2005.

[89] E. P. O'Reilly, A. Lindsay, S. Fahy, S. Tomic, and P. J. Klar, "A tight-binding based analysis of the band anti-crossing model and its application in Ga(In)NAs alloys," in *Dilute Nitride Semiconductors,* M. Henini, Ed., pp. 361–391, Elsevier, 2005.

[90] W. Walukiewicz, K. Alberi, J. Wu, W. Shan, K. M. Yu, and J. W. Ager, *Electronic Band Structure of Highly Mismatched Semiconductor Alloys,* Springer, 2008.

[91] I. Vurgaftman and J. R. Meyer, "Band parameters for nitrogen-containing semiconductors," *Journal of Applied Physics,* vol. 94, no. 6, pp. 3675–3696, 2003.

[92] D. J. Palmer, P. M. Smowton, P. Blood, J. Y. Yeh, L. J. Mawst, and N. Tansu, "Effect of nitrogen on gain and efficiency in InGaAsN quantum-well lasers," *Applied Physics Letters,* vol. 86, no. 7, Article ID 071121, pp. 1–3, 2005.

[93] W. M. McGee, R. S. Williams, M. J. Ashwin et al., "Structure, morphology, and optical properties of $Ga_xIn_{1-x}N_{0.05}As_{0.95}$ quantum wells: influence of the growth mechanism," *Physical Review B,* vol. 76, no. 8, Article ID 085309, 2007.

[94] J. Miguel-Sánchez, A. Guzmán, J. M. Ulloa, A. Hierro, and E. Muñoz, "Effect of nitrogen ions on the properties of InGaAsN quantum wells grown by plasma-assisted molecular beam epitaxy," *IEE Proceedings: Optoelectronics,* vol. 151, no. 5, pp. 305–308, 2004.

[95] A. Y. Egorov, D. Bernklau, D. Livshits, V. Ustinov, Z. I. Alferov, and H. Riechert, "High power CW operation of InGaAsN lasers at 1.3 µm," *Electronics Letters,* vol. 35, no. 19, pp. 1643–1644, 1999.

[96] O. Ambacher, "Growth and applications of group III-nitrides," *Journal of Physics D,* vol. 31, no. 20, pp. 2653–2710, 1998.

[97] H. Carrère, A. Arnoult, A. Ricard, and E. Bedel-Pereira, "RF plasma investigations for plasma-assisted MBE growth of (Ga,In)(As,N) materials," *Journal of Crystal Growth,* vol. 243, no. 2, pp. 295–301, 2002.

[98] E. M. Pavelescu, T. Hakkarainen, V. D. S. Dhaka et al., "Influence of arsenic pressure on photoluminescence and structural properties of GaInNAs/GaAs quantum wells grown by molecular beam epitaxy," *Journal of Crystal Growth,* vol. 281, no. 2–4, pp. 249–254, 2005.

[99] S. Giet, A. J. Kemp, D. Burns et al., "Comparison of thermal management techniques for semiconductor disk lasers," in *Solid State Lasers XVII: Technology and Devices,* vol. 6871 of *Proceedings of SPIE,* San Jose, Calif, USA, 2008.

[100] A. Härkönen, M. Guina, O. Okhotnikov et al., "1-W antimonide-based vertical external cavity surface emitting laser operating at 2-µm," *Optics Express,* vol. 14, no. 14, pp. 6479–6484, 2006.

[101] J. M. Hopkins, N. Hempler, B. Rösener et al., "High-power, (AlGaIn)(AsSb) semiconductor disk laser at 2.0 µm," *Optics Letters,* vol. 33, no. 2, pp. 201–203, 2008.

[102] J. H. V. Price, T. M. Monro, H. Ebendorff-Heidepriem et al., "Mid-IR supercontinuum generation from nonsilica microstruetured optical fibers," *IEEE Journal on Selected Topics in Quantum Electronics,* vol. 13, no. 3, pp. 738–749, 2007.

[103] A. Härkönen, J. Paajaste, S. Suomalainen et al., "Picosecond passively mode-locked GaSb-based semiconductor disk laser operating at 2 µm," *Optics Letters,* vol. 35, no. 24, pp. 4090–4092, 2010.

[104] A. Härkönen, C. Grebing, J. Paajaste et al., "Modelocked GaSb disk laser producing 384 fs pulses at 2 m wavelength," *Electronics Letters,* vol. 47, no. 7, pp. 454–456, 2011.

[105] L. Shterengas, G. Belenky, T. Hosoda, G. Kipshidze, and S. Suchalkin, "Continuous wave operation of diode lasers at 3.36 µm at 12°C," *Applied Physics Letters,* vol. 93, no. 1, Article ID 011103, 2008.

[106] T. Hosoda, G. Kipshidze, L. Shterengas, and G. Belenky, "Diode lasers emitting near 3.44 μm in continuous-wave regime at 300 K," *Electronics Letters*, vol. 46, no. 21, pp. 1455–1457, 2010.

[107] Y. Y. Lai, J. M. Yarborough, Y. Kaneda et al., "340-W peak power from a GaSb 2-μm optically pumped semiconductor laser (OPSL) grown mismatched on GaAs," *IEEE Photonics Technology Letters*, vol. 22, no. 16, Article ID 5512584, pp. 1253–1255, 2010.

[108] U. Keller, K. J. Weingarten, F. X. Kärtner et al., "Semiconductor saturable absorber mirrors (SESAM's) for femtosecond to nanosecond pulse generation in solid-state lasers," *IEEE Journal on Selected Topics in Quantum Electronics*, vol. 2, no. 3, pp. 435–453, 1996.

[109] R. Koskinen, S. Suomalainen, J. Paajaste et al., "Highly nonlinear GaSb-based saturable absorber mirrors," in *Nonlinear Optics and Applications III*, vol. 7354 of *Proceedings of SPIE*, 2009.

[110] J. Paajaste, S. Suomalainen, R. Koskinen, A. Härkönen, G. Steinmeyer, and M. Guina, "GaSb-based semiconductor saturable absorber mirrors for mode-locking 2 μm semiconductor disk lasers," *Physica Status Solidi (C), Special Issue: 38th International Symposium on Compound Semiconductors (ISCS 2011)*, , vol. 9, no. 2, pp. 294–297, 2012.

[111] L. Cerutti, A. Garnache, A. Ouvrard, and F. Genty, "High temperature continuous wave operation of Sb-based vertical external cavity surface emitting laser near 2.3 μm," *Journal of Crystal Growth*, vol. 268, no. 1-2, pp. 128–134, 2004.

[112] M. Rattunde, N. Schulz, C. Ritzenthaler et al., "High brightness GaSb-based optically pumped semiconductor disk lasers at 2.3 μm," in *Quantum Sensing and Nanophotonic Devices IV*, vol. 6479 of *Proceedings of SPIE*, 2007.

[113] A. Garnache, S. Hoogland, A. C. Tropper, I. Sagnes, G. Saint-Girons, and J. S. Roberts, "Sub-500-fs soliton-like pulse in a passively mode-locked broadband surface-emitting laser with 100 mW average power," *Applied Physics Letters*, vol. 80, no. 21, pp. 3892–3894, 2002.

[114] H. Lindberg, M. Sadeghi, M. Westlund et al., "Mode locking a 1550 nm semiconductor disk laser by using a GaInNAs saturable absorber," *Optics Letters*, vol. 30, no. 20, pp. 2793–2795, 2005.

[115] J. Lindfors, J. Paajaste, R. Koskinen, A. Härkönen, S. Suomalainen, and M. Guina, "Highly selective etch stop layer for GaSb substrate removal," in *Proceedings of the 16th Semiconducting and Insulating Materials Conference (SIMC-XVI '11)*, 2011.

[116] N. Yokouchi, T. Miyamoto, T. Uchida, Y. Inaba, F. Koyama, and K. Iga, "40 angstrom continuous tuning of a GaInAsP/InP vertical-cavity surface-emitting laser using an external mirror," *IEEE Photonics Technology Letters*, vol. 4, no. 7, pp. 701–703, 1992.

[117] P. Kreuter, B. Witzigmann, D. J. H. C. Maas, Y. Barbarin, T. Südmeyer, and U. Keller, "On the design of electrically pumped vertical-external-cavity surface-emitting lasers," *Applied Physics B*, vol. 91, no. 2, pp. 257–264, 2008.

[118] J. R. Orchard, D. T.D. Childs, L. C. Lin, B. J. Stevens, D. M. Williams, and R. A. Hogg, "Design rules and characterisation of electrically pumped vertical external cavity surface emitting lasers," *Japanese Journal of Applied Physics*, vol. 50, no. 4, Article ID 04DG05, 2011.

[119] W. Schwarz, "Cavity optimization of electrically pumped VECSELs," II Annual Report, Institute of Optoelectronics, Ulm University, 2006.

[120] M. Jansen, B. D. Cantos, G. P. Carey et al., "Visible laser and laser array sources for projection displays," in *Liquid Crystal Materials, Devices, and Applications XI*, vol. 6135 of *Proceedings of SPIE*, 2006.

[121] A. Mooradian, S. Antikichev, B. Cantos et al., "High power extended vertical cavity surface emitting diode lasers and arrays and their applications," in *Proceedings of the Micro-Optics Conference*, pp. 1–4, 2005.

[122] J. G. McInerney, A. Mooradian, A. Lewis et al., "High-power surface emitting semiconductor laser with extended vertical compound cavity," *Electronics Letters*, vol. 39, no. 6, pp. 523–525, 2003.

[123] A. Härkönen, A. Bachmann, S. Arafin et al., "2.34 μm electrically-pumped VECSEL with buried tunnel junction," in *Semiconductor Lasers and Laser Dynamics IV*, vol. 7720 of *Proceedings of SPIE*, 2010.

6

Spin-Controlled Vertical-Cavity Surface-Emitting Lasers

Nils C. Gerhardt and Martin R. Hofmann

Photonics and Terahertz Technology, Ruhr University Bochum, 44780 Bochum, Germany

Correspondence should be addressed to Nils C. Gerhardt, nils.gerhardt@rub.de

Academic Editor: Rainer Michalzik

We discuss the concept of spin-controlled vertical-cavity surface-emitting lasers (VCSELs) and analyze it with respect to potential room-temperature applications in spin-optoelectronic devices. Spin-optoelectronics is based on the optical selection rules as they provide a direct connection between the spin polarization of the recombining carriers and the circular polarization of the emitted photons. By means of optical excitation and numerical simulations we show that spin-controlled VCSELs promise to have superior properties to conventional devices such as threshold reduction, spin control of the emission, or even much faster dynamics. Possible concepts for room-temperature electrical spin injection without large external magnetic fields are summarized, and the progress on the field of purely electrically pumped spin-VCSELs is reviewed.

1. Introduction

Concepts for the use of the electron spin as an information carrier have become an important research field called "spintronics." The goal of spintronic research is to exploit the carrier spin degree of freedom additionally to the charge degree of freedom in order to develop novel devices, which offer new functionalities or a better performance as their conventional counterparts. Semiconductor spintronics in general includes the search for alternative device concepts as well as the investigation of the fundamental physical processes as spin injection, spin transport, spin manipulation, and spin detection. This research area was strongly stimulated by the suggestion of the so-called spin transistor by Datta and Das in 1990 [1]. Although such a spin transistor has yet to be realized even about 20 years after its suggestion, a lot of progress has been made in terms of understanding the above-mentioned fundamental physical processes. Moreover, new spintronic device concepts have been developed which might have a more realistic application perspective than the spin transistor. For example, spin-optoelectronic devices might be very promising. In such devices the direct connection between carrier spin momentum and photon spin momentum upon radiative recombination will be utilized in order to generate a spin-controlled net circular polarization degree for the light emission. While spin light-emitting diodes (spin-LEDs)

are already established tools in order to characterize and optimize electrical spin injection [2–9], spin controlled semiconductor lasers (spin-lasers) seem to be more promising for mass applications. Spin-lasers might provide properties superior to those of their conventional counterparts. For example, they promise to have faster modulation dynamics [10–14], to operate with lower threshold [15–20] and to offer a stronger polarization determination than conventional lasers with up to a 100% polarization control [17, 21–26]. However, such spin-optoelectronic device concepts are only attractive for applications if they operate at room temperature and without the need for large external magnetic fields, and if they really provide new or superior properties. Therefore, while earlier reviews have nicely discussed the physical background of spin-optoelectronic devices and the scientific achievements [24, 27, 28], we concentrate on approaches operating at room temperature without the need for superconducting magnets. In particular, we analyze the potential for new and superior performance of spin-controlled lasers.

In this article, we first discuss the fundamentals of spin-optoelectronics and then analyze the concepts for spin injection with respect to their potential for room temperature and low magnetic field operation. Then, we discuss the concepts for spin-controlled semiconductor lasers, namely, spin-controlled vertical-cavity surface-emitting lasers (VCSELs), and analyze which properties are particularly attractive for

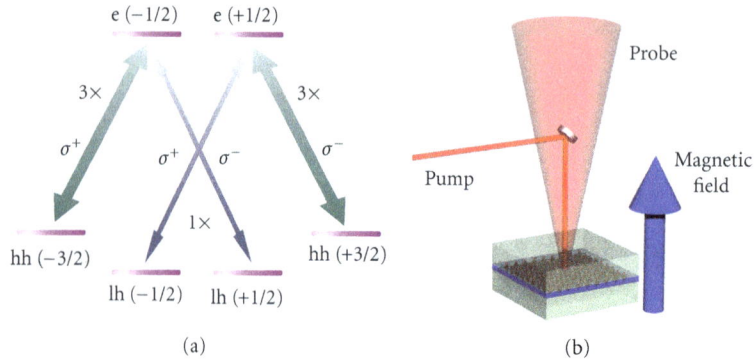

FIGURE 1: (a) Selection rules for optical transitions in direct semiconductor quantum well. The quantum numbers m_j which correspond to the z-projections of the total angular momentum of each Bloch state are printed in brackets. The transition rates and the associated circular polarization states are plotted next to the transition. (b) Vertical geometry for the selection rules in case of quantum well structure.

applications. Finally, we briefly review the state-of-the-art for electrically pumped VCSELs.

2. Fundamentals of Spin-Optoelectronics

2.1. Optical Selection Rules. Spin-optoelectronics is based on the fact that the total spin angular momentum of an electron-hole pair is directly linked to the angular momentum of a photon, which is either absorbed or emitted radiatively. This link is a consequence of the conservation of the angular momentum and is expressed by the so-called optical quantum selection rules for dipole radiation which are shown schematically for direct band-gap semiconductors in Figure 1(a) [29]. This schematic reduces the band structure of a typical direct semiconductor like GaAs in the vicinity of the Γ-point to a 6-level diagram in accordance to usual Bloch states. The Bloch states in Figure 1(a) are denoted by the quantum number m_j which corresponds to the projection of the total angular momentum $\vec{J}$ including orbital and spin momentum onto the positive z axis. The s-like conduction band is represented by two electron (e) levels with opposite spins ($+1/2$ and $-1/2$). The p-like valence band is represented by four states: two heavy hole (hh) states with $m_j = \pm 3/2$, and two light hole (lh) states with $m_j = \pm 1/2$. The split-off band is energetically separated by the spin-orbit splitting energy Δ_{SO} and therefore usually not included in the considerations. However, it should be noted that a sufficiently large spin-orbit splitting is a basic requirement for spin-optoelectronic devices otherwise optically induced spin injection as well as optical detection of carrier spin would not be achievable. In bulk GaAs, the hh and lh states are degenerated at the Γ-point. The energetic splitting of hh and lh states in Figure 1(a) appears because we consider a two-dimensional system, for example, a GaAs quantum well (QW), which is mostly used in spin-optoelectronic devices. Here different confinement energies for heavy and light hole states and possible strain contributions lead to a separation of the hole band states. The projection of the angular momentum of circularly polarized photons of the wave

vector matches $\pm 1\,\hbar$. From this it follows that optical transitions between conduction and valence band states involving circularly polarized light are allowed for $\Delta m_j = \pm 1$ only. In direct bulk semiconductors, this optical selection rule is valid for all directions, but in the case of lower dimensional active regions like quantum wells or quantum dots (QDs) the situation is a little more complex. In narrow QWs, the transitions depicted in Figure 1(a) are only valid for a vertical geometry, where the carrier spin orientation as well as the light emission is perpendicular to the quantum well plane (Figure 1(b)). Sometimes this geometry is denoted as Faraday geometry, even though a magnetic field is not obligatory in this case. The possible transitions are indicated by arrows in Figure 1(a). Due to the different geometries of the wave functions of the hh and lh states, the transitions involving hh and lh states have different probabilities. In detail, the hh transitions are three-times more probable than lh transitions.

As mentioned above, these selection rules directly link the spin polarization of the carriers and the polarization of the emitted or absorbed light. For example, we assume a spin polarization of 100% in the electron $m_j = -1/2$ state. Then, the emission consists of a part with circular right polarization (σ^+, $-1/2$ to $-3/2$) and a part with circular left polarization (σ^-, $-1/2$ to $+1/2$), whereas the right polarized part is three-times stronger than the left one. The electron spin polarization is defined as [27]:

$$P_n = \frac{n_+ - n_-}{n_+ + n_-}. \tag{1}$$

Here, $n_\pm$ are the densities of electrons in the $+1/2$ and $-1/2$ electron states, respectively. If $I(\sigma^+)$ and $I(\sigma^-)$ are the intensities for the right and left circularly polarized light fields, the circular polarization degree can by described as [27]:

$$P_{circ} = \frac{I(\sigma^+) - I(\sigma^-)}{I(\sigma^+) + I(\sigma^-)}. \tag{2}$$

In the following we assume that each hole state is sufficiently populated and thus the electron densities are the only limiting factors for optical transitions. Then the equation can be reformed to [27]:

$$P_{\text{circ}} = \frac{I(\sigma^+) - I(\sigma^-)}{I(\sigma^+) + I(\sigma^-)} = \frac{(n_+ + 3n_-) - (3n_+ + n_-)}{(n_+ + 3n_-) + (3n_+ + n_-)}$$
$$= -\frac{P_n}{2}. \tag{3}$$

From this it follows that the circular polarization degree (CPD) of the light field emitted from an electron spin-polarized QW is always $-1/2$-times the electron spin-polarization degree. In our example with $P_n = -1$ the resulting light field would be right circularly polarized with a CPD of 50%. This correlation is valid as long as we consider both lh and hh transitions. However, the energetic separation of hh and lh states even allows for higher polarization degrees: If the emission or absorption is energy selective, that is, if only the hh-electron transition takes place, a CPD of up to 1 can be obtained for an electron spin-polarization degree of -1.

The situation would change completely, if split-off band related transitions were involved. If we assume, for example, an excitation with right circular polarized light with a photon energy larger than the band-gap energy E_g plus the spin-orbit splitting energy Δ_{so}, no net spin-polarization degree would be achieved due to the additional split-off transitions. Since split-off-transitions have a transition rate of 2, they equalize the summarized transition rates for right and left circularly polarized light [27]. Correspondingly, in order to achieve a spin-polarization degree for the electrons (and holes) using optical pumping, the excitation photon energy has to be less than $E_g + \Delta_{\text{so}}$. This indicates that spin-orbit coupling is principally necessary for spin-optoelectronics. However, we will see in next chapter that the spin-orbit interaction is also the origin for the most relevant spin relaxation mechanisms which are often obstructive for the development of spin-optoelectronic devices.

These selection rules are the basis for spin-optoelectronics because they directly link optical and carrier spin polarizations. In detail, for spin-optoelectronic devices with electrical spin injection (spin-LEDs or spin-VCSELs) they enable an estimate of the spin injection efficiency from the measured light polarization degree of the optical emission. On the other hand, they can be used also to inject carrier spin polarizations optically, by means of circularly polarized light. This is an important tool in order to investigate spin-dependent effects fundamentally, if electrical spin injection is not available.

2.2. Spin Transport and Spin Relaxation. Unfortunately the carrier spin in a semiconductor is not permanently stable like the electron charge but relaxes to equilibrium within a relatively short time called spin relaxation time. This is often a fundamental challenge for the development of spin-optoelectronic devices especially in case of the spin transport. The spin relaxation is due to many reasons like the Elliot-Yafet (EY) [30, 31], the D'yakonov-Perel (DP) [32], and the Bir-Aronov-Pikus (BAP) [33] mechanisms, and others like the hyperfine interaction. These relaxation mechanisms can basically be described as a result of the interaction between the spin magnetic moments and fluctuating effective magnetic fields, originating mainly from the spin-orbit coupling. This interaction forces a group of aligned electron spins into equilibrium of both allowed spin states. Typical spin relaxation times vary over a huge range from 10 ps up to 100 ns. They are depending on various band structure and environment parameters like lattice symmetry, spin-orbit interaction, confinement, carrier, dopant, defect densities, temperature, and others. A detailed recapitulation of the different spin relaxation mechanisms and their dependencies is beyond the scope of this article. However, a nice overview can be found, for example, in [27]. Here we will concentrate on the fundamentals of spin relaxation important for the development of spin-optoelectronic devices operating at higher temperatures.

Usually, the spin relaxation time strongly decreases with temperature. In semiconductors without inversion symmetry like (100) GaAs and at low hole densities, the DP mechanism is the dominant spin relaxation mechanism at elevated temperature. Here the spin states in the conduction band for $k \neq 0$ are no longer degenerated, resulting in an effective magnetic field $B_{\text{eff}} = f(k)$. Consequently, momentum scattering processes lead to magnetic fields fluctuating in time and inducing a spin dephasing and relaxation process [34]. Typical spin relaxation times for (100) GaAs bulk vary from approximately 100 ns at low temperatures [35] to some tens of ps at room temperature [36]. The confinement energy in low-dimensional active regions is an important factor, too. In (100)-GaAs-QWs at room temperature, the spin relaxation rate exhibits a quadratic increase with increasing confinement energy [37]. Accordingly, typical relaxation times for (100)-GaAs-QWs at room temperature are in the regime of several tens of picoseconds for electron spins. Anyway low-dimensional structures can also provide longer spin relaxation times, because their higher degree of spatial confinement limits the carrier motion and possibly reduces relaxation mechanisms like the DP mechanism. This is the reason for the long spin lifetimes, predicted for quantum dots, which are consequently a very promising material system for spin-optoelectronics [24, 38]. Other possibilities in order to obtain long spin lifetimes are to make use of materials with inversion symmetry like Si or (110)-GaAs, because here the usually dominant DP mechanism is suppressed. Especially silicon, which provides electron spin relaxations times up to 7 ns at RT [38] is a very interesting material system, because of its compatibility with the highly developed complementary metal oxide semiconductor (CMOS) technology. Unfortunately, because of its indirect nature, the development of sufficient spin-optoelectronic devices based on silicon remains a fundamental challenge.

Up to now we have concentrated on the electron spin relaxation only. Hole spins usually relax much faster than electron spins. Typical values at room temperature are in the range of 100 fs [39]. Consequently it is often adequate to assume a statistic contribution of the hole spin even after optical spin injection. Nevertheless there are some concepts for hole spin injection and transport but they suffer from a rather low efficiency [3, 40].

However, even the spin relaxation time of the electrons is typically shorter than the transport time of the carriers through a spin-optoelectronic device, for example, a spin-LED. The selection rules provide a direct link between the polarization of the spin-LED emission and the spin polarization of the carriers when they recombine. However, the longer the time is between the generation of the spin-polarized carriers (by optical absorption or electrical spin injection) the more spins relax before they recombine. This relaxation takes place both on the transport path and within the active region where the carriers recombine radiatively. Consequently, the light polarization emitted by a spin-LED only provides a lower estimate of the spin injection. Presuming a sufficient number of holes for recombination, for example, in a p-doped semiconductor, the spin relaxation within the active region can be accounted using [34]

$$P_{\text{circ, eff}} = \frac{P_{\text{circ}}}{1 + (\tau/\tau_s)}. \qquad (4)$$

Here τ_s is the spin lifetime, τ the electron lifetime, and $P_{\text{circ, eff}}$ the effectively measured degree of circular polarization. This has important consequences: the impact of spin relaxation in the active medium is not determined by the spin lifetime alone, but by the electron-to-spin-lifetime ratio τ/τ_s [41] which has to be minimized. This can be accomplished either by a long spin relaxation time or by a short electron lifetime. Furthermore, to ensure a minimized spin relaxation during transport, all transport path lengths in spin-optoelectronic devices should be kept as short as possible. Typical electron spin relaxation lengths for a drift related transport in vertical geometry in n-doped GaAs at room temperature are theoretically predicted to be in the range of 25–50 nm [42]. Recently, these predictions could be verified experimentally using a series of spin-LEDs with varying spin-injection transport path length in vertical drift-based transport geometry by Soldat et al. [43]. The results demonstrate an exponential decrease of the circular polarization degree and thus of the spin-polarization degree in the active region with increasing injection path length. A spin relaxation length of 26 nm at room temperature in undoped GaAs was determined, which corresponds to the lower bound of the theoretically predicted values.

Consequently, the development of electrically pumped spin-optoelectronic devices at room temperature is a huge challenge, because in standard optoelectronic devices in particular lateral carrier transport path lengths easily reach values of several micrometers. In the following section, we briefly review the concepts for electrical spin injection into semiconductors.

2.3. Electrical Spin Injection into Semiconductors. On the long run, spin-optoelectronic devices will only be application relevant if the spin injection can be performed electrically and when the devices operate at room temperature and without the need for high external magnetic fields which would require superconducting magnets. As we will discuss below, this implies that many of the spin injection concepts reported in the literature will never be usable in practical devices. Firstly, we briefly describe the state-of-the-art and the

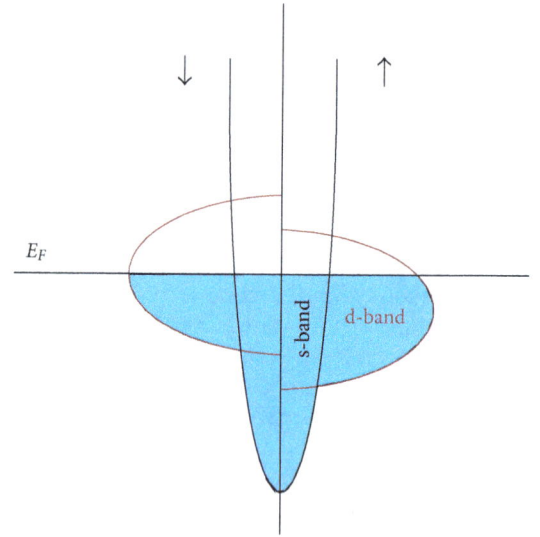

FIGURE 2: Schematics of the exchange field induced spin-splitting of the d-like density of states in a ferromagnetic metal. E_F denotes the Fermi level (Illustration analogous to [44]).

different concepts for spin injection into semiconductors. Again, we are concentrating on concepts which are relevant for the development of room temperature devices.

In general, the alignment of spins upon injection into a semiconductor implies the presences of a magnetic layer somewhere in the vicinity of the contact of the spin-optoelectronic device. The straightforward approach would be to use ferromagnetic contacts, for example, iron contacts. In a ferromagnetic metal like iron, the density of states at the Fermi level has both s- and d-character. The exchange interaction in the ferromagnet leads to a spin splitting of the d-states and therefore to a different density of states for spin-up and spin-down states at the Fermi level [44]. The density of states of a ferromagnetic metal is schematically depicted in Figure 2. Since s-electrons have significantly smaller effective masses than electrons in d-states, the current flow in the metal is dominated by the s-electrons. However, due to the splitting of the d-like density of states, spin-up and spin-down electrons have different probabilities for scattering into the d-states which results in different mobilities for spin-up and spin-down s-electrons. Consequently, the current will be dominated by s-electrons in a spin-state with less d-like density of states at the Fermi level. This can principally be used for spin injection in a nonmagnetic semiconductor [44].

But with a normal ohmic contact between the magnetic iron contact and the semiconductor, the large conductivity mismatch [45] leads to nearly negligible spin injection efficiency. Two different solutions of this problem have been suggested: the first is injection by tunnel contacts from a ferromagnetic metal layer into the semiconductor [46, 47] and the second is to use diluted magnetic semiconductors. Spin injection using dilute magnetic semiconductors is either possible with spin injection out of ferromagnetic semiconductors [3] or with spin alignment with paramagnetic semiconductor layers [2]. Though the highest spin injection efficiencies were realized with diluted magnetic semiconductors, these approaches suffer from fundamental drawbacks. Spin aligners for example, a paramagnetic BeMnZnSe layer in the n-doped region of a GaAs/Al-GaAs LED structure, require large external magnetic fields in the order of a few Tesla for spin alignment. Such high fields can only be provided by superconducting magnets. The ferromagnetic semiconductors used for spin injection so far, have Curie temperatures far below room temperature. Consequently, such concepts require cryogenic cooling. Various candidates for ferromagnetic semiconductors with Curie temperatures above room temperature have been suggested and discussed. The most promising materials are GaMnN [48, 49] and MnAs clusters in GaAs environment [50] due to their compatibility to existing optoelectronic semiconductor technology. Materials like ZnCrTe [51], Cr-doped In_2O_3 [52], $CdMnGeP_2$ [53], or ZnMnO [54] offer high Curie temperatures but are far from being compatible with established optoelectronic technology, yet. Actually, efficient spin injection from ferromagnetic semiconductors at room temperature has yet to be demonstrated and it is not clear whether this demonstration will happen at all.

Many ferromagnetic metals, in contrast, have Curie temperatures far above room temperature. But other problems appear for spin injection from ferromagnetic metals. As mentioned above, the conductivity mismatch between metal and semiconductor prevents spin injection via ohmic contacts. This problem can be solved using tunnel contacts either via Schottky contacts or via isolating tunnel barriers as, for example, MgO. In tunnel contacts the tunnel rates for the electrons are proportional to the product of the densities of states of the materials on both sides of the tunnel barriers [44]. Due to the spin splitting of the d-states in ferromagnetic metals discussed above, this enables a robust spin injection from the metal into the semiconductor circumventing the problem of conductivity mismatch.

Spin injection at room temperature has indeed been successfully realized both with Schottky barriers [4, 7] and with isolating tunnel barriers [8, 55]. While the first approaches reached only spin injection efficiencies of a few percent [4], the record value for spin injection from ferromagnetic metals into semiconductors at room temperature is 32% using MgO tunnel barriers [8]. However, as mentioned above, the optical selection rules usually require an orientation of the injected spins perpendicular to the semiconductor surface. Most ferromagnetic contacts, in contrast, have an easy magnetization axis and thus a spontaneous remanent magnetization in the film plane, that is, parallel to the surface. Accordingly, large magnetic fields in the order of 2 Tesla have to be applied in order to turn the magnetization into the required perpendicular orientation. Similar to the spin aligner concept, this induces the need for a superconducting magnet which is not attractive for device applications. This problem can be solved using ferromagnetic materials with perpendicular magnetization even without external magnetic field. Possible candidates include Fe/Tb multilayers [6, 9, 56–58] and alloys, FePt [59], and PtCo [60]. Indeed, room temperature spin injection has successfully been demonstrated in remanence and at room

temperature with Fe/Tb-contacts [9, 58] and FePt [59]. The polarization degrees of the spin-LED emission and thus the injection efficiencies were in the few percent region in all cases. One reason for this is that the magnetization orientation of the ferromagnetic contacts is not completely perpendicular to the surface. Typical values for the angle between magnetization direction and surface normal are $\sim 30°$ for FeTb alloys and $\sim 40°$ for FeTb multilayers [57, 58]. Thus, further material optimization is required to ensure perfect vertical alignment of the ferromagnetic contacts in remanence. Then, it can be expected that this efficiency can be increased up to about 30% combining the vertically magnetized ferromagnetic contacts with the optimized injector structure of Jiang et al. [8]. Thus, from the actual point of view, the optimum room-temperature spin-LED would look like it is schematically shown in Figure 3.

However, even this optimized spin-LED will probably not become relevant for applications other than the characterization and optimization of spin injection contacts. This is because LEDs are generally too slow for information technology. Moreover, the injection efficiencies that could be obtained even in the best case are still rather low. In contrast, spin-controlled lasers might offer a much higher application potential. In the following section, we will analyze this potential.

3. Spin-Controlled Vertical-Cavity Surface-Emitting Lasers

The first step on the way to a spin-polarized laser is to identify a qualified laser concept. Principally a choice has to be made, whether an edge- or a vertical-emitting geometry suites best. Figure 4 compares schematically the structures of a vertical-cavity surface-emitting laser (VCSEL) and of a conventional edge-emitting laser. On the first glance, an edge-emitting laser might be more promising. It is obvious from the comparison of both laser structures with the spin-LED in Figure 3 that it would be rather easy to transfer the injector concept of a spin-LED to an edge-emitter while electrical spin injection into the VCSEL is much more complicated due to the larger vertical path length. In addition, an edge-emitting concept would allow for an easy remanent spin injection utilizing the natural in-plane magnetization of the ferromagnetic layers due to the shape anisotropy. But the requirement for the Faraday geometry in spin-optoelectronics implies that lasers with vertical architecture such as VCSELs are the first candidates for spin-controlled lasers. At least for narrow quantum wells the angular momentum of the holes lies in the quantum well plane, and the connection between carrier and photon spin is only straightforward for the vertical geometry as discussed in Section 2.1. In principle, an edge-emitting concept utilizing a bulk active region might still be a possibility. Recent experimental results have demonstrated spin injection in an edge-emitting LED utilizing bulk-like wide GaAs QWs [61]. In a wide QW, the heavy-hole angular momentum can principally lie in the QW plane comparable to the case in bulk semiconductors. However, due to the restriction on bulk-like active systems, an edge-emitting concept for room temperature operation would be

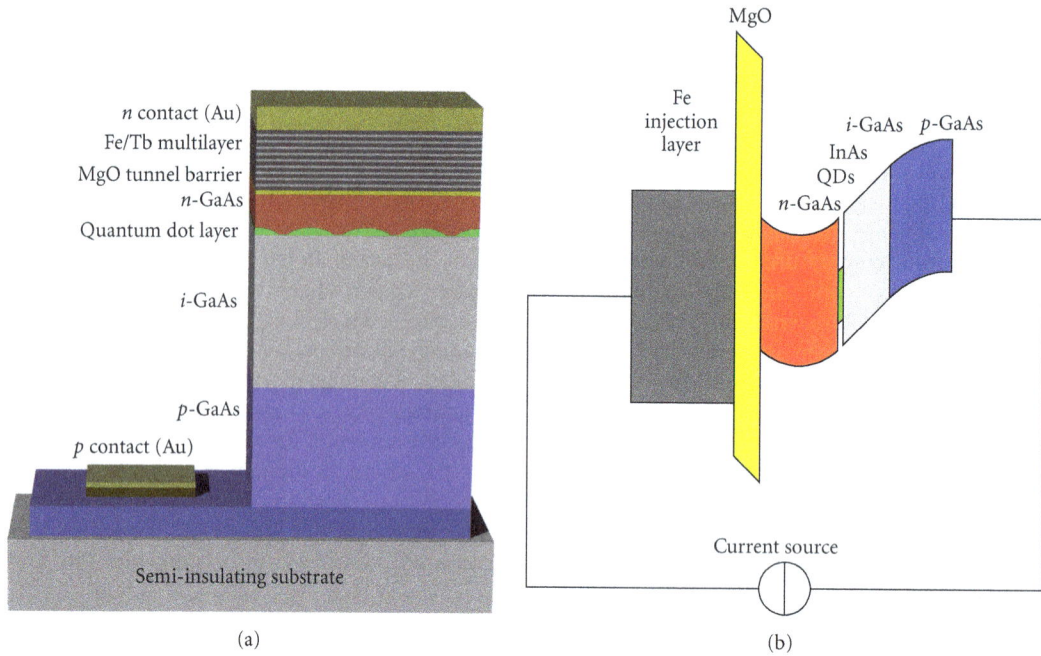

FIGURE 3: Optimized structure for a room temperature spin-LED. The design combines a Fe/Tb-multilayer contact structure with spontaneous perpendicular magnetization, a highly efficient MgO tunnel barrier for spin injection, a minimized electron path length, and InAs QDs with enhanced spin lifetime at room temperature (a), (b) depicts the schematic energy diagram of the spin-LED in growth direction. Spin-polarized electrons will be injected from the lowest Fe layer of the Fe/Tb multilayer contact into n-GaAs via the MgO tunnel barrier and recombine with unpolarized holes in the QDs.

a tough challenge and it remains questionable whether such a concept can be competitive to conventional laser devices. Additionally, waveguide effects have a relevant impact on the polarization state of an edge-emitting laser, usually leading to a linearly polarized laser emission parallel or perpendicular to the waveguide plane. With respect to this, VCSELs have a big advantage, because generally a VCSEL is a laterally isotropic device with nearly perfect circular symmetry, which leads to weak pinning of the polarization state.

Altogether, because of the above-mentioned fundamental disadvantages of edge-emitting concepts a VCSEL seems to be the most qualified concept for a spin-polarized laser at room temperature. Thus, the challenge of a more complicated spin injection concept has to be faced.

As a consequence of the long vertical spin transport path, room temperature spin-VCSELs with electrical spin injection are not available yet. Thus, at this stage, the potential of spin-VCSELs at room temperature has to be analyzed theoretically or with alternative experimental techniques. Since the selection rules shown in Figure 1(a) enable controlled optical spin excitation, most experimental work on this field has been done on spin-VCSELs with optical spin injection. In the following section, we review theoretical and experimental work on room-temperature spin-VCSELs in order to work out the specific advantages spin-VCSELs might deliver for applications.

3.1. Basics and Properties of Spin-VCSELs. The idea that spin-controlled lasers might offer a much higher potential than spin-LEDs arises from the fact that lasers show a dynamical

behavior much different from that of LEDs. One important example affects the influence of the spin relaxation in the active region. In spin-LEDs the effective circular polarization degree will be reduced. This reduction is usually accounted for by the factor $1 + (\tau/\tau_s)$ [34] including the electron-to-spin-lifetime ratio (see Section 2.2). At room temperature the ratio is typically very large due to a small spin lifetime, which leads to a small circular polarization degree. In spin lasers, in contrast, the electron lifetime will be reduced significantly due to the strong stimulated emission, resulting in a vanishing electron lifetime to spin lifetime ratio and a correction factor of approximately 1. Accordingly the spin relaxation rate in the active medium should be less important in spin-VCSELs in comparison to spin-LEDs. This is a first fundamental advantage. Nevertheless the spin relaxation during carrier transport remains still an issue.

The dynamic behavior of spin-polarized laser can be investigated theoretically using a spin-dependent rate-equation model. Several different models have been used in the literature for this purpose [12–15, 19, 24, 62, 63]. In the following a common dynamic spin-flip model (SFM) will be used, originally developed by San Miguel et al. in order to describe the polarization switching and bistability in conventional VCSEL devices [64–66]. The SFM is based on a four-energy-level approximation and takes only transitions between electron and heavy-hole states into account. It describes the polarization dynamics for the two hh-related circularly polarized transitions, considering two distinguished carrier densities for spin-up and spin-down carriers [65]. These two carrier reservoirs are coupled by the spin

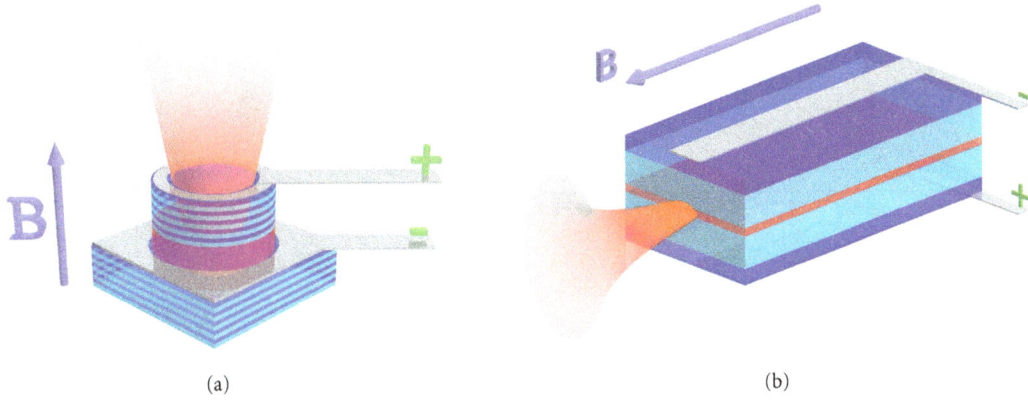

Figure 4: Comparison between spin-VCSEL (a) and spin-edge emitter (b).

relaxation rate γ_s. The spin relaxation rate γ_s describes all kinds of microscopic spin relaxation processes mentioned in Section 2.2 by means of a single phenomenological parameter [66]. The circularly polarized light fields $E_\pm$ are coupled by the cavity anisotropies birefringence (γ_p) and dichroism (γ_a). The four coupled rate equations are as follows:

$$\dot{E}_\pm = \kappa(1 + i\alpha)(N \pm m_z - 1)E_\pm - \left(\gamma_a + i\gamma_p\right)E_\mp$$
$$+ \xi_\pm\sqrt{\beta\gamma(N \pm m_z)},$$
$$\dot{N} = \gamma[\eta_+ + \eta_- - (1 + I_+ + I_-)N - (I_+ - I_-)m_z],$$
$$\dot{m}_z = \gamma(\eta_+ - \eta_-) - [\gamma_s + \gamma(I_+ + I_-)]m_z - \gamma(I_+ - I_-)N.$$
$$(5)$$

In the SFM the light field is coupled to two population inversion variables [66]. The first variable N is the total carrier population, that is, the sum of the populations of spin-up and spin-down states, normalized to the population at laser threshold. Its decay rate is γ. The second variable m_z is the so-called carrier spin magnetization which represents the normalized value of the population difference between spin-up and spin-down states [63]. The intensity $I_\pm$ of the circularly polarized optical laser modes can be described by the complex amplitudes $E_\pm$ of the circularly polarized light fields via $I_\pm = |E_\pm|^2$. κ is the cavity decay rate which can be related to the photon lifetime using $1/2\kappa$ [66]. α is the linewidth enhancement factor. The influence of the spontaneous emission to the laser mode is considered using the spontaneous emission factor β and the spontaneous emission noise terms $\xi_\pm$ which are usually described by complex Gaussian shaped distributions. Optical as well as electrical pumping can be modelled using the pump terms $\eta_\pm$. The optical gain is implemented in the model by using a simple linear dependence of the population inversion. Since the optical gain for the circularly polarized light intensities $I_\pm$ is proportional to $(N \pm m_z - 1)$, the gain values for I_+ and I_- are unequal in case of a carrier spin polarization. This is one of the fundamental concepts of the spin-polarized VCSEL and will be described later in detail. Gain compression, frequency dependencies or temperature effects are not included in this model. However, since a VCSEL operates spectrally single-mode with a usually

small detuning between the cavity mode and the gain mode, the reduction of heavy-hole transitions and neglecting the frequency dependencies are sufficient to describe the main features of conventional and spin-polarized VCSELs. We will use this model in the following in order to discuss the advantages and properties of spin-VCSELs in comparison to spin-LEDs and conventional lasers.

One important difference between a spin-laser and a spin-LED is the nonlinearity of a laser at the laser threshold which enables a kind of amplification of spin information with a spin-controlled carrier injection. Figure 5 shows schematically the gain spectra of a laser with conventional unpolarized pumping in comparison to the case with spin-polarized pumping. Note that only the gain differences at the cavity energy E_{cavity} are relevant for the dynamics so that the spectral dependence of the gain is not considered in our simple model.

The anisotropic pumping leads to a small spin polarization of the carriers in the active region. We assume here a small excess in the occupation of the $e(-1/2)$ state (compare Figure 1) with respect to the $e(+1/2)$ state. This excess is in the regime of a few percent only which is, in principle, accessible by electrical spin injection. This small excess occupation of one spin band leads to a higher inversion in the $e(-1/2)$ state and, accordingly, the corresponding σ^+-transition from $e(-1/2)$ to the $hh(-3/2)$ state sees a higher gain than the σ^--transition from $e(+1/2)$ to $hh(+3/2)$. The transitions to the lh states do not play a role here because they are usually not in resonance with the VCSEL cavity mode and have a lower transition strength. In other words, the spin anisotropic pumping leads directly to a gain anisotropy at the photon energy of the cavity resonance E_{cavity} for the circularly polarized laser modes. We further consider a polarization independent loss level indicated by the horizontal line in Figure 5. In the situation shown in Figure 5(b), the laser is just above threshold for σ^+ emission and still below threshold for σ^- emission, which results in a nearly 100% right circularly polarized laser emission. This behavior has been investigated experimentally by Ando et al. for the first time [21]. They demonstrated that the laser polarization in optically pumped VCSEL structures at room temperature can be

(a) Unpolarized inversion

(b) Spin-polarized inversion

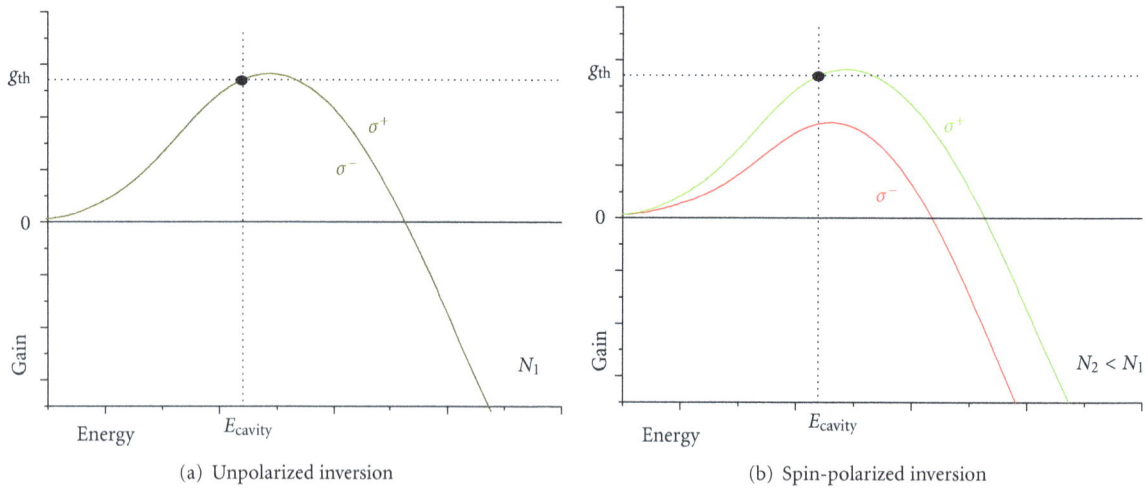

FIGURE 5: Schematical illustration of the optical gain as a function of photon energy in a VCSEL with conventional unpolarized pumping (a) and for a small spin polarization of the carriers in the active region (b). The spin-polarized pumping leads to separation of the gain spectra for the σ^+ and σ^- transition. This results in a gain anisotropy at the cavity resonance energy E_{cavity}, which marks the relevant photon energy for the laser emission.

controlled via optical spin injection. In an ideal case, even a small excess in the spin polarizations of the electrons in the active region should be efficient in order to generate a 100% σ^+ polarization of the optical emission. Accordingly, it can be expected that a spin-VCSEL, in contrast to a spin-LED, provides a kind of spin amplification, that is, the polarization degree of the optical emission becomes higher than the spin-polarization degree in the active region in the vicinity of the laser threshold. This expectation was confirmed by Hövel et al. [22, 23]. They showed that in an optically pumped spin-VCSEL the circular polarization degree can indeed become higher than the input polarization degree.

In these experiments, a Ti : sapphire laser was taken for optical pumping an InGaAs/GaAs-QW-VCSEL-structure at room temperature. The spin injection was obtained by pulsed excitation with a pulse length of 80 fs. Figure 6(a) shows the measured circular polarization degree of the VCSEL emission as a function of the injected spin polarization of the carriers in the active region. To circumvent the stopband of the VCSEL Bragg mirrors, the excitation was performed with some excess energy leading to excitation of heavy hole- and light hole-transitions in the GaAs barriers. According to the selection rules in Figure 1(a) the maximum spin-polarization degree of the excited electrons is 50%. Accordingly, a circular polarization of 100% leads to an injected carrier spin polarization of 50% at most, if we neglect any kind of spin relaxation in the barriers and in the quantum wells. The results in Figure 6 confirm the expectation of spin amplification with a spin-VCSEL. A circular polarization degree of 100% is already obtained with a spin polarization of 30%, and a spin-polarization of 13% still provides a polarization degree of 50%. The results are in a good agreement with theoretical calculations based on the SFM mentioned above. The simulations also shown in Figure 6(a) were obtained for a spin relaxation time of 40 ps

in the active region. Later experimental and numerical work confirmed that spin amplification also works for continuous wave excitation [23]. Nevertheless it has to be noted that both effects, the possibility to control the light polarization by the carrier spin and the amplification of spin information are restricted to a pump region near the laser threshold. In a simple steady-state picture this can easily be understood taking the clamping of the carrier density at the laser threshold into account. If we consider the situation in Figure 5(b), the laser emission with a 100% σ^+ polarization state leads to a clamping of the carrier density in the $e(-1/2)$ electron spin band but not in the $e(+1/2)$ electron spin band. Accordingly, if the carrier density will be increased and the spin-polarization degree is less than 100%, the gain anisotropy will be reduced and the σ^- gain spectrum reaches threshold for a high carrier density, too. Thus the σ^- polarization starts to emit and diminishes the spin control and amplification effects.

Accordingly, a spin-VCSEL has two threshold carrier densities N_{th1} and N_{th2} for the two circularly polarized laser modes. Their difference depends on the spin-polarization degree of the carriers in the active medium. If the device operates in a carrier density regime N with $N_{th2} < N < N_{th1}$, predominantly the electrons with the correct spin-polarization participate in the laser process. From this it follows that less injected carriers are sufficient to reach threshold in a spin-VCSEL in comparison to a conventional VCSEL without spin-control. This spin polarization induced threshold reduction can be seen from the data depicted in Figure 6(b). The laser threshold is significantly reduced for a carrier spin polarization of 50% in comparison to unpolarized pumping. Threshold reduction in spin-VCSELs has motivated a lot of work in the field of spin-lasers, because the direct pumping of only one spin state basically allows reducing the threshold by up to 50% [15]. However this implies 100% spin-polarization degree in the active medium and a sufficiently long spin lifetime, taking simple rate equations into

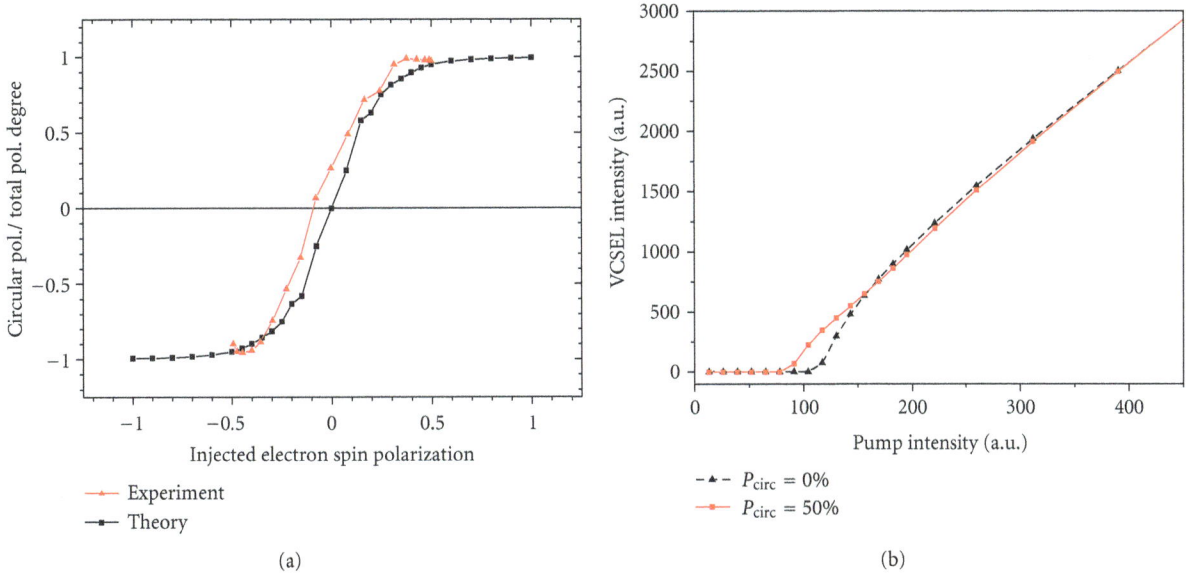

FIGURE 6: Circular polarization degree as a function of the carrier spin polarization in the active region (a). The experiments have been obtained at room temperature for pulsed excitations and show a good agreement with theoretical results based on the SFM for a spin relaxations time of 40 ps. (b) shows the calculated laser characteristics for a carrier spin polarization of 0% and 50%.

account. This threshold reduction was indeed observed. At low temperatures, Rudolph et al. have reported a threshold reduction of 23% in optically pumped devices [15], while at room temperature still a reduction of 2.5% was observable [16]. Recent theoretical work about threshold reduction even predicts threshold reductions significantly higher than 50% depending on several device parameters like nonradiative recombination [18], hole spin relaxation [67], and valence-band mixing [20] Anyway, the best result for electrically pumped devices so far is a threshold reduction of 14% at 200 K [68]. However, even though threshold reduction is an important property of spin-VCSEL, its efficiency depends strongly on the spin lifetime and it is still questionable whether in terms of threshold reduction a spin-VCSEL will be ever competitive with optimized conventional devices at room temperature. This spin-induced threshold reduction will probably be overcompensated by the additional effort needed for spin injection by using ferromagnetic contacts, for example.

A closer look onto the experimental details of the work done on purely optically pumped spin-VCSELs indicates problems that might occur in electrically pumped devices, too. In order to obtain the effects like spin amplification and threshold reduction great care had to be taken to ensure a perfectly circularly symmetric pump spot. Otherwise additional anisotropic carrier density and temperature effects introduce parasitic anisotropies into the cavity, which lead to a coupling of both circular polarized laser modes and result in linearly polarized laser emission. Such cavity anisotropies like birefringence and dichroism are known to have an enormous impact on the polarization dynamics of electrically pumped VCSEL [69–71]. The anisotropies, caused for example, by the internal electric fields and anisotropic strain induce a pinning of the polarization mode to a certain

linearly polarized state and thus are the origin of chaotic polarization behavior in electrically pumped VCSEL devices [64, 70–74]. Accordingly, for the development of electrically pumped spin-VCSELs the role of cavity anisotropies on the laser dynamics is an important issue and might in the worst case be stronger than any spin-induced effects. Hövel et al. have investigated whether the influence of cavity anisotropies can be overcompensated by the spin in electrically pumped devices [25]. The experiments were performed using a special hybrid pumping scheme for a conventional electrically pumped VCSEL structure. The VCSEL was pumped electrically with a continuous spin unpolarized current in the vicinity of the electrical threshold. Additionally, a short circularly polarized light pulse with 3 ps pulse length was used to inject a small amount of spin-polarized carriers in the active region. The experimental setup shown in Figure 7 was used for these experiments. Just, instead of the streak camera shown in Figure 7, a time integrating photodetector was used to analyze the time averaged output of the VCSEL. The results indeed confirmed that spin control and spin amplification are also feasible in electrically pumped VCSEL devices. In other words, the spin effects are strong enough to overcompensate the above-mentioned cavity anisotropies at least for time-integrated measurements [25].

So far, we have only discussed time averaged stationary effects. But, besides spin amplification, spin control, and threshold reduction in continuously operating spin-VCSELs, dynamical effects might be even more promising. For example, it was predicted that spin-VCSELs might be considerably faster than their conventional counterparts. First fundamental investigations by Hallstein et al. confirmed this potential [10]. They investigated an optically pumped VCSEL structure at low temperature and in a high magnetic field and observed a fast modulation of the VCSEL polarization

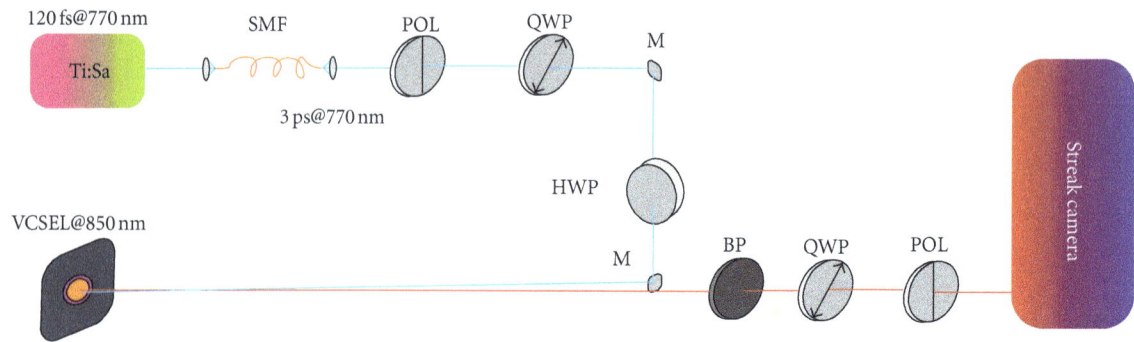

FIGURE 7: Setup for hybride excitation and spin control of a commercial VCSEL device. (SMF) single mode fiber, (POL) linear polarizer, (QWP) quarter wave plate, (M) mirror, (HWP) half wave plate, and (BP) bandpass.

due to spin precession of the carriers in the magnetic field with modulation frequencies up to 120 GHz [10, 11]. But also under realistic device conditions, strong indications for improved dynamical performance of spin-VCSELs have been found. Li et al. investigated an electrically pumped commercial VCSEL device with additional optical spin injection at room temperature [14]. Now, the entire experimental arrangement shown in Figure 7 was used for these experiments. Instead of the time integrated detection used by Hövel et al. [25], a time and polarization resolved analysis of the VCSEL emission was performed with a streak camera synchronized to the exciting Ti : sapphire laser. The time-resolution of the setup was approximately 3 ps. Again, the above-mentioned hybrid excitation scheme for the electrically pumped VCSEL was used and the VCSEL was operated in the vicinity of the threshold.

The results for a pulsed spin injection into the $e(-1/2)$ electron spin band are presented in Figure 8. While the intensity dynamics exhibits a typical short pulse response, followed by some relaxation oscillations, the dynamics of the circular polarization degree perform a very fast oscillation within the first VCSEL pulse. The corresponding oscillation frequency is in the range of 10 GHz and thus much faster than the relaxation oscillation in the device for the same pump conditions. Accordingly, even in electrically pumped devices at room temperature, the combination of the spin dynamics with the photon dynamics in a laser cavity obviously leads to an improved speed of spin-VCSEL devices in comparison to conventional devices. The results are in a good agreement with simulations utilizing the rate equation spin-flip-model discussed before (Figures 8(c) and 8(d)) [14, 75]. A detailed analysis based on the SFM additionally revealed that the observed dynamics are a consequence of an interplay between the spin dynamics of the carriers and the birefringence in the laser cavity which can be described as follows: in case of a zero spin polarization as in a conventional VCSEL, only one linearly polarized mode is lasing. In case of spin injection, we have an imbalance of the spin band populations, which results in laser emission with a nonzero circular polarization degree. This corresponds to a simultaneous emission of two orthogonal linearly polarized laser modes. Due to the birefringence in the cavity, their frequencies are different [70]. The resulting beating of both

modes leads to an oscillation of the circular polarization degree [76]. The polarization oscillations feed back into the carrier spin dynamics and can stabilize the dynamics which potentially results in a long oscillation lifetime. The damping of polarization oscillations depends on the effective dichroism in the cavity and the oscillations are sustained the longer, the smaller the dichroism is. Since the effective dichroism and thus the damping of the oscillations can be controlled by the current, this allows for both very long and very short oscillation lifetimes. This is potentially interesting to stabilize spin information, as well as for the generation of short polarization bursts which are interesting for information transmission. This concept has been verified experimentally very recently by Gerhardt et al. [76]. They demonstrated polarization oscillations with a frequency of 11.6 GHz for a device with a modulation bandwidth of only less than 4 GHz at room temperature. The oscillation lifetimes could be controlled by the current in the vicinity of a polarization switching point significantly above threshold, where the effective dichroism is minimized. It should be added that while the damping of the oscillations is current dependent the oscillations themselves are not restricted to any current region. In comparison to the other spin effects in VCSELs this is an important advantage for applications and will be discussed later. At the polarization switching point, oscillation lifetimes of at least 5 ns could be demonstrated which is 200-times longer than the estimated spin lifetime in the device [76]. The oscillation frequency is determined by the linear birefringence and small corrections due to nonlinear spin effects only and is principally independent of the carrier dynamics. Hence, by tuning the birefringence, for example, by applying additional strain, the oscillation frequency can possibly be enhanced significantly. It is not restricted by conventional relaxation oscillations. Since strain-induced birefringence values of 80 GHz have already been reported in the literature [77] this concept has a strong potential for future ultrahigh bandwidth lasers in the 100 GHz region [75, 76].

The ability for enhance speed and modulation bandwidth is a very promising advantage of spin-VCSELs, which is attractive for a lot of applications, for example, high-speed optical data communication technology. Accordingly, a lot of work has been concentrated on this issue, and

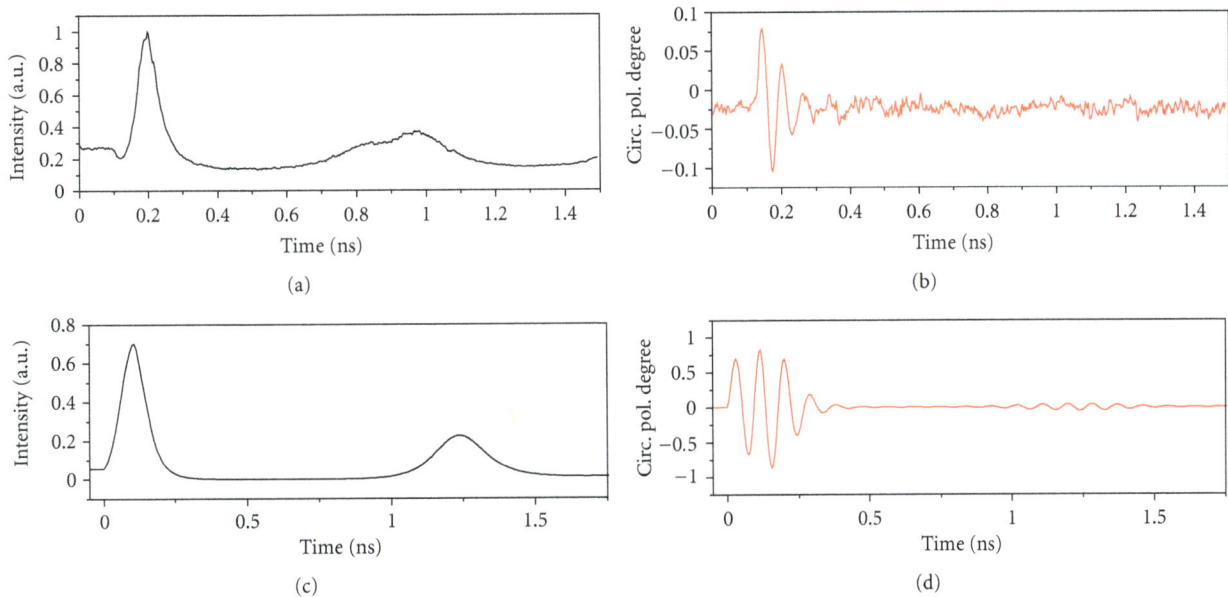

FIGURE 8: Dynamics of the total intensity (a) and the circular polarization degree (b) of an electrically pumped VCSEL after additional spin injection using a short right circularly polarized light pulse. Theoretical calculations based on the SFM are depicted ((c), (d)).

several other concepts have been presented, recently. Lee et al. investigated the small-signal modulation properties of both circularly polarized laser modes in a spin-polarized VCSEL, theoretically [13]. They predict that the modulation bandwidth can be enhanced for the favored circularly polarized laser mode as compared to conventional devices, due to spin injection. This concept is directly correlated to the threshold reduction for this laser mode and accordingly restricted to this current region close to threshold. In another recent publication Saha et al. have studied both the small-signal and the large signal modulation properties using a rate equation model [12]. The results for small-signal modulation support the reported prediction of Lee et al. and emphasize that the laser dynamics can be improved, when only one circularly polarized mode is operating. Additionally, they demonstrate experimentally using an electrically pumped InAs-QD-based spin-VCSEL that the amplification of spin information and the polarization control can significantly be enhanced for pulsed operation (see Section 3.3).

However, even though a lot of progress has been made in emphasizing the advantages of spin-polarized laser devices in comparison to conventional ones and a lot of promising properties have already been identified, a device with superior performance in comparison with conventional devices is still missing. One important challenge which has to be solved is the realization of efficient electrically pumped spin-VCSELs at room temperature. The current state of technology for electrically pumped VCSELs will be discussed in the next section. However, another important problem is that most concepts for spin-VCSELs are restricted to an operation near the laser threshold. This can be an important drawback on the way to realistic applications, because here the efficiency and the power of a laser are low and the laser dynamics is inherently slow. Accordingly, the search for other

spin-dependent properties which are not restricted to this current region is one of the most relevant tasks in the near future. Anyway, the usable current region can be significantly enhanced by improving the spin polarization degree. Here the development of spin-VCSELs using (110) GaAs quantum wells might deliver an important progress due to the long spin relaxations time in this material. Recently the first optically pumped spin-VCSEL grown on (110) GaAs substrate and operating at room temperature could be demonstrated by Iba et al. [26]. They reported circularly polarized lasing with a circular polarization degree of 0.96 at room temperature using pulsed excitation. The spin lifetime in the active region was estimated to be 0.7 ns which is significantly larger than in conventional (100) GaAs QWs. First transient investigations using a (110) GaAs-based VCSEL with an (110) InGaAs active region operating at 77 K additionally demonstrate optically induced switching of the circular polarization mode in the GHz range [78]. Even though these results are very promising, the realization of exceptional (110) based electrically pumped devices is still a big challenge. This is because "standard" all electrically pumped spin-VCSELs on conventional (100) oriented substrates already induce severe technological difficulties which have not yet been solved completely. We will review this work in the next section.

3.2. Electrically Pumped Spin-VCSELs. All electrically pumped spin-polarized VCSEL devices published so far have been developed at the University of Michigan, USA. The first device was presented in 2005 by Holub et al. [40]. Here a GaMnAs spin aligner, located intracavity underneath the top mirror was used for electrical spin injection. Because GaMnAs is an intrinsically p-doped material, the spin-VCSEL concept based on hole spin injection, accepting the associated ultrafast spin relaxation. Five $In_{0.2}Ga_{0.8}As$

quantum wells served as active gain media. Polarization measurements as a function of the magnetic field revealed a maximum circular polarization degree 4.6% at 80 K. The concept to implement the spin-injection contact directly into the cavity of the VCSEL, which displays a direct transfer of a spin-LED concept to spin-VCSELs, has the advantage that the lateral transport length equals zero whereas the vertical transport length could be minimized to ~0.25 μm [40]. Unfortunately, the intracavity spin-aligner layer has a critical impact on the cavity quality and induces additional magnetic circular dichroism (MCD). The latter has a potentially strong influence on the polarization state of the emission, depending on the number of reflection within the laser cavity [24]. Accordingly, due to the impact of the MCD and the utilized hole spin injection, the results have been critically discussed in the literature [79, 80]. However, a small amount of circular polarization degree of approximately 1% has finally been stated to be due to the injection of spin polarized holes [24, 80], thus demonstrating the first realization of purely electrically pumped spin-polarized VCSEL.

Taking the difficulties due to hole spin injection and MCD in the cavity into account, the next concept for an electrically pumped spin-VCSEL consequently based on electron spin injection contacts. This concept was realized by Holub et al. in 2007 [17] using a Fe/$Al_{0.1}Ga_{0.9}$As Schottky tunnel barrier for electron spin injection in combination with an n-doped intracavity contact. The VCSEL design is depicted in the inset of Figure 9(b). Again, an active region containing five compressively strained $In_{0.2}Ga_{0.8}$As QWs was used for the laser operation. One important advantage of this concept is the utilization of a magnetic layer free cavity, reducing the impact of MCD to a minimum. However this led to a significantly increased average electron spin transport length of ~4.5 μm for a circular post VCSEL with a mesa diameter of 15 μm. Nevertheless laser operation with a maximum CPD of 23% at 50 K and for an external magnetic field of 2.2 T could be realized with this device (Figure 9(a)). Additionally, threshold reduction in electrically pumped spin-VCSELs could be demonstrated for the first time, showing a maximum value of 11% at 50 K (Figure 9(b)) [17, 81]. Using this value, a cavity spin polarization degree of 16.8% was estimated for the barrier layers using a spin-dependent rate equation model comparable to the model used by Rudolph et al. [15, 16]. This value is significantly higher than the carrier spin polarization in the InGaAs-QWs estimated to be 7%. However, the circular polarization degree of the laser is stated to be dominantly determined by the cavity spin degree, while the spin polarization in the active region is less important due to the short carrier lifetime. Comparing the 16.8% cavity spin polarization with the maximum value of 23% for the CPD, the results nevertheless demonstrate the predicted amplification of the spin information.

In order to optimize the spin-VCSEL concepts for an operation at elevated temperatures, the next concept published by Basu et al. in 2008 based on InAs/GaAs quantum dots as active gain medium [68]. Electron spin injection was obtained using a MnAs/$Al_{0.1}Ga_{0.9}$As Schottky tunnel contact in a comparable VCSEL design comparable to that

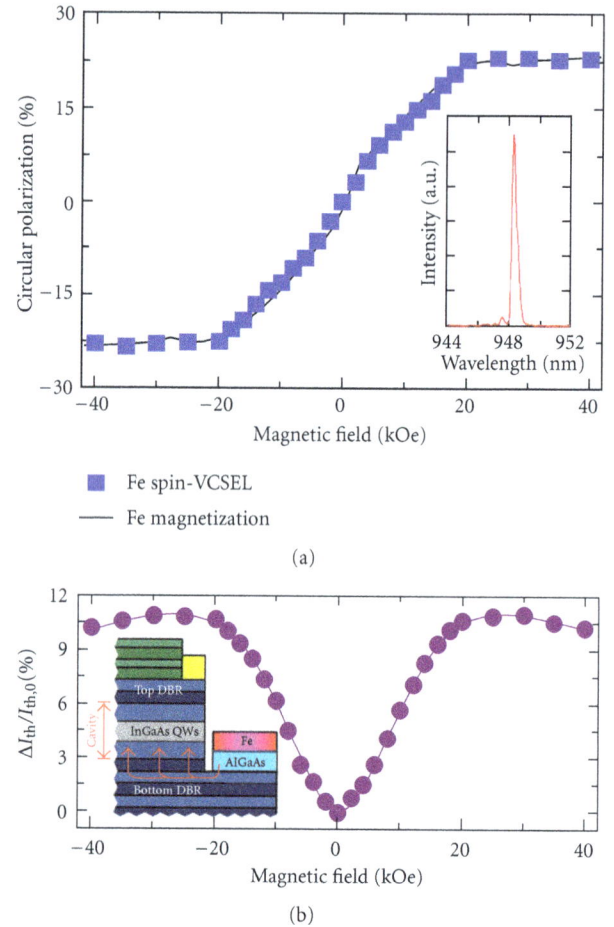

(a)

(b)

FIGURE 9: Circular polarization degree (a) and threshold current reduction (b) as a function of the magnetic field in an electrically pumped spin-VCSEL at 50 K. The VCSEL design is displayed in the inset of (b). The spin-VCSEL based on electron spin injection utilizing a Fe/AlGaAs Schottky tunnel contact design. (from [17]).

displayed in the inset of Figure 9(b)). The active region contains 10 layers of self-organized QDs grown by molecular beam epitaxy (MBE) with a QD density of 3×10^{10} cm^{-2}. The utilization of InAs-QDs represented an important progress, allowing to increase the operation temperature of the laser up to 200 K. Here a CPD of 8% and a maximum threshold reduction of 14% could be demonstrated for external magnetic fields of approximately 2 T, using a device with 15 μm mesa diameter. A following publication based on comparable devices showed values of ~15% CPD and ~8% threshold reduction at 200 K and demonstrates the electrically controlled modulation of the output polarization in the vicinity of the threshold for the first time [19]. The issues of polarization control and high-frequency dynamics in such devices have been addressed recently in 2010 by Saha et al. [12]. They investigated the transient characteristics of an electrically pumped spin-VCSEL structure, comparable to the devices in [19, 68] but with mesa diameters down to 10 μm. The operation temperature could be increased to 230 K and an average CPD of up to 55% could be realized for pulsed bias conditions with a pulse length of 3 ns. These results were obtained for an estimated spin polarization in the active region of ∼6%, demonstrating a significant amplification process for the spin information, due to the stimulated emission in spin-VCSELs.

Though impressive progress has been made concerning the development of electrically pumped spin-VCSELs, a device operating at room temperature is still missing. However, the realization of this goal is absolutely necessary to develop spin-VCSELs in order to benefit from the potential for superior performance in comparison to conventional devices in the near future. Thus, additional intensive research effort is required in this field in the next years. But the already demonstrated results are encouraging and raise the hope that this goal can be reached soon.

4. Conclusions and Outlook

In this paper we review the state-of-the-art of spin-controlled VCSELS with a particular focus on the most promising concepts for real devices. After discussing the fundamentals of spin-optoelectronics we discuss the concepts for electrical spin injection into semiconductor light-emitting diodes (LEDs). However, spin-controlled lasers are generally more attractive for applications than spin-controlled LEDs. Lasers have much faster dynamics than LEDs and the nonlinearity of the laser at threshold potentially enables a strong amplification of spin-dependent effects. Among the different concepts for semiconductor lasers, VCSELS are most attractive for spin-control because of their vertical device architecture and circular symmetry. Fundamental studies confirm that spin-VCSELs promise to have lower thresholds than their conventional counterparts and that they enable spin-control of the output polarization. But their most promising advantage is that they might be much faster than their conventional counterparts. For practical applications, the interaction of the spin effects with cavity anisotropies like birefringence and dichroism might enable enormously high modulation frequencies. Recently, Li et al. and Gerhardt et al. [14, 76] have reported spin-induced oscillations much faster than the relaxation oscillation frequency which, in conventional devices, roughly determines the upper modulation frequency limit. Since the damping of the spin induced oscillations and thus their lifetime can be tuned by the current, this concept is interesting for many applications like high-bandwidth data communication or spin information storage. However, in order to realize a device with superior performance, this concept has to be further analyzed in detail. In particular, a clever cavity design with careful engineering of the birefringence might open the door to modulation frequencies significantly above 100 GHz. While most spin-induced effects work only in the vicinity of threshold, such polarization oscillations can be utilized at higher pump levels. This is potentially an important breakthrough, because the usual restriction to an operation near the laser threshold is a major drawback for applications.

The greatest challenge in the field of spin-VCSELs still remains the realization of room temperature operation with pure electrical spin injection. Further massive effort will have to be invested into the engineering of appropriate spin injectors, injection paths in the semiconductor, and of materials with weak spin relaxation such as (110) GaAs, for example. Additionally a successful integration of efficient spin injection contacts with perpendicular magnetization, providing low switching fields would be an important issue towards realistic applications. However, there has been considerable progress in this area of electrically pumped VCSELs in the past few years. Together with the above-mentioned high potential of spin-VCSELs it can thus be expected that spin-controlled VCSELs remain a scientifically stimulating research area with growing application potential within the next decade.

Acknowledgments

The authors thank the German science foundation for financial support within the SFB 491. Helpful discussions with and technical support from T. Ackemann, C. Brenner, M. Li, and H. Höpfner are greatfully acknowledged.

References

[1] S. Datta and B. Das, "Electronic analog of the electro-optic modulator," *Applied Physics Letters*, vol. 56, no. 7, pp. 665–667, 1990.

[2] R. Flederling, M. Kelm, G. Reuscher et al., "Injection and detection of a spin-polarized current in a light-emitting diode," *Nature*, vol. 402, no. 6763, pp. 787–790, 1999.

[3] Y. Ohno, D. K. Young, B. Beschoten, F. Matsukura, H. Ohno, and D. D. Awschalom, "Electrical spin injection in a ferromagnetic semiconductor heterostructure," *Nature*, vol. 402, no. 6763, pp. 790–792, 1999.

[4] H. J. Zhu, M. Ramsteiner, H. Kostial, M. Wassermeier, H. P. Schönherr, and K. H. Ploog, "Room-temperature spin injection from Fe into GaAs," *Physical Review Letters*, vol. 87, no. 1, Article ID 016601, 4 pages, 2001.

[5] A. T. Hanbicki, B. T. Jonker, G. Itskos, G. Kioseoglou, and A. Petrou, "Efficient electrical spin injection from a magnetic metal/tunnel barrier contact into a semiconductor," *Applied Physics Letters*, vol. 80, no. 7, pp. 1240–1242, 2002.

[6] N. C. Gerhardt, S. Hövel, C. Brenner et al., "Electron spin injection into GaAs from ferromagnetic contacts in remanence," *Applied Physics Letters*, vol. 87, no. 3, Article ID 32502, 3 pages, 2005.

[7] C. Adelmann, X. Lou, J. Strand, C. J. Palmstrøm, and P. A. Crowell, "Spin injection and relaxation in ferromagnet-semiconductor heterostructures," *Physical Review B*, vol. 71, no. 12, Article ID 121301, 4 pages, 2005.

[8] X. Jiang, R. Wang, R. M. Shelby et al., "Highly spin-polarized room-temperature tunnel injector for semiconductor spintronics using MgO(100)," *Physical Review Letters*, vol. 94, no. 5, Article ID 056601, 2005.

[9] S. Hövel, N. C. Gerhardt, M. R. Hofmann et al., "Room temperature electrical spin injection in remanence," *Applied Physics Letters*, vol. 93, no. 2, Article ID 021117, 2008.

[10] S. Hallstein, J. D. Berger, M. Hilpert et al., "Manifestation of coherent spin precession in stimulated semiconductor emission dynamics," *Physical Review B*, vol. 56, no. 12, pp. R7076–R7079, 1997.

[11] M. Oestreich, J. Hübner, D. Hägele et al., "Spintronics: spin electronics and optoelectronics in semiconductors," in *Advances in Solid State Physics*, B. Kramer, Ed., pp. 173–186, Springer, Berlin, Germany, 2001.

[12] D. Saha, D. Basu, and P. Bhattacharya, "High-frequency dynamics of spin-polarized carriers and photons in a laser," *Physical Review B*, vol. 82, no. 20, Article ID 205309, 2010.

[13] J. Lee, W. Falls, R. Oszwałdowski, and I. Žutić, "Spin modulation in semiconductor lasers," *Applied Physics Letters*, vol. 97, no. 4, Article ID 041116, 2010.

[14] M. Y. Li, H. Jähme, H. Soldat, N. C. Gerhardt, M. R. Hofmann, and T. Ackemann, "Birefringence controlled room-temperature picosecond spin dynamics close to the threshold of vertical-cavity surface-emitting laser devices," *Applied Physics Letters*, vol. 97, no. 19, Article ID 191114, 2010.

[15] J. Rudolph, D. Hägele, H. M. Gibbs, G. Khitrova, and M. Oestreich, "Laser threshold reduction in a spintronic device," *Applied Physics Letters*, vol. 82, no. 25, pp. 4516–4518, 2003.

[16] J. Rudolph, S. Döhrmann, D. Hägele, M. Oestreich, and W. Stolz, "Room-temperature threshold reduction in vertical-cavity surface-emitting lasers by injection of spin-polarized electrons," *Applied Physics Letters*, vol. 87, no. 24, Article ID 241117, pp. 1–3, 2005.

[17] M. Holub, J. Shin, D. Saha, and P. Bhattacharya, "Electrical spin injection and threshold reduction in a semiconductor laser," *Physical Review Letters*, vol. 98, no. 14, Article ID 146603, 2007.

[18] I. Vurgaftman, M. Holub, B. T. Jonker, and J. R. Meyer, "Estimating threshold reduction for spin-injected semiconductor lasers," *Applied Physics Letters*, vol. 93, no. 3, Article ID 031102, 2008.

[19] D. Basu, D. Saha, and P. Bhattacharya, "Optical polarization modulation and gain anisotropy in an electrically injected spin laser," *Physical Review Letters*, vol. 102, no. 9, Article ID 093904, 2009.

[20] M. Holub and B. T. Jonker, "Threshold current reduction in spin-polarized lasers: role of strain and valence-band mixing," *Physical Review B*, vol. 83, no. 12, Article ID 125309, 2011.

[21] H. Ando, T. Sogawa, and H. Gotoh, "Photon-spin controlled lasing oscillation in surface-emitting lasers," *Applied Physics Letters*, vol. 73, no. 5, pp. 566–568, 1998.

[22] S. Hövel, N. Gerhardt, M. Hofmann, J. Yang, D. Reuter, and A. Wieck, "Spin controlled optically pumped vertical cavity surface emitting laser," *Electronics Letters*, vol. 41, no. 5, pp. 251–253, 2005.

[23] N. Gerhardt, S. Hövel, M. Hofmann, J. Yang, D. Reuter, and A. Wieck, "Enhancement of spin information with vertical cavity surface emitting lasers," *Electronics Letters*, vol. 42, no. 2, pp. 88–89, 2006.

[24] M. Holub and P. Bhattacharya, "Spin-polarized light-emitting diodes and lasers," *Journal of Physics D*, vol. 40, no. 11, article R01, pp. R179–R203, 2007.

[25] S. Hövel, A. Bischoff, N. C. Gerhardt et al., "Optical spin manipulation of electrically pumped vertical-cavity surface-emitting lasers," *Applied Physics Letters*, vol. 92, no. 4, Article ID 041118, 2008.

[26] S. Iba, S. Koh, K. Ikeda, and H. Kawaguchi, "Room temperature circularly polarized lasing in an optically spin injected vertical-cavity surface-emitting laser with (110) GaAs quantum wells," *Applied Physics Letters*, vol. 98, no. 8, Article ID 81113, 2011.

[27] I. Žutić, J. Fabian, and S. D. Sarma, "Spintronics: fundamentals and applications," *Reviews of Modern Physics*, vol. 76, no. 2, pp. 323–410, 2004.

[28] M. Oestreich, M. Bender, J. Hübner et al., "Spin injection, spin transport and spin coherence," *Semiconductor Science and Technology*, vol. 17, no. 4, pp. 285–297, 2002.

[29] F. Meier and B. P. Zakharchenya, *Optical Orientation. Modern Problems in Condensed Matter Sciences*, vol. 8, North-Holland-Elsevier Science, New York, NY, USA, 1984.

[30] R. J. Elliott, "Theory of the effect of spin-Orbit coupling on magnetic resonance in some semiconductors," *Physical Review*, vol. 96, no. 2, pp. 266–279, 1954.

[31] Y. Yafet, "Solid state physics," in *Advances in Research and Applications*, F. Seitz and D. Turnbull, Eds., pp. 2–96, Academic Press, 1963.

[32] M. I. D'Yakonov and V. I. Perel, "Optical orientation in a system of electrons and lattice nuclei in semiconductors. Theory," *Soviet Physics*, vol. 38, pp. 177–183, 1974.

[33] G. Bir, A. Aronov, and G. Pikus, "Spin relaxation of electrons due to scattering by holes," *Soviet Physics*, vol. 42, pp. 705–712, 1976.

[34] M. Dyakonov, *Spin Physics in Semiconductors*, Springer, 2008.

[35] R. I. Dzhioev, K. V. Kavokin, V. L. Korenev et al., "Low-temperature spin relaxation in n-type GaAs," *Physical Review B*, vol. 66, no. 24, Article ID 245204, 7 pages, 2002.

[36] A. V. Kimel, F. Bentivegna, V. N. Gridnev, V. V. Pavlov, R. V. Pisarev, and T. Rasing, "Room-temperature ultrafast carrier and spin dynamics in GaAs probed by the photoinduced magneto-optical Kerr effect," *Physical Review B*, vol. 63, no. 23, Article ID 235201, 8 pages, 2001.

[37] A. Malinowski, R. S. Britton, T. Grevatt, R. T. Harley, D. A. Ritchie, and M. Y. Simmons, "Spin relaxation in GaAs/AlM$_x$Ga$_{1-x}$As quantum wells," *Physical Review B*, vol. 62, no. 19, pp. 13034–13039, 2000.

[38] J. Fabian, A. Matos-Abiague, C. Ertler, P. Stano, and I. Žutić, "Semiconductor spintronics," *Acta Physica Slovaca*, vol. 57, no. 4-5, pp. 565–907, 2007.

[39] D. J. Hilton and C. L. Tang, "Optical orientation and femtosecond relaxation of spin-polarized holes in GaAs," *Physical Review Letters*, vol. 89, no. 14, Article ID 146601, 4 pages, 2002.

[40] M. Holub, J. Shin, S. Chakrabarti, and P. Bhattacharya, "Electrically injected spin-polarized vertical-cavity surface-emitting lasers," *Applied Physics Letters*, vol. 87, no. 9, Article ID 91108, pp. 1–3, 2005.

[41] M. Ramsteiner, H. Y. Hao, A. Kawaharazuka et al., "Electrical spin injection from ferromagnetic MnAs metal layers into GaAs," *Physical Review B*, vol. 66, no. 8, Article ID 081304, 4 pages, 2002.

[42] S. Saikin, M. Shen, and M. C. Cheng, "Spin dynamics in a compound semiconductor spintronic structure with a Schottky barrier," *Journal of Physics Condensed Matter*, vol. 18, no. 5, pp. 1535–1544, 2006.

[43] H. Soldat, M. Li, N. C. Gerhardt et al., "Room temperature spin relaxation length in spin light-emitting diodes," *Applied Physics Letters*, vol. 99, no. 5, Article ID 051102, 2011.

[44] J. F. Gregg, I. Petej, E. Jouguelet, and C. Dennis, "Spin electronics—a review," *Journal of Physics D*, vol. 35, no. 18, pp. R121–R155, 2002.

[45] G. Schmidt, D. Ferrand, L. W. Molenkamp, A. T. Filip, and B. J. Van Wees, "Fundamental obstacle for electrical spin injection from a ferromagnetic metal into a diffusive semiconductor," *Physical Review B*, vol. 62, no. 8, pp. R4790–R4793, 2000.

[46] E. I. Rashba, "Theory of electrical spin injection: tunnel contacts as a solution of the conductivity mismatch problem," *Physical Review B*, vol. 62, no. 24, Article ID R16267, pp. R16267–R16270, 2000.

[47] A. Fert and H. Jaffrès, "Conditions for efficient spin injection from a ferromagnetic metal into a semiconductor," *Physical Review B*, vol. 64, no. 18, Article ID 184420, 9 pages, 2001.

[48] T. Dietl, H. Ohno, F. Matsukura, J. Cibert, and D. Ferrand, "Zener model description of ferromagnetism in zinc-blende magnetic semiconductors," *Science*, vol. 287, no. 5455, pp. 1019–1022, 2000.

[49] M. L. Reed, N. A. El-Masry, H. H. Stadelmaier et al., "Room temperature ferromagnetic properties of (Ga, Mn)N," *Applied Physics Letters*, vol. 79, no. 21, pp. 3473–3475, 2001.

[50] M. Tanaka, "Ferromagnet (MnAs)/III-V semiconductor hybrid structures," *Semiconductor Science and Technology*, vol. 17, no. 4, pp. 327–341, 2002.

[51] H. Saito, S. Yamagata, and K. Ando, "Room-temperature ferromagnetism in a II-VI diluted magnetic semiconductor Zn1-xCrxTe," *Physical Review Letters*, vol. 90, no. 20, Article ID 207202, 4 pages, 2003.

[52] J. Philip, A. Punnoose, B. I. Kim et al., "Carrier-controlled ferromagnetism in transparent oxide semiconductors," *Nature Materials*, vol. 5, no. 4, pp. 298–304, 2006.

[53] G. A. Medvedkin, T. Ishibashi, T. Nishi, K. Hayata, Y. Hasegawa, and K. Sato, "Room temperature ferromagnetism in novel diluted magnetic semiconductor Cd1-xMnxGeP2," *Japanese Journal of Applied Physics*, vol. 39, no. 10 A, pp. L949–L951, 2000.

[54] T. Dietl and H. Ohno, "Ferromagnetic III-V and II-VI semiconductors," *MRS Bulletin*, vol. 28, no. 10, pp. 714–719, 2003.

[55] T. Manago and H. Akinaga, "Spin-polarized light-emitting diode using metal/insulator/semiconductor structures," *Applied Physics Letters*, vol. 81, no. 4, pp. 694–696, 2002.

[56] N. C. Gerhardt, S. Hövel, C. Brenner et al., "Spin injection light-emitting diode with vertically magnetized ferromagnetic metal contacts," *Journal of Applied Physics*, vol. 99, no. 7, Article ID 073907, 2006.

[57] E. Schuster, R. A. Brand, F. Stromberg et al., "Epitaxial growth and interfacial magnetism of spin aligner for remanent spin injection: [Fe/Tb]n /Fe/MgO/GaAs -light emitting diode as a prototype system," *Journal of Applied Physics*, vol. 108, no. 6, Article ID 063902, 2010.

[58] A. Ludwig, R. Roescu, A. K. Rai et al., "Electrical spin injection in InAs quantum dots at room temperature and adjustment of the emission wavelength for spintronic applications," *Journal of Crystal Growth*, vol. 323, no. 1, pp. 376–379, 2011.

[59] A. Sinsarp, T. Manago, F. Takano, and H. Akinaga, "Electrical spin injection from out-of-plane magnetized FePt/MgO tunneling junction into GaAs at room temperature," *Japanese Journal of Applied Physics Part 2*, vol. 46, no. 1–3, pp. L4–L6, 2007.

[60] L. Grenet, M. Jamet, P. Nó et al., "Spin injection in silicon at zero magnetic field," *Applied Physics Letters*, vol. 94, no. 3, Article ID 032502, 2009.

[61] O. M. J. Van't Erve, G. Kioseoglou, A. T. Hanbicki, C. H. Li, and B. T. Jonker, "Remanent electrical spin injection from Fe into AlGaAs/GaAs light emitting diodes," *Applied Physics Letters*, vol. 89, no. 7, Article ID 072505, 2006.

[62] M. J. Adams and D. Alexandropoulos, "Parametric analysis of spin-polarized VCSELs," *IEEE Journal of Quantum Electronics*, vol. 45, no. 6, pp. 744–749, 2009.

[63] R. Oszwałdowski, C. Gøthgen, and I. Žutić, "Theory of quantum dot spin lasers," *Physical Review B*, vol. 82, Article ID 85316, 2010.

[64] M. San Miguel, Q. Feng, and J. V. Moloney, "Light-polarization dynamics in surface-emitting semiconductor lasers," *Physical Review A*, vol. 52, no. 2, pp. 1728–1739, 1995.

[65] A. Gahl, S. Balle, and M. San Miguel, "Polarization dynamics of optically pumped VCSEL's," *IEEE Journal of Quantum Electronics*, vol. 35, no. 3, pp. 342–351, 1999.

[66] J. Martin-Regalado, F. Prati, M. San Miguel, and N. B. Abraham, "Polarization properties of vertical-cavity surface-emitting lasers," *IEEE Journal of Quantum Electronics*, vol. 33, no. 5, pp. 765–783, 1997.

[67] C. Gøthgen, R. Oszwadowski, A. Petrou, and I. Žutić, "Analytical model of spin-polarized semiconductor lasers," *Applied Physics Letters*, vol. 93, no. 4, Article ID 042513, 2008.

[68] D. Basu, D. Saha, C. C. Wu, M. Holub, Z. Mi, and P. Bhattacharya, "Electrically injected InAsGaAs quantum dot spin laser operating at 200 K," *Applied Physics Letters*, vol. 92, no. 9, Article ID 091119, 2008.

[69] M. Travagnin, M. P. Van Exter, A. K. Jansen Van Doorn, and J. P. Woerdman, "Role of optical anisotropies in the polarization properties of surface-emitting semiconductor lasers," *Physical Review A*, vol. 54, no. 2, pp. 1647–1660, 1996.

[70] M. P. Van Exter, M. B. Willemsen, and J. P. Woerdman, "Polarization fluctuations in vertical-cavity semiconductor lasers," *Physical Review A*, vol. 58, no. 5, pp. 4191–4205, 1998.

[71] M. Sondermann, M. Weinkath, and T. Ackemann, "Polarization switching to the gain disfavored mode in vertical-cavity surface-emitting lasers," *IEEE Journal of Quantum Electronics*, vol. 40, no. 2, pp. 97–104, 2004.

[72] M. B. Willemsen, M. P. Van Exter, and J. P. Woerdman, "Anatomy of a polarization switch of a vertical-cavity semiconductor laser," *Physical Review Letters*, vol. 84, no. 19, pp. 4337–4340, 2000.

[73] E. L. Blansett, M. G. Raymer, G. Khitrova et al., "Ultrafast polarization dynamics and noise in pulsed vertical-cavity surface-emitting lasers," *Optics Express*, vol. 9, no. 6, pp. 312–318, 2001.

[74] T. Ackemann and M. Sondermann, "Characteristics of polarization switching from the low to the high frequency mode in vertical-cavity surface-emitting lasers," *Applied Physics Letters*, vol. 78, no. 23, pp. 3574–3576, 2001.

[75] R. Al-Seyab, D. Alexandropoulos, I. D. Henning, and M. J. Adams, "Instabilities in spin-polarized vertical-cavity surface-emitting lasers," *IEEE Photonics Journal*, vol. 3, no. 5, pp. 799–809, 2011.

[76] N. C. Gerhardt, M. Y. Li, H. Jähme, H. Höpfner, T. Ackemann, and M. R. Hofmann, "Ultrafast spin-induced polarization oscillations with tunable lifetime in vertical-cavity surface-emitting lasers," *Applied Physics Letters*, vol. 99, no. 15, Article ID 151107, 2011.

[77] K. Panajotov, B. Nagler, G. Verschaffelt et al., "Impact of in-plane anisotropic strain on the polarization behavior of vertical-cavity surface-emitting lasers," *Applied Physics Letters*, vol. 77, no. 11, pp. 1590–1592, 2000.

[78] K. Ikeda, T. Fujimoto, H. Fujino, and T. Katayama, "Switching of lasing circular polarizations in a (110)-VCSEL," *IEEE Photonics Technology Letters*, vol. 21, no. 18, pp. 1350–1352, 2009.

[79] D. Hägele, M. Oestreich, M. Holub, and P. Bhattacharya, "Comment on "electrically injected spin-polarized vertical-cavity surface-emitting lasers" [Applied Physics Letters vol. 87, article 091108, 2005]," *Applied Physics Letters*, vol. 88, no. 5, Article ID 56101, p. 1, 2006.

[80] M. Holub and P. Bhattacharya, "Response to Comment on "Electrically injected spin-polarized vertical-cavity surface-emitting lasers" [Applied Physics Letters, vol. 87, article 091108, 2005]," *Applied Physics Letters*, vol. 88, no. 5, Article ID 56102, p. 1, 2006.

[81] M. Holub, P. Bhattacharya, J. Shin, and D. Saha, "Electron spin injection from a regrown Fe layer in a spin-polarized vertical-cavity surface-emitting laser," *Journal of Crystal Growth*, vol. 301-302, pp. 602–606, 2007.

7

Nanoscale Biomolecular Detection Limit for Gold Nanoparticles Based on Near-Infrared Response

Mario D'Acunto,[1,2] **Davide Moroni,**[2] **and Ovidio Salvetti**[2]

[1] *Istituto di Struttura della Materia, Consiglio Nazionale delle Ricerche, Via Fosso del Cavaliere 100, 00133 Roma, Italy*
[2] *Istituto di Scienze e Tecnologia dell'Informazione, Consiglio Nazionale delle Ricerche, Via Moruzzi 1, 56124 Pisa, Italy*

Correspondence should be addressed to Mario D'Acunto, mario.dacunto@ism.cnr.it

Academic Editor: Carlo Corsi

Gold nanoparticles have been widely used during the past few years in various technical and biomedical applications. In particular, the resonance optical properties of nanometer-sized particles have been employed to design biochips and biosensors used as analytical tools. The optical properties of nonfunctionalized gold nanoparticles and core-gold nanoshells play a crucial role for the design of biosensors where gold surface is used as a sensing component. Gold nanoparticles exhibit excellent optical tunability at visible and near-infrared frequencies leading to sharp peaks in their spectral extinction. In this paper, we study how the optical properties of gold nanoparticles and core-gold nanoshells are changed as a function of different sizes, shapes, composition, and biomolecular coating with characteristic shifts towards the near-infrared region. We show that the optical tenability can be carefully tailored for particle sizes falling in the range 100–150 nm. The results should improve the design of sensors working at the detection limit.

1. Introduction

The development of biosensors devices requires sophisticated approaches to detect and to identify the analytes [1–3]. Because the sensor signal-to-noise ratio increases with decreasing size for many devices, many researchers are expending considerable effort for miniaturizing sensing devices down to nanoscale size [3–7]. In the past decade several authors have analyzed the effect of flow, size, and adsorption isotherms on biomolecular adsorption and in a limited way the effects at nanometer length scales. The role of nanoscale size can be shown examining a simple sensor geometry (e.g., hemisphere) that absorbs the analyte [6, 8, 9]. The maximum number of molecules, $N(t)$, that can accumulate on a sensor due to irreversible adsorption may be determined by

$$N(t) = \int_0^t F(\tau)d\tau = \int_0^t \int_A f \, d\sigma \, d\tau, \qquad (1)$$

where F is the total flux (molecule s^{-1}), f is the flux at the sensors (molecules s^{-1}m^{-2}), σ is the unit area, A is the sensor area, and t is time. Time dependence in (1) involves a crucial question for clinical applications; that is, if any analyte molecule that contacts the sensor surface can be detected, what is the minimum detectable concentration for a given accumulation time? Sheehan and Whitman [6] showed that DNA microarrays with a 200 μm diameter hemisphere can detect $\sim$1 fM in $\sim$1 min, which is actually very close to the actual DNA detection limit, $\sim$20 fM. Analogously, from (1) is possible to deduce the accumulation events

$$N(t) = 4DN_A c_0 a t, \qquad (2)$$

where N_A is Avogadro's number, c_0 is the initial solution concentration, a is the radius of the sensing surface, and D is a diffusion constant. It is remarkable that (2) is linear in both radius and time. This implies that when reducing the size of the absorbing surface, the time for a critical accumulation of events increases consequently. The advent of nanotubes and nanowires as biosensors introduces new geometries into the sensing activity with some changes to accumulation events as in (2).

The use of nanotechnologies for diagnostic applications meets the rigorous demands of clinical standards sensitivity with cost effectiveness. Today, main nanodiagnostic tools include quantum dots (QDs), gold nanoparticles, and cantilevers [10]. The effectiveness of nanoparticles as biomedical imaging contrast and therapeutic agents depends on their optical properties. Biosensing applications based on surface plasmon resonance shifts need strong resonance in the wavelength sensitivity range of the instrument as well as narrow optical resonance line widths. For actual *in vivo* imaging and therapeutic applications, the optical resonance of the nanoparticles is strongly desired to be in the near-infrared (NIR) region of the biological water window, where the tissue transmissivity is the highest [11–15].

The optical properties of gold nanoparticles in the visible and near-infrared (vis-NIR) domains are governed by the collective response of conduction electrons. These form an electron gas that moves away from its equilibrium position when perturbed by an external light field, thus creating induced surface polarization changes that act as a restoring force on the electron gas. This results in a collective oscillatory motion of the electrons similar to the vibrations of a plasma and characterized by a dominant resonance band lying in the vis-NIR for gold and called plasmon excitations. Thus the surface of gold nanoparticles can be used as a sensing element because when biomolecules attach to such surface the binding event can be revealed by optical changes. In addition, it is possible to calculate the maximum number of possible binding events and the consequent optical response when the shape and size of the gold particles are known.

In this paper, we will calculate and simulate the optical tunability of gold nanoparticles and core-gold nanoshells (optical changes with nanoparticles sizes) using the classical Mie theory and discrete dipole approximation (DDA). We will show that when biomolecules bind to gold surface optical shifts toward the near-infrared region are produced, and wavelengths changes can be accurately quantified. The accurate knowledge of the optical tunability could allow for improving sensing performance, sensitivity, and figure of merit of biosensors operating at detection limit in short time as requested in clinical applications.

2. Nanoscale-Sized Systems for Detection Limit

The increased demand for sensitivity requires that a diagnostically significant interaction occurs between analyte molecules and signal-generating particles, thus enabling detection of a single analyte molecule. Nanotechnology has enabled one-to-one interaction between analytes and signal-generating particles such as QDs and gold nanoparticles. Here, we review briefly the most important nanoscale tools for detection limit, such as QDs, cantilevers, and gold nanoparticles [10, 16].

QDs are semiconductor nanocrystals, characterized by strong light absorbance, that can be used as fluorescent labels for biomolecules. A typical QD has a diameter of 2–10 nm and is usually composed of a core consisting of a semiconductor material enclosed in a shell of another semiconductor material with a larger spectral bandgap. When a QD absorbs a photon with energy higher than the bandgap energy of the composing semiconductor, an exciton (electron-hole pair) is created. As a result, a broadband absorption spectrum occurs because of the increased probability of absorption at shorter wavelengths. The recomposition of the exciton, generally characterized by a long lifetime >10 ns, to a lower energy state leads to emission of a photon with a narrow symmetric band [17–19]. QDs are reported to emit with lifetimes of 5–40 ns, whereas conventional dyes emit in a time less than 2 ns. This characteristic produces a strong and stable fluorescence signal. In addition, due to quantum confinement effect a direct relationship exists between the QD size and the values of the quantized energy levels. Methods for tracking and detecting QDs are numerous and include fluorometry and several types of microscopy, such as fluorescence, confocal, total internal reflection, wide-field epifluorescence, near-field optical microscopy, and multiphoton microscopy. The choice of detection techniques depends on the emission of wavelengths; for example, QDs composed by CdSe/NzS emit in the 530–630 nm range and InP and InAs QDs emit in the NIR range, while PbS emits in the 850–950 nm range and PbSe QD in the mid-infrared range.

Another nanoscale detection technique is based on cantilevers [20–22]. Cantilevers are small beams similar to those used in atomic force microscopy and they operate detection by use of nanomechanical deflections. An instructive example of operating cantilevers is the detection of DNA hybridization. The cantilever surface holds a particular DNA sequence capable of binding to a specific target. When the hybridization occurs with the cantilever single-stranded DNA, mechanical stress produced by the rebinding process underlying the hybridization deflects the cantilever, with a measurable deflection by the optical lever method proportional to the amount of DNA hybridized. This technique can be used as a microarray, allowing for multiple analyses. However, this method currently needs further development to solve the problem of nonspecific binding [23].

Gold nanoparticles and gold nanoshells provide great sensitivity for the detection of DNA antibodies and proteins [24–28]. The instrumental platform for the detection of biomolecules including gold particles is based on surface plasmon resonance (SPR). SPR is an optical technique that measures the refractive index of very thin layers of material adsorbed on a metal. It offers real-time in situ analysis of dynamic surface events and is capable of defining rates of adsorption and desorption for surface interactions. Plasmon-plasmon resonance, resulting from the interaction of locally adjacent gold nanoparticle labels that have bound to a target, produces changes in optical properties that can be used for detection. It is known that the characteristic red color of gold colloid changes to a bluish-purple color on colloid aggregation because of this effect. Analogously, the reduction of gold particle sizes involves shifts toward the NIR region [29]. Raman spectroscopy is a favored detection method using silver in the visualization process. In this case, gold nanoparticles can be coated with silver shells;

silver-coated gold particles less than 100 nm in size have strong light-scattering properties and can easily be detected by optical microscopies operating at the detection limit for oligonucleotides down to ~10 fM, that is, nearly 50-fold lower than conventional fluorophore-based methods. Gold nanoshells could allow direct, rapid, and economically feasible analysis of whole blood samples. Generally, a nanoshell consists of concentric spherical nanoparticles with a dielectric core, meanly consisting of gold sulfide or silica, surrounded by a thin gold shell. Variations in the relative thickness of the core and outer shell allow the optical resonance of gold to go into the midinfrared region. Other relative surface properties can be used to focus the absorption wavelength range into the near-infrared, just above the absorption of hemoglobin and below that of water [30]. This characteristic aids in avoiding interference from hemoglobin giving the possibility to do a direct analysis of whole blood. An additional advantage of gold nanoshells is the excellent biocompatibility.

The optical tunability of gold nanoparticles and core-gold nanoshells resides in plasmon resonance properties. It is well known that the plasmon resonance of metal nanoparticles is strongly sensitive to the nanoparticles' size and shape and the dielectric properties of the surrounding environments. We will simulate the optical tunability of gold nanoparticles and gold nanoshells (optical changes with nanoparticles' sizes and surface biomolecular coating) using the classical Mie theory and DDA-modified approach. Finally, we will obtain the wavelength shift as a function of different size dimension, environments, and biomolecular coating layer of the gold nanoparticles. The accurate knowledge of the optical tunability and wavelength shifts should allow for development of biosensors operating at detection limit in short time as requested in clinical applications.

The design of biosensors working at the detection limit must attempt to compare sensing performance, sensitivity, and figure of merit (FOM) [31]. The sensitivity S is defined as the ratio of the resonant wavelength shift $\partial\lambda_{res}$ to the variation of the surrounding refractive index ∂n_s, while the FOM is defined as the ratio of the refractive index sensitivity to the resonance width $\Delta\lambda$:

$$S = \frac{\partial\lambda_{res}}{\partial n_s}, \qquad FOM = \frac{S}{\Delta\lambda}. \qquad (3)$$

Resolution is typically defined as the minimum detection limit. In addition, sensitivity and thence resolution can be modified by size and geometry of the gold nanoparticles and nanostructure environment. In general, the adsorbed mass on the sensor surface can be approximated by the De Freijter's formula, which is based on the refractive index change [32]

$$\gamma = \frac{d\Delta n}{\partial n/\partial c}, \qquad (4)$$

where d is the dimension of the adsorbed species (biomolecule) falling in the range of nanometers, Δn is the refractive index difference between the medium and the species, and $\partial n/\partial c$ is the correspondent biomolecular refractive increment. Analogously, the spectral response of the refractometric nanoplasmonic sensors can be described by

$$\Delta\lambda = m(n_{eff} - n_{medium}), \qquad (5)$$

where m is the refractive index sensitivity, expressed in plasmon resonance peak shift per refractive index unit (RIU), and n_{eff} is the effective refractive index of the adsorbate layer. In the simplest approximation, the surface plasmon resonance-induced evanescent field decays exponentially from the surface as $E(z) = \exp(-z/l_d)$ with a decay length l_d. The effective refractive index of the adsorbate layer is described as

$$n_{eff} = \frac{2}{l_d}\int_0^d n(z)E^2(z)dz. \qquad (6)$$

Using (5)-(6) into (4) we obtain a relationship of the surface plasmon resonance peak shift with the surface coverage of the adsorbate:

$$\gamma(t) = \frac{d\Delta\lambda(t)}{m(1 - \exp(-2d/l_d))\partial n/\partial c}. \qquad (7)$$

In the next section, we will provide the accurate calculation of wavelength shifts due to increased size of gold nanoparticles and coating of gold surface in core-gold nanoshells. Equation (7) establishes that the accurate knowledge of such shifts gives the possibility to design gold nanoparticles based sensors working at the biomolecular detection limit.

3. Optical-Infrared Response of Gold Nanoparticles and Core-Gold Nanoshells: Calculation and Simulation Methods

The complexity of the electromagnetic field in the presence of arbitrarily shaped nanoparticles is such that Maxwell's equations must be solved using numerical methods [33]. The far-field performance of a nanoparticle is summarized in the wavelength-dependent absorption and scattering cross sections, and as a consequence, the optical properties of gold nanospheres and silica-gold nanoshells will be quantified in terms of their calculated absorption and scattering efficiency. The light intensity transmitted through a dilute dispersion I (assuming no more than one photon-particle collision per photon) is $I = I_0\exp[-(\sigma_{abs} + \sigma_{sc})NL]$, where I_0 is the incident light intensity, N is the number of particles per unit volume, L is the path length inside the dispersion, and σ_{abs} and σ_{sc} are the wavelength-dependent absorption and scattering cross sections, respectively. The optical modeling of nanoparticles thus relies on the solution of Maxwell's equations for each specific geometry and set of illumination conditions, assuming a local dielectric function, $\varepsilon(\omega)$, and description of the materials involved. When dipolar contribution in plasmon excitation is prominent, then the far-field cross sections can be obtained from the polarizability α using the following expressions [34]:

$$\sigma_{abs} + \sigma_{sc} = \frac{2\pi}{\lambda\sqrt{\varepsilon_m}}Im\{\alpha\}, \qquad \sigma_{sc} = \frac{8\pi^3}{3\lambda^4}|\alpha|^2, \qquad (8)$$

where ε_m is the environmental permittivity of the medium outside the particle and λ is the light wavelength.

To simulate the optical properties of gold nanoparticles, many methods can be found. Mie theory is one of such methods [35]. Mie scattering theory begins with Maxwell's equations and the necessary boundary conditions, which are then transformed into spherical polar coordinates with a solution of the wave vector equation emerging. The coefficients for scattering arise and the theory is extended to the far-field solution based on a plane constructed from the incident and scattered waves. At this point, the Stokes parameters that are the measurable quantities and the formulation of matrix plane can be quantified. From Mie theory, once scattering matrices have been derived, information about the direction and polarization dependence of the scattered light can be extracted, so that absorption, scattering, and extinction cross sections for any arbitrary spherical particle with dielectric function ε can be calculated. Since extincted power is the sum of the scattered and absorbed power, the absorption cross section is simply $\sigma_{abs} = \sigma_{ext} - \sigma_{sca}$, while the scattering and extinction cross sections can be calculated from the total cross section of a spherical particle, that can be written as

$$\sigma = \frac{\lambda_m^2}{2\pi} \sum_{l=1}^{l=\infty} (2l+1)\left[\mathrm{Im}\left\{\varphi_l^E\right\} + \mathrm{Im}\left\{\varphi_l^M\right\} \right], \quad (9)$$

where $\varphi_l^{E,M}$ are the electric and magnetic scattering coefficients, respectively, and l is the orbital momentum number (e.g., $l = 1$ for a dipole). These coefficients admit analytical expressions in terms of spherical Bessel and Hankel functions for a homogeneous sphere [33]

$$\varphi_l^E =$$

$$\frac{-\varepsilon_m j_l(\rho_m)\left[j_l(\rho) + \rho j_l'(\rho) \right] + \varepsilon\left[j_l(\rho_m) + \rho_m j_l'(\rho_m) \right] j_l(\rho)}{\varepsilon_m h_l(\rho_m)\left[j_l(\rho) + \rho j_l'(\rho) \right] - \varepsilon\left[h_l(\rho_m) + \rho_m h_l'(\rho_m) \right] j_l(\rho)}$$

$$\varphi_l^M = \frac{-\rho j_l(\rho_m) j_l'(\rho) + \rho_m j_l'(\rho_m) j_l(\rho)}{\rho h_l(\rho_m) j_l'(\rho) - \rho_m h_l'(\rho_m) j_l(\rho)}, \quad (10)$$

where $\rho = (2\pi R/\lambda)\sqrt{\varepsilon}$, $\rho_m = (2\pi R/\lambda)\sqrt{\varepsilon_m}$, and the prime represents the first differentiation with respect to the argument in parentheses. It should be noted that plasmon modes are related to the poles of φ_l^E. In fact, the magnetic contribution to scattering produces only resonances for particle sizes comparable to the wavelength. The multipolar polarizability is proportional to φ_l^E, and in particular, the dipolar polarizability reads $\alpha = (3\lambda^3/4\pi^2)\varphi_{l=1}^E$, where the dipolar electric scattering coefficient can be readily calculated using $j_1(x) = \sin x/x^2 - \cos x/x$ and $h_1(x) = (1/x^2 - i/x)\exp(ix)$, so that (8) can be used instead of (9). The value of α obtained is really accurate for describing gold spheres

with diameters up to 200 nm and simplified expression can be found as, for example,

$$\alpha = 3V\varepsilon_m$$

$$\times \frac{1 - 0.1(\varepsilon + \varepsilon_m)\theta^2/4}{(\varepsilon + 2\varepsilon_m)/(\varepsilon - \varepsilon_m) - (0.1\varepsilon + \varepsilon_m)\theta^2/4 - i(2/3)\varepsilon_m^{3/2}\theta^3},$$

$$(11)$$

where V is the particle volume, ε and ε_m are the permittivities of the particle and the surrounding medium, and $\theta = 2\pi R/\lambda$ is the size parameters that recover the electrostatic limit for $\theta = 0$. Analogously, the dipolar polarizability of a coated gold nanosphere is given by

$$\alpha = 3V\varepsilon_m$$

$$\times \frac{(R_0/R_i)^3(2\varepsilon_{coat} + \varepsilon)(\varepsilon_{coat} - \varepsilon_m) - (\varepsilon_{coat} - \varepsilon)(2\varepsilon_{coat} + \varepsilon_m)}{(R_0/R_i)^3(2\varepsilon_{coat} + \varepsilon)(\varepsilon_{coat} + 2\varepsilon_m) - 2(\varepsilon_{coat} - \varepsilon)(\varepsilon_{coat} - \varepsilon_m)},$$

$$(12)$$

where R_i is the internal radius of the core material described by ε, the coating of permittivity ε_{coat} extends up to a radius R_0, and the medium outside the particle has permittivity ε_m. An analogous formula can be derived for a barium titanate core particle with a gold nanoshell. Calculations of the optical absorption and scattering efficiency of gold nanospheres and barium titanate-gold nanoshells will be presented below. The required parameters for the code were the value of the core and shell radii R_1 and R_2, the complex refractive indices for the core, shell, and biomolecular coating, and the surrounding medium n_c, n_s, n_b, and n_m, respectively, Figure 1.

The discrete-dipole approximation (DDA) has been used as a complimentary method to Mie theory [36]. DDA is a flexible and powerful technique for computing scattering and absorption by targets of arbitrary geometry. DDA calculations require choices for the locations and the polarizabilities of the point dipoles that represent the targets. To approximate a certain geometry (e.g., a sphere or a core-shell-environment system, where the shell is made by gold) with a finite number of dipoles, we might consider using some number of closely spaced, weaker dipoles in regions near the target boundaries to do a better job of approximating the boundary geometry. For a core-shell system, we use the following algorithm to generate the dipole array. Given a coordinate reference system, (1) we generate a trial lattice defined by a lattice spacing d and coordinates of the lattice point nearest the origin. (2) All lattice sites are located within the volume V of the core-shell-surrounding system (where the surrounding is considered as a sphere enveloping the core-shell system). (3) Try different values of d and dipoles coordinates and maximize some goodness-of-fit criterion for a list of occupied sites $i = 1,\ldots,N$. Each of these occupied sites represents a cubic subvolume d^3 of material centered on the site. Particular efforts have been addressed to distinguish between lattice sites near the surface and those in the interior. (4) We therefore rescale the array

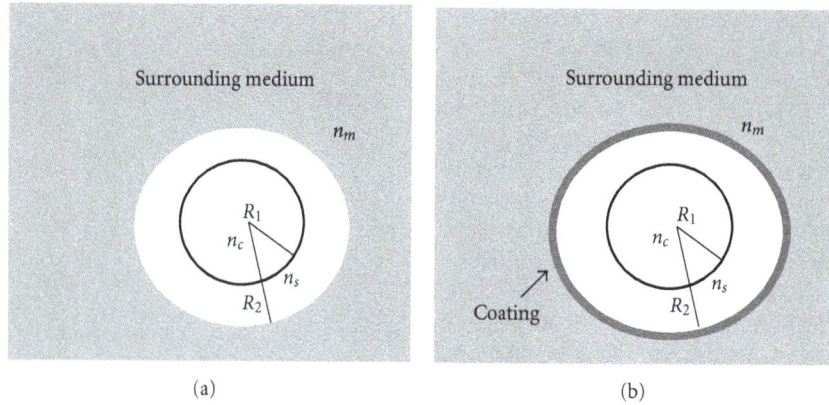

FIGURE 1: (a) Schematic sketch of a gold nanoshell, R_1 and R_2 are the core and shell radii, and the complex refractive indices for the core, shell, and the surrounding medium are n_c, n_s, and n_m, respectively. (b) The same gold nanoshell with a biomolecular absorbed coating layer. Note that the shell radius is defined as the total radius minus the core radius.

by requiring that $d = (V/N)^{1/3}$, so that the volume Nd^3 of the occupied lattice sites is equal to the volume V of the original target. (5) For each occupied N sites, assign a dipole polarizability α_i.

The assignation of dipole polarizabilities is a crucial question in a DDA method. Following the seminal paper of Draine and Flatau [37], the polarizability $\alpha(\omega)$ can be found analytically in the long-wavelength limit $|n|kd \ll 1$ as a series expansion in the powers of kd, with the criterion that $N > (4\pi/3)|n|^3(kr_{eff})^3$ where n is the refractive index and $r_{eff} = (3V/4\pi)^{1/3}$ is the effective radius of the nanoparticles of volume V. As a consequence, for materials with higher refractive indexes the DDA method can overestimate absorption cross sections. A customized code for the DDA calculations of gold nanoshells has been written adapting DDSCAT program that is a freely available code [38, 39].

The finiteness of the speed of light has important consequences that affect the response of gold nanoparticles. First of all, the electromagnetic field cannot penetrate beyond a certain depth inside the metal, the so-called skin depth, which is of the order of 15 nm in the vis-NIR. But more importantly, redshifts take place as the particles size increases, and retardation effects play a significant role when the diameter is a consistent fraction of the mode wavelength λ_m in the surrounding medium, which is related to the free-space wavelength through $\lambda_m = \lambda/\sqrt{\varepsilon_m}$. In particular, opposite charges are separated by roughly one particle diameter in a dipole mode, so that the reaction of one end of the particle to changes produced in the other end takes place with a phase delay of the order of $4\pi R/\lambda_m$, and consequently, the period of one mode oscillation increases to accommodate this delay. Analogously, quadrupoles and higher-order modes produce more nodes in the distribution of the polarization charges induced on the surface of the particle, which reduce the effective interaction distance.

Figure 2 shows the shift towards the NIR region for the efficiency of the absorption calculated using the Mie theory and DDA approach for a gold nanoparticle (radius 120 nm, black line) and a barium titanate ($BaTiO_3$)-gold nanoshell

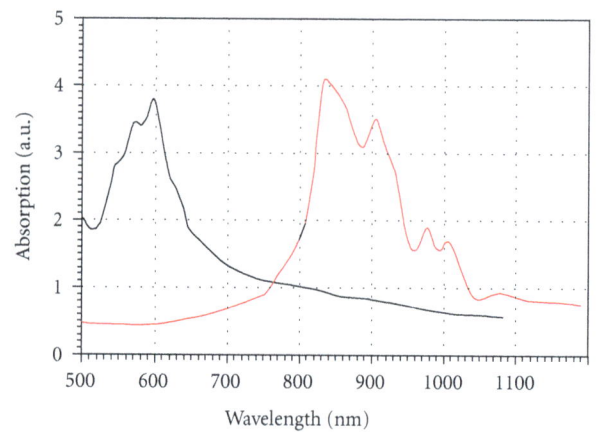

FIGURE 2: Calculated spectra of the efficiency of absorption as a function of the wavelength for gold nanoparticle (120 nm, black line) and barium titanate-gold nanoshell (100 + 20 nm, red line).

(core 100 nm, gold 20 nm, red line). The barium titanate refractive index was taken to follow the dispersion formula as $n^2 - 1 = 4.187\lambda^2/(\lambda^2 - 0.223^2)$. The refractive index of the surrounding medium was considered to be $n_m = 1.34 + 0i$ at all wavelengths, close to the water.

Optical tunability plays a fundamental role for the identification of size, shape, and core-shell composition of the nanoparticles for biomedical sensors and detection limit definition. In order to define the optical tunability, the red shifts must be carefully identified. In Figure 3, the tunability of the extinction cross section of a gold nanoparticle as a function of the diameter size is reported, while in Figure 4, the same quantity for a core-gold nanoshell (core = $BaTiO_3$, gold shell = 20 nm) is reported as a function of different core/gold ratios. In the case of a gold nanosphere, the total extinction cross section increases linearly as the nanosphere size increases, with an initial value of σ_{ext} (nm^2) = 4871 for a gold nanosphere diameter of 40 nm. Approximately, linear behavior of total extinction cross section is waited for core/shells systems, Figure 4.

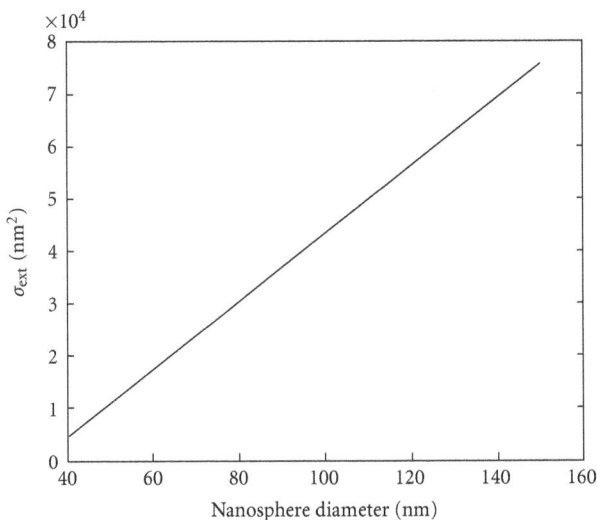

FIGURE 3: Tunability of the extinction cross section of nanoparticles. Variation of the total extinction cross section as a function of the nanosphere diameter size. The extinction cross section changes linearly with the gold nanosphere diameter giving the possibility of direct tuning of optical response with gold nanosphere dimension.

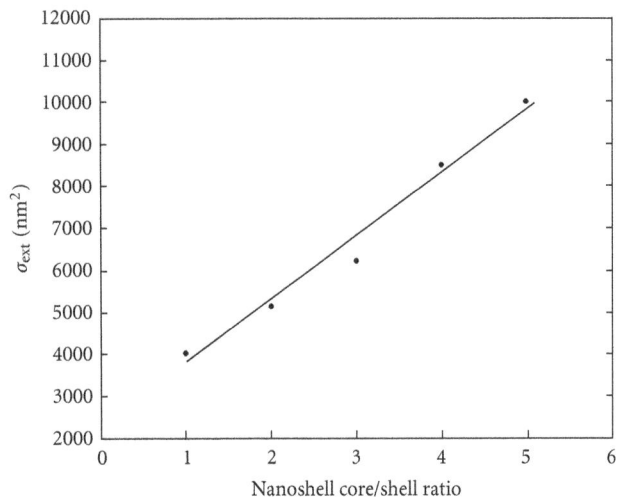

FIGURE 4: Tunability of the extinction cross section as a function of nanoshells core/shell ratios. As in the case of nanospheres, the extinction cross section shows a behavior close to be linear with core/shell ratio increments (gold shell = 20 nm).

The increase in the extinction scattering with the nanoparticle volume has been related to increased radiative damping in larger nanoparticles based on experimental scattering spectra of gold nanospheres and core-gold systems [29–33]. This behavior suggests that larger nanoparticles of high core/shell ratios would be more suitable for biological sensors applications based on light scattering. In the case of nanoshells, the relative core-shell dimensions vary linearly the magnitude of light extinction. This implies that a decrease in the core/shell ratio can be seen to be an effective handle in increasing the scattering contribution to the total extinction. The next calculated quantity is the effect of a biomolecular adsorbed layer on the gold surface for a core/shell system with fixed ratio (80 nm for $BaTiO_3$ and 20 nm for gold surface) on wavelength shifts $\Delta\lambda$ for different layer dimensions. This quantity is particularly important for the definition of sensitivity response and FOM of a biosensor, as described by (3). In Figure 5, the progressive plasmon resonance maximum wavelength red-shift is shown as a function of increased homogeneous layers of adsorbed bovine serum albumin (BSA, 170 mg/mol). The calculation was made, as in the previous case, using (8) and (12), but including in the DDA approach progressive 1-to-3 supplemental layers with a known refractive index [40]. It is interesting to note that when increasing the size of the nanoparticle (from left black line to cyan line on right) an increasing wavelength red-shift is observed. The reduction of the peaks is an artifact effect and the absorption quantity should be normalized including the increasing nanoparticle volume.

Figure 5 represents the most important result of the present paper. In fact, once the red-shift, $\Delta\lambda$, is accurately known, it should be possible to quantify the refractive index changes, γ, (7), to design high sensitive biosensors operating at detection limit with effective optimal $\partial n/\partial c$ factor. The absolute magnitude of the optical cross section provides a limited reliable measure of the optical properties of a population of nanoparticles employed in the biomedical sensing activity or other real-life biomedical applications because broad size nanoparticle population can have a different optical response with respect to few nanoparticles but with higher sizes. The red-shifts as a function of nanoparticles diameter size, shown in Figure 5, are one quantity that can be measured in scanning near optical microscopy (SNOM) technique, measurements that will provide the next development of the present paper.

4. Conclusions

The development of biosensors devices requires sophisticated approaches to detect and to identify the analytes. Because the sensor signal-to-noise ratio increases with decreasing size for many devices, many researchers are expending considerable effort for miniaturizing sensing devices down to nanoscale size. Gold nanoparticles have been widely used during the past few years in various technical and biomedical applications. In particular, the resonance optical properties of nanometer-sized particles have been employed to design biochips and biosensors used as analytical tools. The optical properties of nonfunctionalized gold nanoparticles and core-gold nanoshells play a crucial role for the design of biosensors where gold surface is used as a sensing component. Gold nanoparticles exhibit excellent optical tunability at visible and near-infrared frequencies leading to sharp peaks in their spectral extinction. In this paper, we have simulated how the optical properties of gold nanoparticles are changed as a function of different sizes, shapes, and composition. We have shown that the optical

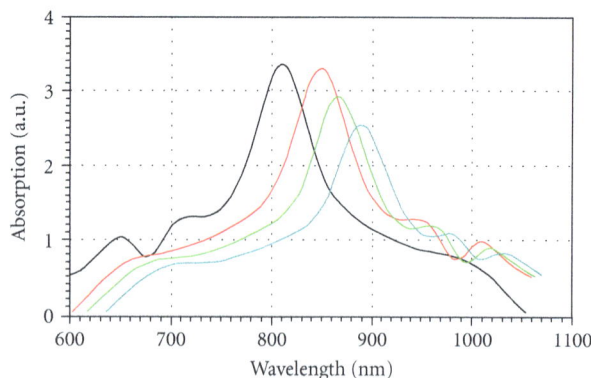

FIGURE 5: Plasmon resonance maximum red-shift for the absorption in a core/shell nanoparticle (BaTiO$_3$ 80 nm, Au 40 nm) with a different number of coating layers of BSA biomolecules (170 mg/mol). Black line represents the uncoated nanoshell, red line represents the coating with a uniform BSA layer, the green line represents two layers and, finally, the cyan line shows the effect of the third layer.

tunability can be carefully tailored for particle sizes falling in the range 100–150 nm. Nanoshells made by a barium titanate core have been considered and the optical response of gold surfaces coated by 1-to-3 layer of BSA was taken in consideration. The simulated results will help experimental optical tests that will be performed using nanoscale optical scanning technology.

Acknowledgment

M. D'Acunto wishes to acknowledge the NanoICT Project for useful support.

References

[1] J. M. Ramsey and A. van der Berg, *Micro Total Analysis Systems*, Kluwer Academic Publisher, Boston, Mass, USA, 2001.

[2] D. R. Reyes, D. Iossifidis, P. A. Auroux, and A. Manz, "Micro total analysis systems. 1. Introduction, theory, and technology," *Analytical Chemistry*, vol. 74, no. 12, pp. 2623–2636, 2002.

[3] P. A. Auroux, D. Iossifidis, D. R. Reyes, and A. Manz, "Micro total analysis systems. 2. Analytical standard operations and applications," *Analytical Chemistry*, vol. 74, no. 12, pp. 2637–2652, 2002.

[4] J. Kong, N. R. Franklin, C. Zhou et al., "Nanotube molecular wires as chemical sensors," *Science*, vol. 287, no. 5453, pp. 622–625, 2000.

[5] Y. Cui, Q. Wei, H. Park, and C. M. Lieber, "Nanowire nanosensors for highly sensitive and selective detection of biological and chemical species," *Science*, vol. 293, no. 5533, pp. 1289–1292, 2001.

[6] P. E. Sheehan and L. J. Whitman, "Detection limits for nanoscale biosensors," *Nano Letters*, vol. 5, no. 4, pp. 803–807, 2005.

[7] E. S. Jeng, A. E. Moll, A. C. Roy, J. B. Gastala, and M. S. Strano, "Detection of DNA hybridization using the near-infrared band-gap fluorescence of single-walled carbon nanotubes," *Nano Letters*, vol. 6, no. 3, pp. 371–375, 2006.

[8] P. R. Nair and M. A. Alam, "Screening-limited response of NanoBiosensors," *Nano Letters*, vol. 8, no. 5, pp. 1281–1285, 2008.

[9] L. Soleymani, Z. Fang, E. H. Sargent, and S. O. Kelley, "Programming the detection limits of biosensors through controlled nanostructuring," *Nature Nanotechnology*, vol. 4, no. 12, pp. 844–848, 2009.

[10] H. M. E. Azzazy, M. M. H. Mansour, and S. C. Kazmierczak, "Nanodiagnostics: a new frontier for clinical laboratory medicine," *Clinical Chemistry*, vol. 52, no. 7, pp. 1238–1246, 2006.

[11] R. Weissleder, C. H. Tung, U. Mahmood, and A. Bogdanov, "*In vivo* imaging of tumors with protease-activated near-infrared fluorescent probes," *Nature Biotechnology*, vol. 17, no. 4, pp. 375–378, 1999.

[12] W. C. W. Chan and S. Nie, "Quantum dot bioconjugates for ultrasensitive nonisotopic detection," *Science*, vol. 281, no. 5385, pp. 2016–2018, 1998.

[13] S. M. Bachilo, M. S. Strano, C. Kittrell, R. H. Hauge, R. E. Smalley, and R. B. Weisman, "Structure-assigned optical spectra of single-walled carbon nanotubes," *Science*, vol. 298, no. 5602, pp. 2361–2366, 2002.

[14] X. Y. Wu, H. J. Liu, J. Q. Liu et al., "Immunofluorescent labeling of cancer marker Her2 and other cellular targets with semiconductor quantum dots," *Nature Biotechnology*, vol. 21, no. 4, pp. 41–43, 2003.

[15] C. L. Amiot, S. Xu, S. Liang, L. Pan, and J. X. Zhao, "Near-infrared fluorescent materials for sensing of biological targets," *Sensors*, vol. 8, no. 5, pp. 3082–3105, 2008.

[16] H. M. E. Azzazy, M. M. H. Mansour, and S. C. Kazmierczak, "From diagnostics to therapy: prospects of quantum dots," *Clinical Biochemistry*, vol. 40, no. 13-14, pp. 917–927, 2007.

[17] X. Michalet, F. F. Pinaud, L. A. Bentolila et al., "Quantum dots for live cells, *in vivo* imaging, and diagnostics," *Science*, vol. 307, no. 5709, pp. 538–544, 2005.

[18] Y. Xing and J. Rao, "Quantum dot bioconjugates for *in vitro* diagnostics & *in vivo* imaging," *Cancer Biomarkers*, vol. 4, no. 6, pp. 307–319, 2008.

[19] Z. Jin and N. Hildebrandt, "Semiconductor quantum dots for *in vitro* diagnostics and cellular imaging," *Trends in Biotechnology*, vol. 30, no. 7, pp. 394–403, 2012.

[20] J. Fritz, M. K. Baller, H. P. Lang et al., "Translating biomolecular recognition into nanomechanics," *Science*, vol. 288, no. 5464, pp. 316–318, 2000.

[21] R. McKendry, J. Zhang, Y. Arntz et al., "Multiple label-free biodetection and quantitative DNA-binding assays on a nanomechanical cantilever array," *Proceedings of the National Academy of Sciences of the United States of America*, vol. 99, no. 15, pp. 9783–9788, 2002.

[22] J. Mertens, C. Rogero, M. Calleja et al., "Label-free detection of DNA hybridization based on hydration-induced tension in nucleic acid films," *Nature Nanotechnology*, vol. 3, no. 5, pp. 301–307, 2008.

[23] R. Datar, S. Kim, S. Jeon et al., "Cantilever sensors: nanomechanical tools for diagnostics," *MRS Bulletin*, vol. 34, no. 6, pp. 449–454, 2009.

[24] C. A. Mirkin, R. L. Letsinger, R. C. Mucic, and J. J. Storhoff, "A DNA-based method for rationally assembling nanoparticles into macroscopic materials," *Nature*, vol. 382, no. 6592, pp. 607–609, 1996.

[25] C. C. You, O. R. Miranda, B. Gider et al., "Detection and identification of proteins using nanoparticle-fluorescent polymer 'chemical nose' sensors," *Nature Nanotechnology*, vol. 2, no. 5, pp. 318–323, 2007.

[26] P. Baptista, E. Pereira, P. Eaton et al., "Gold nanoparticles for the development of clinical diagnosis methods," *Analytical and Bioanalytical Chemistry*, vol. 391, no. 3, pp. 943–950, 2008.

[27] J. A. A. Ho, H. C. Chang, N. Y. Shih et al., "Diagnostic detection of human lung cancer-associated antigen using a gold nanoparticle-based electrochemical immunosensor," *Analytical Chemistry*, vol. 82, no. 14, pp. 5944–5950, 2010.

[28] A. Kumar, B. M. Boruah, and X. J. Liang, "Gold nanoparticles: promising nanomaterials for the diagnosis of cancer and HIV/AIDS," *Journal of Nanomaterials*, vol. 2011, Article ID 202187, 17 pages, 2011.

[29] P. K. Jain, K. S. Lee, I. H. El-Sayed, and M. A. El-Sayed, "Calculated absorption and scattering properties of gold nanoparticles of different size, shape, and composition: applications in biological imaging and biomedicine," *Journal of Physical Chemistry B*, vol. 110, no. 14, pp. 7238–7248, 2006.

[30] J. Wang, M. O'Toole, A. Massey et al., "Highly specific, MIR fluorescent contrast agent with emission controlled by gold nanoparticle," *Advances in Experimental Medicine and Biology*, vol. 701, no. 4, pp. 149–154, 2011.

[31] T. Chung, S. Y. Lee, E. Y. Song, H. Chun, and B. Lee, "Plasmonic nanostructures for nano-scale bio-sensing," *Sensors*, vol. 11, no. 11, pp. 10907–10929, 2011.

[32] S. Chen, M. Svedendahl, M. Käll, L. Gunnarsson, and A. Dmitriev, "Ultrahigh sensitivity made simple: nanoplasmonic label-free biosensing with an extremely low limit-of-detection for bacterial and cancer diagnostics," *Nanotechnology*, vol. 20, no. 43, Article ID 434015, 2009.

[33] V. Myroshnychenko, J. Rodríguez-Fernández, I. Pastoriza-Santos et al., "Modelling the optical response of gold nanoparticles," *Chemical Society Reviews*, vol. 37, no. 9, pp. 1792–1805, 2008.

[34] J. D. Jackson, *Classical Electrodynamics*, Wiley, New York, NY, USA, 1999.

[35] G. Mie, "Beiträge zur optik trüber medien, speziell kolloidaler metallösungen," *Annalen der Physik*, vol. 330, no. 3, pp. 377–445, 1908.

[36] H. De Voe, "Optical properties of molecular aggregates. I. Classical model of electronic absorption and refraction," *The Journal of Chemical Physics*, vol. 41, no. 2, pp. 393–400, 1964.

[37] B. T. Draine and P. J. Flatau, "Discrete-dipole approximation for scattering calculations," *Journal of the Optical Society of America A*, vol. 11, no. 4, pp. 1491–1499, 1994.

[38] http://code.google.com/p/ddscat/.

[39] http://arxiv.org/pdf/1202.3424v3.pdf.

[40] Y. W. Jung, J. J. Yoon, Y. D. Kim, and D. Woo, "Study of the interaction between biomolecule monolayers using total internal reflection ellipsometry," *Journal of the Korean Physical Society*, vol. 58, no. 42, pp. 1031–1034, 2011.

Single Mode Photonic Crystal Vertical Cavity Surface Emitting Lasers

Kent D. Choquette,[1] Dominic F. Siriani,[1] Ansas M. Kasten,[2] Meng Peun Tan,[1] Joshua D. Sulkin,[1] Paul O. Leisher,[3] James J. Raftery Jr.,[4] and Aaron J. Danner[5]

[1] Department of Electrical and Computer Engineering, University of Illinois, Urbana, IL, 61801, USA
[2] Micro and Nano Structures Technologies, GE Global Research, Niskayuna NY 12309, USA
[3] Department of Physics and Optical Engineering, Rose-Hulman Institute of Technology, Terre Haute, IN 47803, USA
[4] Department of Electrical Engineering, United States Military Academy, West Point, NY 10996, USA
[5] Department of Electrical and Computer Engineering, National University of Singapore, Singapore 117576

Correspondence should be addressed to Kent D. Choquette, choquett@illinois.edu

Academic Editor: Krassimir Panajotov

We review the design, fabrication, and performance of photonic crystal vertical cavity surface emitting lasers (VCSELs). Using a periodic pattern of etched holes in the top facet of the VCSEL, the optical cavity can be designed to support the fundamental mode only. The electrical confinement is independently defined by proton implantation or oxide confinement. By control of the refractive index and loss created by the photonic crystal, operation in the Gaussian mode can be insured, independent of the lasing wavelength.

1. Introduction

Vertical cavity surface emitting lasers (VCSELs) have emerged as the commercial laser source of choice for short distance digital fiber optical interconnects and sensing applications. The principle advantage that VCSELs have in many of these applications are their low operating power requirements (only a few mW) as well as their low cost and large volume manufacturing. Compared to edge emitting semiconductor lasers, VCSELs also possess the benefits of a circular output beam, on wafer testing, and the ability to form 2-dimensional arrays. The most common emission wavelength is 850 nm, although wavelengths from 640 to 1300 nm have been demonstrated for VCSELs monolithically grown on GaAs substrates.

Unlike an edge emitting laser, VCSELs have a single longitudinal mode but tend to operate in multiple transverse optical modes. This arises because the optical cavity of the VCSEL is short in the direction of light propagation (typically the cavity is 1 wavelength long, or approx. 265 nm for 850 nm emission), but the transverse cavity width defined by a selectively oxidized or ion implanted aperture [1]

is much greater (typically a few to tens of microns in diameter). Edge emitting semiconductor lasers have much longer cavities (several hundreds of microns in length) supporting numerous longitudinal modes, but with a cavity cross-section that supports a few or a single transverse/lateral mode. The number of laser emission modes will influence the spectral width of the laser emission, while the near field and far field beam profile is determined by the transverse mode profiles. For a typical multimode 850 nm VCSEL, the emission bandwidth can be roughly 3 nm, while the far field is often a ring shape, due to the higher order mode operation. The multimode VCSEL bandwidth can be a limiting factor for high speed digital modulation through optical fiber due to spectral dispersion, while the smallest focused spot size will come from operation only in the fundamental Gaussian mode.

Many approaches to achieve single fundamental mode operation in VCSELs have been reported. The simplest approach is reducing the oxide [2] or implant [3] cavity diameter, but this comes at the expense of high current density and inferior laser reliability. Other approaches include a hybrid combination of oxide/implant apertures [4], surface

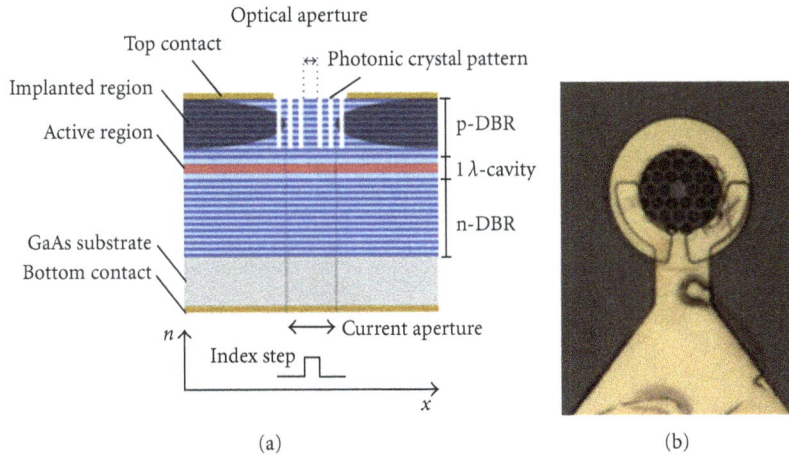

FIGURE 1: (a) Cross-section sketch and (b) top view image of planar ion implanted photonic crystal VCSEL.

relief etched structures [5], increasing the optical cavity length [6], etched holey patterns [7, 8], and etched photonic crystal VCSELs [9–14].

In most cases, the technique for achieving single mode operation requires an added epitaxial semiconductor layer or etched feature that has a stringent dimension control related to the wavelength. All of these approaches favor single fundamental mode operation by either increasing the gain for the fundamental mode, increasing the loss of higher order modes, or by engineering the transverse index profile of the VCSEL to support only the lowest order mode.

In this review, we will focus on index-guided single mode photonic crystal (PhC) VCSELs. In these devices, a 2-dimensional periodic pattern of holes is etched into the top distributed Bragg reflector mirror (DBR). The lasing cavity is defined by the absence of one or more holes as depicted in Figure 1. Unlike in-plane 2-dimensional photonic crystal structures where the defect mode often lies within a frequency bandgap, the type of out-of-plane defect mode in the PhC VCSEL is not created by a photonic band gap. Instead, the PhC VCSEL out-of-plane mode is confined similarly to the case in a solid-core photonic crystal fiber, which guides by way of total internal reflection [15]. As we will show, the unetched defect has a higher refractive index as compared to the surrounding etched photonic crystal region.

The refractive index and optical loss arising from the PhC region can be engineered via the photonic crystal parameters of hole arrangement, period, diameter, and etch depth. A key feature is the periodicity of the etched pattern of holes which allows us to accurately calculate the effective refractive index. The modified index is not a geometric average of the low index etched regions within the semiconductor, but is actually much less [12]. Note that nonperiodic but symmetric etched features, such as wedges or rings, can also be used to engineer the refractive index [16, 17]. However, this approach does rely upon geometrically averaging the index and thus is very sensitive to feature size and fabrication error. Moreover, etched features that are continuous, such as rings, will lead to degraded electrical injection.

The index confinement and optical loss can be exploited to create fundamental mode photonic crystal VCSELs. As shown in the comparison of Figure 2, conventional multimode VCSELs can be altered to show single mode lasing emission more than 30 dB above the nonlasing modes. Moreover, insuring single mode behavior can be maintained independent of the lasing wavelength or crystal design [18], precluding the need for accurate and/or wavelength specific features. Finally, the electrical diameter can be increased independent to the optical cavity to enable lower current density operation [13, 19] and thus potentially long laser lifetime [20].

In Section 2, we review the fabrication details for both oxide-confined and ion implanted PhC VCSELs. In Section 3, a description of the design approach which incorporates both index and loss confinement effects from the photonic crystal is introduced. In Section 4, the performance of various PhC VCSELs are presented. Finally, Section 5 summaries this review.

2. Fabrication

The epitaxial materials used are conventional all-semiconductor VCSELs consisting of upper p-type and lower n-type DBR mirrors which surround a quantum well active region [1]. In our studies, we have fabricated PhC VCSELs grown on GaAs substrates which emit at 670, 780, 850, 980, or 1300 nm [18, 21]. The device fabrication for oxide-confinement or ion implantation follows the typical steps as detailed below, with the addition of the patterning and anisotropic etching of the photonic crystal. In our PhC VCSELs, the optical cavity confinement is determined by the photonic crystal parameters of hole diameter (a), period (b), and hole depth, whereas the electrical confinement is provided by the proton-implanted or oxide aperture. This allows for precise engineering of the index guidance effect of the photonic crystal independent of the electrical confinement of the VCSEL.

Implant-confined photonic crystal VCSELs are fabricated as follows [1]. Ohmic ring contacts (Ti/Au) for the top

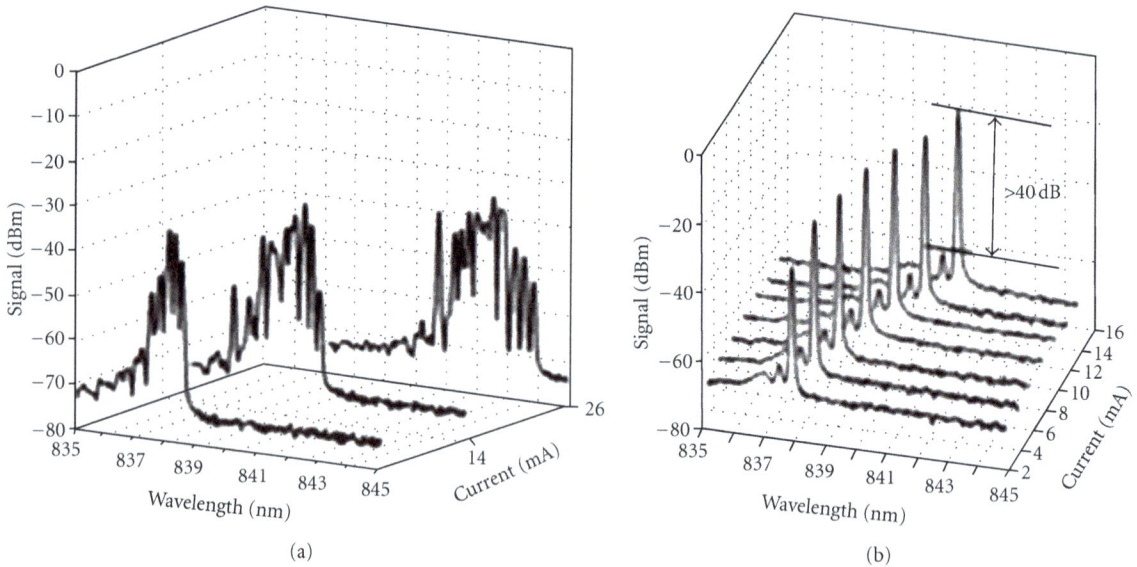

FIGURE 2: Optical spectrum of (a) oxide-confined and (b) photonic crystal VCSEL.

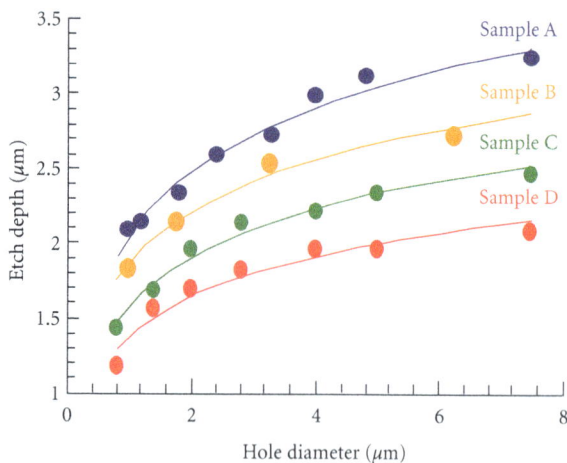

FIGURE 3: Etch depth versus hole diameter for b/a = 0.4, 0.5, 0.6, and 0.7.

p-type DBR are patterned using optical photolithography, deposited by electron-beam evaporation, and formed using conventional liftoff. A backside contact (AuGe/Ni/Au) is deposited to form an ohmic contact to the n-type substrate. A thin protective layer of SiO_2 is deposited, and thick (12 μm) photoresist pillars are patterned in the center of each top ring contact with diameter varying from 6 to 20 μm. The exposed SiO_2 is etched using CF_4 reactive ion etching (RIE). These pillars serve to mask and define the apertures for the implant process. Protons are next implanted at typically 340 keV (the implant energy will depend on the intended implant projected range [1]) with a dose of 4 $\times$ 10^{14} cm^{-2} and afterwards the resist pattern is removed. Another larger resist mask is used for a second multiple implantation step to create a stacked implant from the implant aperture to the

surface for electrical isolation between devices in a planar device topology. The implant isolation can be done either before or after the photonic crystal etch.

The photonic crystal patterns are defined by optical lithography in an SiO_2 mask. Note that the periodic index variation of the hole pattern needs not be of the same size as the light wavelength, since the light is propagating parallel to the hole patterns [12]; hence micron-sized holes are sufficient, which allows for optical photolithographic patterning. The photonic crystal holes are etched into the top DBR using $SiCl_4$/Ar inductively coupled plasma RIE. The etch rate of the holes will depend on their diameter, b, due to size-dependent etch effects [22]. The etch depth for the samples A, B, C, and D, which have a b/a = 0.4, 0.5, 0.6, and 0.7, respectively, is shown in Figure 3 as a function of photonic crystal hole diameter (b). The holes were inspected in a scanning electron microscope at an angle of 35° off-normal, and the etch depth was calculated by counting the number of DBR periods etched and multiplying by the thickness of one period. For a given etching time, smaller diameter holes etched to a shallower depth. The etch depth data can be empirically fit with a logarithmic function. The hole etch depth is limited by the large aspect ratio, but as we show below etching to and/or through the active region is undesirable. However, an important consequence is that the optical mode only experiences the photonic crystal in part of the longitudinal cavity; we will show below that this creates optical loss and confinement. After etching the photonic crystal pattern, the oxide mask is removed by CF_4 RIE, and the samples are subjected to 30 sec rapid thermal annealing at 325°C for ohmic contact formation. Figure 1(a) illustrates a schematic of a completed implant-confined photonic crystal VCSEL and Figure 1(b) shows a top view of a planar implant PhC VCSEL.

For oxide-confined VCSELs, a high aluminum content layer ($Al_{0.98}Ga_{0.02}As$) is placed above the active region for

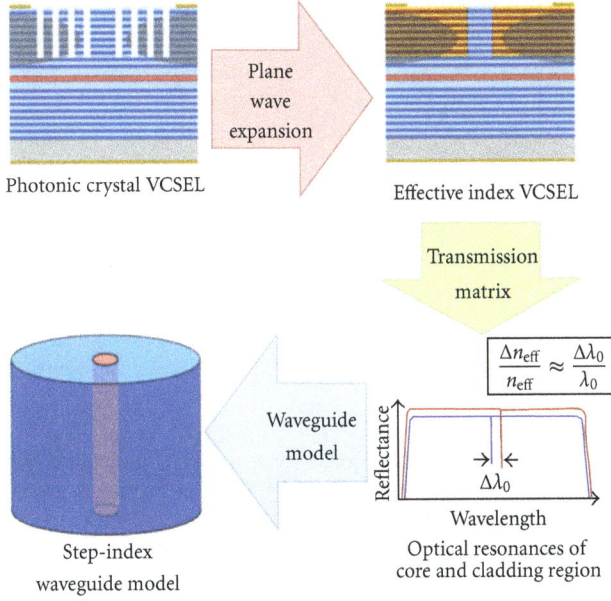

FIGURE 4: Design approach for photonic crystal VCSELs.

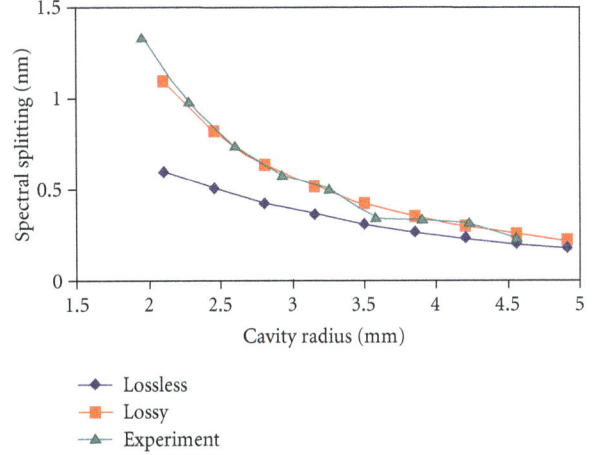

FIGURE 5: Spectral splitting between fundamental and first higher order mode versus cavity radius.

fabrication of an oxide aperture [1, 23]. After top and bottom contact metals are defined and deposited, photonic crystal patterns and an oxidation mesa or trench are defined by standard optical lithography and are etched by ICP-RIE. The hole pattern and the oxidation trench are simultaneously patterned by optical lithography but can be etched separately or simultaneously [24]. This results in self-alignment of the PhC defect to the oxide aperture. After removal of the etch mask, selective oxidation at 420°C in a steam environment creates the oxide aperture.

The index step of the PhC defect cavity can be engineered by carefully designing the periodic hole pattern as well as by controlling the etch depth of the air holes. However, to simply achieve single fundamental mode operation, PhC parameters such as lattice constant, hole diameter, or etching depths of the air holes do not need to be stringently controlled [18]. The large tolerances in the fabrication process makes PhC VCSELs suitable for mass-production and thus an excellent candidate for low cost/high volume consumer applications.

3. Design

The design of the photonic crystal has been developed to incorporate both the index confinement [10] as well as the confinement that arises due to optical loss [25, 26]. The photonic crystal has a reduced effective refractive index, due to the periodic refractive index variation (see Figure 1). The basic design approach is schematically shown in Figure 4. The VCSEL is modeled using a simplified step-index optical fiber waveguide model: the defect in the photonic crystal is considered to be the core of the fiber with index n_{core}, and the photonic crystal region surrounding is taken as the reduced-index cladding region with n_{clad}. More computationally intensive and rigorous methods, such

as finite-difference time-domain, finite element, or vectorial and three-dimensional calculations [27–29] can also be employed.

The waveguiding and optical loss effects are incorporated by using a complex refractive index in the PhC clad region. The real part of the effective refractive index of the PhC cladding is found from the band diagram calculated using the plane wave expansion method. The slope of the band for out-of-plane propagation determines the effective refractive index for each high and low index layer of the DBR mirror. This effective homogeneous index is used to replace the DBR layers that are penetrated by the photonic crystal, thereby accounting for the finite etch depth. The index step difference between the core and clad regions can be found using the difference in the calculated resonance wavelengths [30] of a matrix calculation [26]. This procedure intrinsically accounts for the finite etch depth of the holes.

The optical loss introduced by the finite etch depth of the holes is accounted for using a complex refractive index in the cladding region [31]:

$$n_{clad} = n'_{PhC} + i n''_{loss}. \tag{1}$$

To determine the magnitude of the imaginary component, we rely upon the confinement induced by the loss. This can be inferred from the cold cavity spectral splitting between the modes observed below threshold, as discussed below. From the Helmholtz equation:

$$\nabla^2 U + n^2(r) k_0^2 U = 0, \tag{2}$$

where U is the field in three spatial dimensions, n is the radial-dependent refractive index profile, and k_0 is the free-space wave number, we assume separable solutions of the form:

$$U(r, \phi, z) = u(r) e^{-im\phi} e^{-ik_z z}, \tag{3}$$

where u is the radial field profile, m is an integer, and k_z is an effective propagation constant. The propagation constant k_z

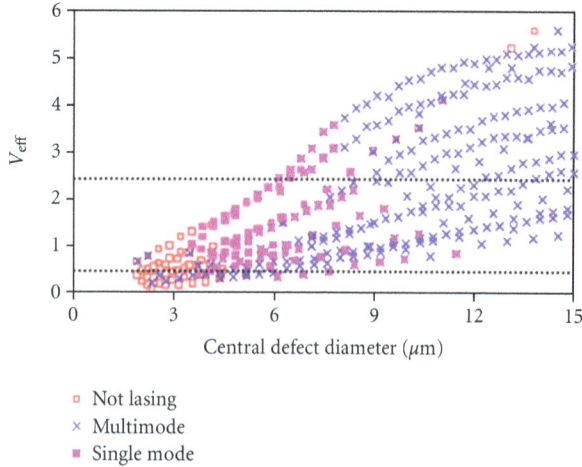

FIGURE 6: Normalized frequency and modal properties of oxide-confined 850 nm VCSELs.

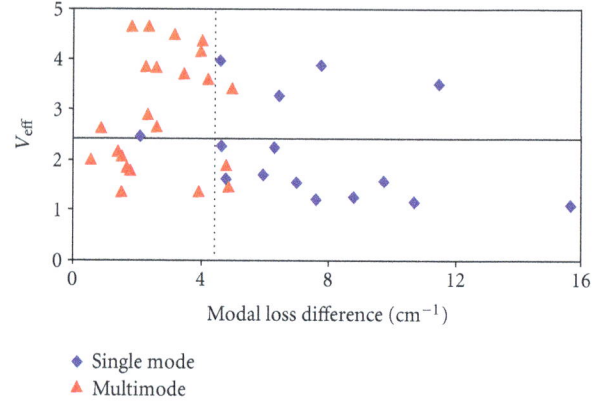

FIGURE 7: Normalized frequency versus modal loss difference between fundamental and first higher order mode of oxide-confined 850 nm VCSELs.

is set by Fabry-Perot cavity. Inserting the solutions in (3) into (2) gives

$$\frac{d^2u}{dr^2} + \frac{1}{r}\frac{du}{dr} + \left(n^2(r)k_0^2 - k_z^2 - \frac{m^2}{r^2}\right)u(r) = 0. \quad (4)$$

Solving the eigenvalue equation (4) using finite differences produces eigenvectors u and eigenvalues k_0, which are set of solutions for the resonant modes of the waveguide whose wave numbers are k_0. The resonant wavelength is

$$\lambda_0 = \frac{2\pi c}{\text{Re}\{\omega_0\}} = \frac{2\pi}{\text{Re}\{k_0\}}, \quad (5)$$

and the loss experienced by the mode is

$$\alpha_i = \text{Im}\{k_0\}. \quad (6)$$

We compare the calculated resonances (fundamental and first higher order mode) to spectral measurements from a fabricated PhC VCSEL with injection current less than threshold [25]. Figure 5 shows spectral splitting between the fundamental and first higher order mode as a function of cavity radius. The lossless (lossy) curve corresponds to a real (complex) PhC refractive index used to solve (4). Note that the splitting increases for reduced cavity size but is greater for a nonzero n'_{loss}. By varying this parameter, we can match the experimental curve shown in Figure 5. Hence, the effect of optical loss from the photonic crystal is to increase the cavity confinement.

For a lossless waveguide, the modal properties can be quantified from the normalized frequency parameter, V_{eff}:

$$V_{\text{eff}} = \frac{2\pi R}{\lambda}\sqrt{n_{\text{core}}^2 - n_{\text{clad}}^2}, \quad (7)$$

where $R = a - b/2$ is the core radius and λ is the free-space optical wavelength. For a lossless guide, the single-mode cutoff is $V_{\text{eff}} < 2.405$ [32]. Note that in (7), the finite etch depth dependence is accounted in the determination of the purely real n_{clad}. In the next section, we compare the

calculated values of V_{eff} and the extracted values of optical loss, with the modal characteristics of PhC VCSELs.

4. Performance

The characteristics of PhC VCSELs were measured using on-wafer probing at room temperature. For continuous wave light versus current measurements, the input current was varied and both the device voltage and the output from a silicon photodetector were measured using a semiconductor parameter analyzer. The spectral characteristics were measured using an optical spectrum analyzer with resolution bandwidth of 0.06 nm. At maximum power, each VCSEL was categorized as either not lasing, multimode lasing, or single mode lasing, where single mode is defined as a single spectral peak at the fundamental (lowest energy) wavelength with greater than 30 dB side mode suppression ratio. Note that VCSELs that lased single mode at threshold with higher order mode lasing at maximum power are defined as multimode.

In Figure 6 we show the calculated V_{eff} and the observed modal properties for approximately 1500 different photonic crystal designs with differing period, b/a ratios, and etch depth [33]. For many of these 850 nm oxide-confined VCSELs with $V_{\text{eff}} < 2.4$, single mode lasing is observed, particularly for cavity diameters less than 6 μm. However, multi-mode VCSELs with $V_{\text{eff}} < 2.4$ can be seen, as well as single mode PhC VCSELs with $V_{\text{eff}} > 2.4$. Hence, the loss induced confinement is a critical design parameter.

In Figure 7, we show a smaller population of approximately 40 oxide-confined 850 nm photonic crystal VCSELs with differing designs where V_{eff} and the modal loss difference between the fundamental and the first higher order mode are determined. The single mode cutoff is shown by a horizontal line in Figure 7. Again, we observe that the normalized frequency parameter is not a good determination for single mode VCSELs. However, if optical loss is considered, a cutoff can be seen between single and multimode lasing at approximately 5 cm^{-1} of loss difference

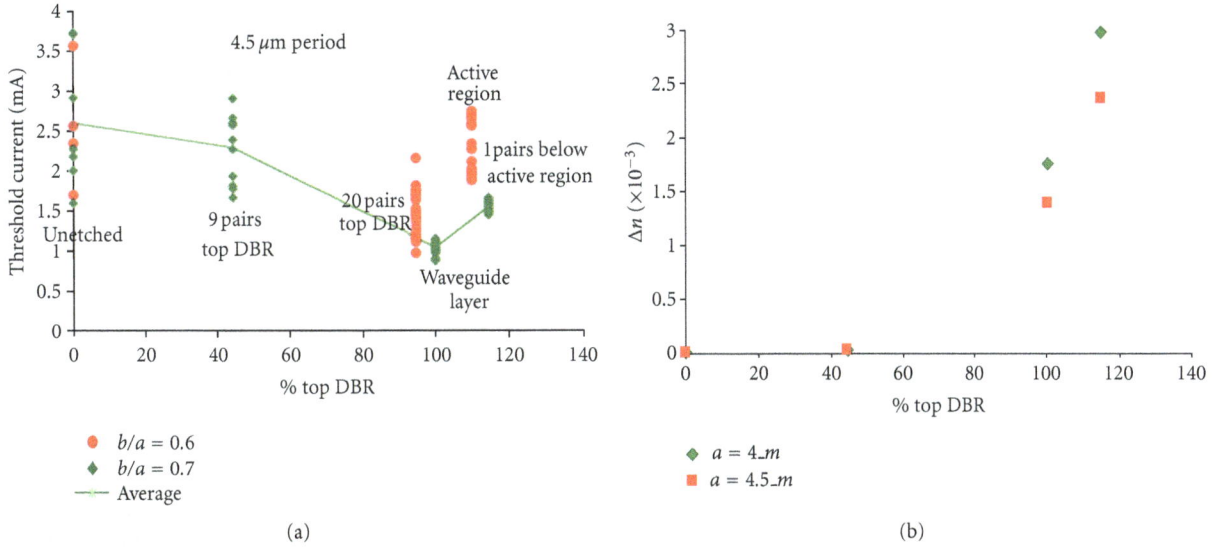

FIGURE 8: (a) Threshold current and (b) index contrast versus etch depth into top DBR for ion implanted 850 nm VCSELs.

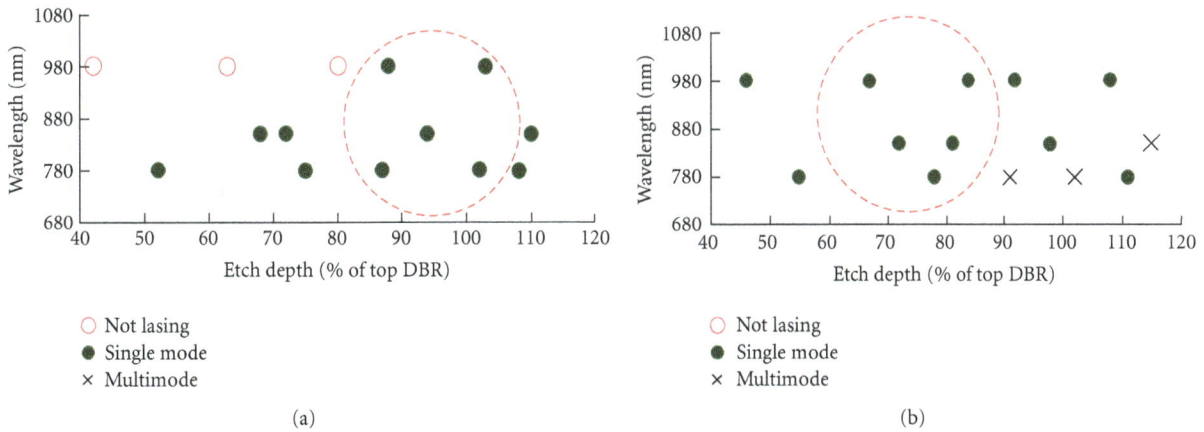

FIGURE 9: Modal properties of oxide-confined PhC VCSELs; designs that are independent of wavelength are shown in dashed circles: (a) b/a = 0.7, a = 4 μm, oxide aperture = 12 μm and (b) b/a = 0.7, a = 4.5 μm, oxide aperture = 12 μm [18].

(dotted vertical line in Figure 7). This condition corresponds to the point at which the loss to higher order mode is too great for gain to compensate and thus lasing to occur.

The effect of the etch depth on confinement is explicitly shown for implanted 850 nm photonic crystal VCSELs in Figure 8(a). Here, we calculate the difference between the core and clad index as a function of the percent etch depth into the top DBR for two b/a ratios. The confinement increases as the hole depth increases. However, etching through the active region leads to increased threshold current, as seen in Figure 8(b). Multiple VCSELs with two b/a ratios etched at different depths into the top DBR are shown. The highest threshold currents are for lasers with no etching, which arises from the diffraction loss inherent to ion implanted VCSELs [34]. The threshold current decreases with increased etch depth (increased confinement), reaching the lowest values when the photonic crystal is etched precisely through the

top DBR. However, when the VCSELs are etched through the active region, the threshold current increases, presumably due to nonradiative recombination occurring at the edges of the penetrating holes into the quantum wells.

A notable feature of photonic crystal designs is that the designs are wavelength agnostic. In Figure 9, we present the modal properties of 780, 850, and 980 nm VCSELs, all with the same photonic crystal pattern [18]. In this study, single mode operation is defined as >35 dB side mode suppression, and all lasers emit >1 mW of single mode output. In other studies, we have achieved as high as 3 mW of fundamental mode emission [13].

For Gaussian fundamental mode emission, we expect to see a Gaussian far field mode with a minimum of divergence. The far field divergence for visible implanted PhC VCSELs is plotted in Figure 10 as a function of their measured side mode suppression ratio (SMSR). The insets in Figure 10 are false color representation of the far field intensity

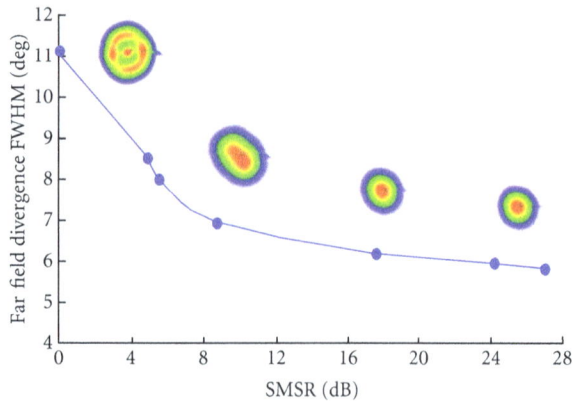

FIGURE 10: Far field divergence of implanted 670 nm PhC VCSELs as a function of side mode suppression ratio; the insets show representative far field patterns.

patterns. For multimode VCSELs (0 dB SMSR), a ring-like far field is observed. For increasing single mode operation, the divergence angle monotonically decreases.

5. Summary

Implant and oxide-confined photonic crystal VCSELs emitting from 670 to 1300 nm were designed, fabricated, and characterized to parametrically study the effects of photonic crystal design and etch depth on single mode operation. A lossy model for the photonic crystal waveguide used for transverse confinement in VCSELs has been developed. We have found a wide range of designs and fabrication approaches to achieve reproducible single fundamental mode operation. Moreover, these designs are found to be independent of wavelength and lead to 1–3 mW of output power in single aperture devices.

These developments in the control of transverse modes can significantly advance the use of VCSELs for a number of applications. With the methods developed here, photonic crystal VCSELs can be designed to operate only on a single mode while maintaining a larger aperture size and higher efficiency. Thus, these advancements could enable a single mode VCSEL with high output powers and greater reliability for sensing and long-haul communication applications.

Acknowledgments

The authors thank many former University of Illinois students for extensive VCSEL characterization. This work was partially supported by the National Science Foundation under Awards nos. 03-35082, DMI 0328162, and NDSEG Graduate Fellowship.

References

[1] K. D. Choquette and K. M. Geib, "Fabrication and performance of vertical cavity surface emitting lasers," in *Vertical Cavity Surface Emitting Lasers*, C. Wilmsen, H. Temkin, L. Coldren et al., Eds., chapter 5, Cambridge University Press, Cambridge, UK, 1999.

[2] C. Jung, R. Jager, M. Grabherr et al., "4.8 mW single-mode oxide confined top surface emitting vertical-cavity laser diodes," *Electronics Letters*, vol. 33, no. 21, pp. 1790–1791, 1997.

[3] R. A. Morgan, G. D. Guth, M. W. Focht et al., "Transverse mode control of vertical-cavity top-surface-emitting lasers," *IEEE Photonics Technology Letters*, vol. 5, no. 4, pp. 374–377, 1993.

[4] E. W. Young, K. D. Choquette, S. L. Chuang, K. M. Geib, A. J. Fischer, and A. A. Allerman, "Single-transverse-mode vertical-cavity lasers under continuous and pulsed operation," *IEEE Photonics Technology Letters*, vol. 13, no. 9, pp. 927–929, 2001.

[5] H. Martinsson, J. A. Vukušić, M. Grabherr et al., "Transverse mode selection in large-area oxide-confined vertical-cavity surface-emitting lasers using a shallow surface relief," *IEEE Photonics Technology Letters*, vol. 11, no. 12, pp. 1536–1538, 1999.

[6] H. J. Unold, S. W. Z. Mahmoud, R. Jager, M. Kicherer, M. C. Riedl, and K. J. Ebeling, "Improving single-mode VCSEL performance by introducing a long monolithic cavity," *IEEE Photonics Technology Letters*, vol. 12, no. 8, pp. 939–941, 2000.

[7] A. Furukawa, S. Sasaki, M. Hoshi, A. Matsuzono, K. Moritoh, and T. Baba, "High-power single-mode vertical-cavity surface-emitting lasers with triangular holey structure," *Applied Physics Letters*, vol. 85, no. 22, pp. 5161–5163, 2004.

[8] P. O. Leisher, A. J. Danner, J. J. Raftery Jr., and K. D. Choquette, "Proton implanted singlemode holey vertical-cavity surface-emitting lasers," *Electronics Letters*, vol. 41, no. 18, pp. 1010–1011, 2005.

[9] D. S. Song, S. H. Kim, H. G. Park, C. K. Kim, and Y. H. Lee, "Single-fundamental-mode photonic-crystal vertical-cavity surface-emitting lasers," *Applied Physics Letters*, vol. 80, no. 21, p. 3901, 2002.

[10] N. Yokouchi, A. J. Danner, and K. D. Choquette, "Two-dimensional photonic crystal confined vertical-cavity surface-emitting lasers," *IEEE Journal on Selected Topics in Quantum Electronics*, vol. 9, no. 5, pp. 1439–1445, 2003.

[11] A. J. Danner, J. J. Raftery Jr., T. Kim, P. O. Leisher, A. V. Giannopoulos, and K. D. Choquette, "Progress in photonic crystal vertical cavity lasers," *IEICE Transactions on Electronics*, vol. E88-C, no. 5, pp. 944–949, 2005.

[12] N. Yokouchi, A. J. Danner, and K. D. Choquette, "Etching depth dependence of the effective refractive index in two-dimensional photonic-crystal-patterned vertical-cavity surface-emitting laser structures," *Applied Physics Letters*, vol. 82, no. 9, pp. 1344–1346, 2003.

[13] A. J. Danner, T. S. Kim, and K. D. Choquette, "Single fundamental mode photonic crystal vertical cavity laser with improved output power," *Electronics Letters*, vol. 41, no. 6, pp. 325–326, 2005.

[14] H. P. D. Yang, F. I. Lai, Y. H. Chang et al., "Singlemode (SMSR<40dB) proton-implanted phptonic crystal vertical-cavity surface-emitting lasers," *Electronics Letters*, vol. 41, no. 6, pp. 326–328, 2005.

[15] T. A. Birks, J. C. Knight, and P. S. J. Russell, "Endlessly single-mode photonic crystal fiber," *Optics Letters*, vol. 22, no. 13, pp. 961–963, 1997.

[16] A. Furukawa, S. Sasaki, M. Hoshi, A. Matsuzono, K. Moritoh, and T. Baba, "High-power single-mode vertical-cavity surface-emitting lasers with triangular holey structure," *Applied Physics Letters*, vol. 85, no. 22, pp. 5161–5163, 2004.

[17] P. O. Leisher, A. J. Danner, J. J. Raftery, D. Siriani, and K. D. Choquette, "Loss and index guiding in single-mode proton-implanted holey vertical-cavity surface-emitting lasers," *IEEE Journal of Quantum Electronics*, vol. 42, no. 10, pp. 1091–1096, 2006.

[18] A. M. Kasten, M. P. Tan, J. D. Sulkin, P. O. Leisher, and K. D. Choquette, "Photonic crystal vertical cavity lasers with wavelength-independent single-mode behavior," *IEEE Photonics Technology Letters*, vol. 20, no. 23, pp. 2010–2012, 2008.

[19] C. Chen, Z. Tian, K. D. Choquette, and D. V. Plant, "25-Gb/s direct modulation of implant confined holey vertical-cavity surface-emitting lasers," *IEEE Photonics Technology Letters*, vol. 22, no. 7, pp. 465–467, 2010.

[20] A. M. Kasten, J. D. Sulkin, P. O. Leisher, D. K. McElfresh, D. Vacar, and K. D. Choquette, "Manufacturable photonic crystal single-mode and fluidic vertical-cavity surface-emitting lasers," *IEEE Journal on Selected Topics in Quantum Electronics*, vol. 14, no. 4, pp. 1123–1131, 2008.

[21] P. O. Leisher, A. J. Danner, and K. D. Choquette, "Single-mode 1.3-μm photonic crystal vertical-cavity surface-emitting laser," *IEEE Photonics Technology Letters*, vol. 18, no. 20, pp. 2156–2158, 2006.

[22] R. A. Gottscho and C. W. Jurgensen, "Microscopic uniformity in plasma etching," *Journal of Vacuum Science & Technology B*, vol. 10, p. 2133, 1992.

[23] K. D. Choquette, R. P. Schneider Jr., K. L. Lear, and K. M. Geib, "Low threshold voltage vertical-cavity lasers fabricated by selective oxidation," *Electronics Letters*, vol. 30, no. 24, pp. 2043–2044, 1994.

[24] M. S. Alias, S. Shaari, P. O. Leisher, and K. D. Choquette, "Single transverse mode control of VCSEL by photonic crystal and trench patterning," *Photonics and Nanostructures*, vol. 8, no. 1, pp. 38–46, 2010.

[25] D. F. Siriani, P. O. Leisher, and K. D. Choquette, "Loss-induced confinement in photonic crystal vertical-cavity surface-emitting lasers," *IEEE Journal of Quantum Electronics*, vol. 45, no. 7, pp. 762–768, 2009.

[26] D. F. Siriani, M. P. Tan, A. M. Kasten et al., "Mode control in photonic crystal vertical-cavity surface-emitting lasers and coherent arrays," *IEEE Journal on Selected Topics in Quantum Electronics*, vol. 15, no. 3, Article ID 4781555, pp. 909–917, 2009.

[27] P. Bienstman, R. Baets, J. Vukusic et al., "Comparison of optical VCSEL models on the simulation of oxide-confined devices," *IEEE Journal of Quantum Electronics*, vol. 37, no. 12, pp. 1618–1631, 2001.

[28] T. Czyszanowski, M. Dems, and K. Panajotov, "Single mode condition and modes discrimination in photonic-crystal 1.3 μm AlInGaAs/InP VCSEL," *Optics Express*, vol. 15, no. 9, pp. 5604–5609, 2007.

[29] T. Czyszanowski, M. Dems, and K. Panajotov, "Optimal parameters of photonic-crystal vertical-cavity surface-emitting diode lasers," *Journal of Lightwave Technology*, vol. 25, no. 9, pp. 2331–2336, 2007.

[30] G. R. Hadley, "Effective index model for vertical-cavity surface-emitting lasers," *Optics Letters*, vol. 20, no. 13, pp. 1483–1485, 1995.

[31] A. E. Siegman, "Propagating modes in gain-guided optical fibers," *Journal of the Optical Society of America A*, vol. 20, no. 8, pp. 1617–1628, 2003.

[32] R. S. Quimby, *Photonics and Lasers*, John Wiley & Sons, New Jersey, NJ, USA, 2006.

[33] A. J. Danner, J. J. Raftery Jr., P. O. Leisher, and K. D. Choquette, "Single mode photonic crystal vertical cavity lasers," *Applied Physics Letters*, vol. 88, no. 9, Article ID 091114, 2006.

[34] G. Hasnain, K. Tai, L. Yang et al., "Performance of gain-guided surface emitting lasers with semiconductor distributed Bragg reflectors," *IEEE Journal of Quantum Electronics*, vol. 27, no. 6, pp. 1377–1385, 1991.

9

End-to-End Image Simulator for Optical Imaging Systems: Equations and Simulation Examples

Peter Coppo,[1] **Leandro Chiarantini,**[1] **and Luciano Alparone**[2]

[1] *Selex Galileo, Via A. Einstein, 35, Florence, 50013 Campi Bisenzio, Italy*
[2] *Department of Electronics & Telecommunications, University of Florence, Via S. Marta 3, 50139 Florence, Italy*

Correspondence should be addressed to Peter Coppo; peter.coppo@selexgalileo.com

Academic Editor: Marija Strojnik

The theoretical description of a simplified end-to-end software tool for simulation of data produced by optical instruments, starting from either synthetic or airborne hyperspectral data, is described and some simulation examples of hyperspectral and panchromatic images for existing and future design instruments are also reported. High spatial/spectral resolution images with low intrinsic noise and the sensor/mission specifications are used as inputs for the simulations. The examples reported in this paper show the capabilities of the tool for simulating target detection scenarios, data quality assessment with respect to classification performance and class discrimination, impact of optical design on image quality, and 3D modelling of optical performances. The simulator is conceived as a tool (during phase 0/A) for the specification and early development of new Earth observation optical instruments, whose compliance to user's requirements is achieved through a process of cost/performance trade-off. The Selex Galileo simulator, as compared with other existing image simulators for phase C/D projects of space-borne instruments, implements all modules necessary for a complete panchromatic and hyper spectral image simulation, and it allows excellent flexibility and expandability for new integrated functions because of the adopted IDL-ENVI software environment.

1. Introduction

Hyper-spectral imaging has dramatically changed the rationale of remote sensing of the Earth relying on spectral diversity.

Since the pioneering Hyperion mission launched in 2001 [1], hyper spectral imaging airborne and satellite sensors have shown their utility by obtaining calibrated data for determining a wide variety of bio- and geophysical products from the collected imagery.

However, all sensors have their own set of performance characteristics, response functions, noise statistics, and so on, which determine and can challenge the validity of the generated data products. Through simulation of the sensor response, the utility of a new sensor design can be ascertained prior to construction, by running algorithms on simulated remote sensing data sets. In the case of existing well-characterised sensors the generation of simulated data assists in debugging sensor problems and provides a better understanding of a particular sensor's performance in new operational environments.

In this paper, an end-to-end Selex Galileo (SG) simulation tool developed in the ENVI-IDL [2] environment for the generation of simulated data from airborne/space-borne optical and infrared instruments, starting from high resolution imagery is presented.

High resolution hyper-spectral data from airborne campaigns can be typically used as input for space-borne sensors simulations. As an alternative, the input images can be completely synthesized by modelling the geometrical and spectral characteristics of the observed targets. The simulator is based on six different modules describing the reflectance scenario, the atmospheric conditions, the instrument models and the atmospheric inversion model.

The core modules aim to simulate instrument performances (spectral, spatial, and radiometric) from a variety of sensor parameters including optics, detector, scanning, and electronics characteristics. The Atmospheric module is based on the standard Modtran [3] model, whereas the scenario simulation module aims at associating a spectral signature to each pixel of a synthetic thematic map, whenever a high resolution image taken by an airborne instrument is not available.

Compared to a detailed instrument simulator, typically developed for the realization and commissioning phases (B/C phases) of a spaceborne/airborne payload, the proposed simplified end-to-end simulator is conceived as a tool (phase 0/A) to enable the rapid dimensioning of a new optical instrument and to trace the link between user and instrument requirements. SG simulator (SG_SIM) pursues a similar philosophy as other approaches useful for 0/A phases (e.g., SENSOR, MODO, CAMEO, and PICASSO), and it includes all main functions (implemented in the IDL-ENVI SW environment) necessary for a complete hyper spectral image simulation, which are not often simultaneously present in the others.

For instance, in comparison to SENSOR [4] the control of spectral mixing and the generation of synthetic scenes are also considered, whereas in comparison to US simulators, for example CAMEO [5] and PICASSO [6–8]), the extension to the MWIR/LWIR spectral bands, and a 3D reflectance rendering are missing.

After a detailed theoretical description of SG_SIM model equations and its key concepts (Section 2), some simulation examples for satellite and airborne hyper spectral and panchromatic data study cases are reported (Section 3).

2. Simulator Equations Description

The flow diagram of the software tool is shown in Figure 1. The input data can be either airborne reflectance images at high spatial, spectral, and radiometric resolution or synthetic reflectance maps, coming from a thematic map and a reflectance data base, and specifications for the instrument to be simulated (e.g., spatial and spectral response, sampling, transfer function, noise model, viewing geometry, and quantisation).

The simulation procedure consists of four different processing steps. First the at-sensor radiance images are obtained by using the Atmospheric Modtran code, then the signal is spatially, spectrally, and radiometric degraded by applying the specific instrument response models to generate the instrument simulated radiance image.

2.1. Atmospheric Simulation. The Atmospheric Module ingests as input a reflectance image taken at high spatial and spectral resolution which is then transformed into sensor radiance images by using the atmospheric radiances and transmittances generated by the Modtran code.

A preliminary simplified atmospheric model has been used. It considers Lambertian surface scattering, near-nadir observation, no adjacency effects, and a flat Earth. The input spectral radiance $L(\lambda, h)$ for an observation sensor at altitude h is obtained on the basis of the following relationship, derived from the radiative transfer model depicted in

Figure 2. The radiance is described from the following:

$$L(\lambda, h) = I_{\text{TOA-SUN}}(\lambda) \cdot \frac{\cos(\theta_{\text{sun}})}{d^2} \cdot \tau_\downarrow(\lambda) \cdot \frac{\rho(\lambda)}{\pi} \cdot \tau_\uparrow(\lambda, h)$$
$$+ L_{\uparrow\text{ATM}}(\lambda, h),$$

$$(1)$$

with

(i) $I_{\text{TOA-sun}}(\lambda)$ = Top of atmosphere sun irradiance (W/m^2/μm);

(ii) $\rho(\lambda)$ = Earth surface reflectance;

(iii) θ_{sun} = Sun observation angle (function of latitude, longitude, day of the year, and time);

(iv) d^2 = Earth-Sun distance normalised to mean (depending from day of the year);

(v) $\tau_\downarrow(\lambda)$ = Total downwards atmosphere transmission;

(vi) $\tau_\uparrow(\lambda, h)$ = Total upwards atmosphere transmission from ground to the observation altitude h;

(vii) $L_{\uparrow ATM}(\lambda, h)$ = Scattered atmosphere radiance (W/m^2/sr/μm) from ground to the observation altitude h;

(viii) $L(\lambda, h)$ = Total Atmosphere radiance (W/m^2/sr/μm) which represents the input to the instrument at altitude h.

The downwards/upwards atmospheric transmittances $\tau_\downarrow(\lambda)$, $\tau_\uparrow(\lambda, h)$ and the atmospheric radiance $L_{\uparrow ATM}(\lambda, h)$ depend on the concentration of all atmospheric gases and the aerosols distribution. The simulator allows the control of the major variable atmospheric gases (i.e., the columnar water vapour and CO_2 contents), the aerosols visibility at a certain observation altitude h, and the aerosols profile. These parameters can be controlled by means of Modtran code inputs, while the other parameters are considered constant. A dedicated graphical interface is used to create the Modtran input charts.

Generally the surface reflectance's images $\rho(\lambda)$, used as input to the simulator, come from a data base of experimental airborne or ground truth data acquired with other spectrometers, and they are affected by the spectral response of those instruments used for the database acquisition.

The radiances $L(\lambda, h)$ are generated from the Modtran code at the maximum spectral resolution (1 cm^{-1}) and are convolved with the spectral response (SR) of the instrument used to generate the data base. This spectral response (SR) is approximated with a Gaussian function with the centre wavelength λ_c and the Full Width at Half Maximum (FWHM) equal to $\Delta\lambda/4$, where the integral is performed in a $\Delta\lambda$ spectral

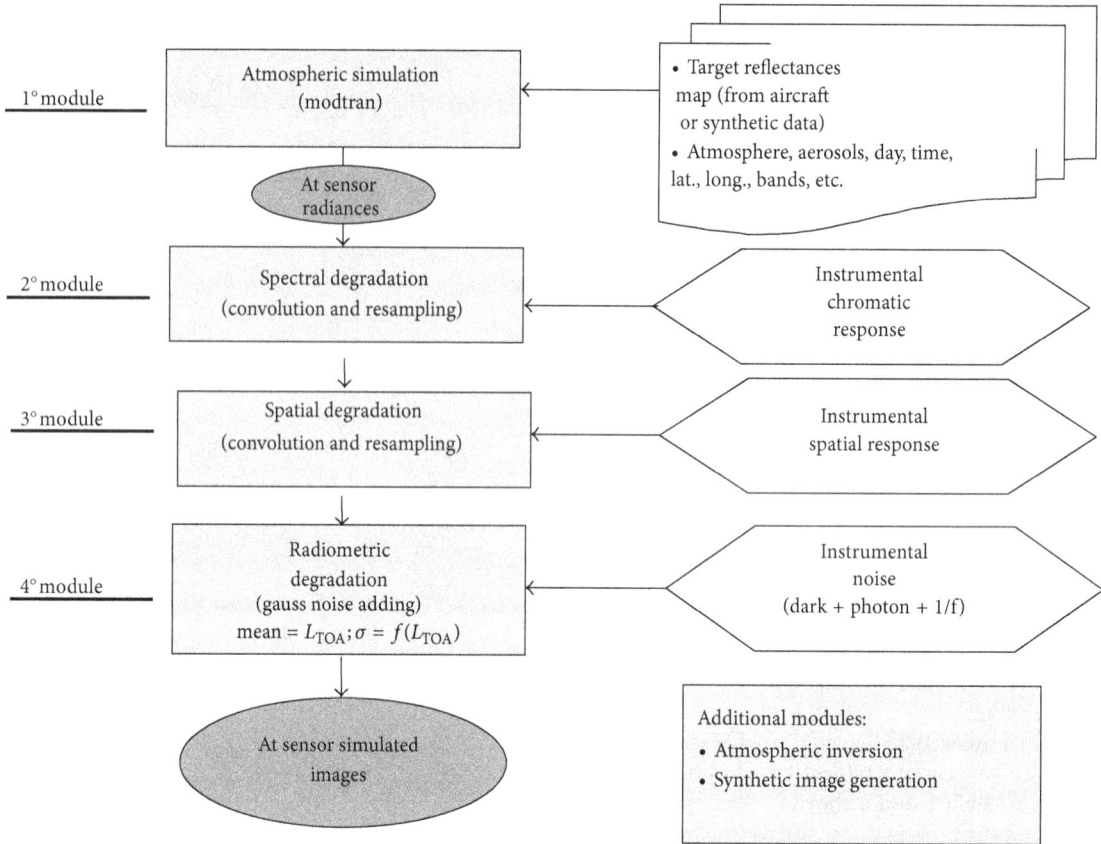

FIGURE 1: Airborne/space-borne optical sensor data simulator (flow diagram).

range centred in λ_c, and can be written as in the following:

$$
L_{1i}(\lambda, h) = \int_{\lambda_c - \Delta\lambda/2}^{\lambda_c + \Delta\lambda/2} \frac{L_i(\lambda, h) \cdot \mathrm{SR}_i[\lambda_c(i) - \lambda] \cdot d\lambda}{\Delta\lambda}
$$

$$
= \int_{\lambda_c - \Delta\lambda/2}^{\lambda_c + \Delta\lambda/2} \left[I_{\mathrm{TOA\text{-}SUN}}(\lambda) \cdot \frac{\cos(\theta_{\mathrm{sun}})}{d^2} \cdot \tau_\downarrow(\lambda) \right.
$$
$$
\left. \cdot \frac{\rho_i(\lambda_c(i))}{\pi} \cdot \tau_\uparrow(\lambda, h) + L_{\uparrow\mathrm{ATM}}(\lambda, h) \right]
$$
$$
\cdot \frac{\mathrm{SR}_i[\lambda_c(i) - \lambda] \cdot d\lambda}{\Delta\lambda}
$$

$$
\cong \int_{\lambda_c - \Delta\lambda/2}^{\lambda_c + \Delta\lambda/2} \left[I_{\mathrm{TOA\text{-}SUN}}(\lambda) \cdot \frac{\cos(\theta_{\mathrm{sun}})}{d^2} \cdot \tau_\downarrow(\lambda) \right.
$$
$$
\left. \cdot \frac{\overline{\rho}_i(\lambda_c, \Delta\lambda)}{\pi} \cdot \tau_\uparrow(\lambda, h) + L_{\uparrow\mathrm{ATM}}(\lambda, h) \right]
$$
$$
\cdot \frac{\mathrm{SR}_i[\lambda_c(i) - \lambda] \cdot d\lambda}{\Delta\lambda}
$$

$$
= \int_{\lambda_c - \Delta\lambda/2}^{\lambda_c + \Delta\lambda/2} \frac{\overline{L}_i(\lambda, h) \cdot \mathrm{SR}_i[\lambda_c(i) - \lambda] \cdot d\lambda}{\Delta\lambda}.
$$

$$(2)$$

With

(i) $L_i(\lambda, h) = [I_{\mathrm{TOA\text{-}SUN}}(\lambda) \cdot \cos(\theta_{\mathrm{sun}})/d^2 \cdot \tau_\downarrow(\lambda) \cdot \rho_i(\lambda_c(i))/ \pi \cdot \tau_\uparrow(\lambda, h) + L_{\uparrow\mathrm{ATM}}(\lambda, h)]$ the output spectral radiance obtained from Modtran by using the real surface reflectivity $\rho_i(\lambda)$ for each spectral pixel i;

(ii) $\mathrm{SR}_i[\lambda, \lambda_c(i)]$ the normalised spectral response of instrument used to generate the data base for the ith spectral channel (with $\lambda_c(i)$ the central wavelength) as a function of wavelength λ. Each $\mathrm{SR}_i[\lambda, \lambda_c(i)]$ has been simulated with a Gaussian response centred at λ_c and FWHM equal to $\Delta\lambda/4$.

(iii) $\overline{L}_i(\lambda, h) = [I_{\mathrm{TOA\text{-}SUN}}(\lambda) \cdot \cos(\theta_{\mathrm{sun}})/d^2 \cdot \tau_\downarrow(\lambda) \cdot \overline{\rho}(\lambda_c(i), \Delta\lambda)/\pi \cdot \tau_\uparrow(\lambda, h) + L_{\uparrow\mathrm{ATM}}(\lambda, h)]$ the output spectral radiance from Modtran by using the data base reflectivity value.

(iv) $\overline{\rho}_i(\lambda_c(i), \Delta\lambda) = (\int_{\lambda c - \Delta\lambda/2}^{\lambda c + \Delta\lambda/2} \rho_i(\lambda_c(i)) \cdot \mathrm{SR}_i[\lambda_c(i) - \lambda] \cdot d\lambda)/ \Delta\lambda$ the weighting mean of the Earth surface reflectivity within the $\mathrm{SR}(\lambda)$ spectral response, which represents the data-base reflectivity value.

(v) $L_{1i}(\lambda, h)(\mathrm{W/m^2/sr/\mu m})$ the mean spectral radiance within the $\mathrm{SR}(\lambda)$ spectral response.

The high resolution at sensor radiance $L_1(\lambda, h)$ is simulated with Modtran code for different values of the surface reflectivity and a 3D Look-Up-Table (reflectivity, radiance,

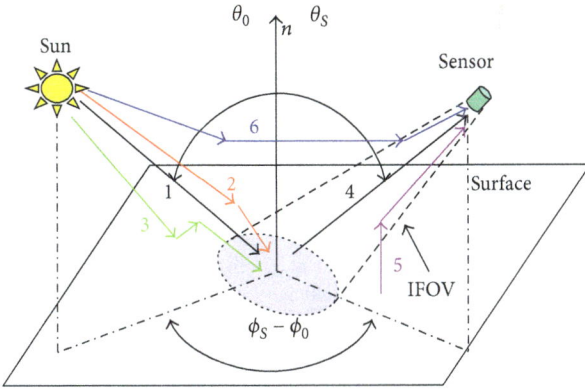

FIGURE 2: General atmospheric model scheme: the simplified atmospheric simulator takes into account in first approximation the 1, 2, 4, and 6 paths. The adjacent effects described from 3 and 5 paths will be updated in the next version of simulator.

and wavelength) is generated. Finally, for each wavelength, the simulation module determines the best linear fit between radiance and surface reflectivity, which is applied to all input reflectivity image pixels $\rho_i(\Delta\lambda)$ to generate the at sensor radiances $L_{1i}(\lambda, h)$.

2.2. Spectral Degradation. The second processing block applies a spectral degradation where the at-sensor radiance image is further spectrally degraded to the spectral channels and response of the airborne/satellite instrument to be simulated by means of a spectral interpolation and a convolution with the Instrument Chromatic Response (ICR). The $ICR_i[\lambda, \lambda_c(i)]$ represents the normalized to maximum instrument response for the ith spectral channel (with $\lambda_c(i)$ defined as the central wavelength) to a spatially uniform monochromatic source as a function of wavelength λ. The at-sensor radiance $L_2(W/m^2/sr)$ is obtained from the following:

$$L_2\left(\lambda_c(i), h\right) = \int L_1(\lambda, h) \cdot ICR_i\left[\lambda_c(i) - \lambda\right] \cdot d\lambda. \quad (3)$$

2.3. Spatial Degradation. The spatial degradation module ingests the at-sensor radiance image and degrades it to the required spatial sampling. This process is applied by means of a convolution between the input image and the Instrument Spatial Response (ISR) of the optical sensor to be simulated followed by a resampling process (decimation) (Figure 3). The ISR is defined as the response of the overall instrument in a given spatial pixel to a spectrally uniform point source as a function of its position in space. The spatially degraded radiance image (L_3) is described by the following:

$$L_3\left(\lambda_c(i), h, x, y\right) = \iint_{image} L_2\left(\lambda_c(i), h, x', y'\right)$$
$$\cdot ISR\left(x - x', y - y'\right) \cdot dx' \cdot dy'. \quad (4)$$

The ISR is calculated as the Inverse Fourier Transform of the Modulation Transfer Function (MTF), which assumes

the overall system is a linear shift invariant system. Then a "cascade model" for the system MTF is applied on the hypothesis of independent subsystems.

The hypothesis of independent subsystems is exact for many instruments, while the use of MTF, without taking into account the phase effects is valid only as a first approximation in incoherent imaging systems using well-corrected optics [10].

The "cascade model" (Figure 4) takes into account the hypothesis of separability of spatial frequency variables. Due to the properties of the Fourier Transform, the separability in the frequency domain corresponds to separability in the space domain. The along-track and across-track MTFs are calculated starting from a theoretical formulation and the Inverse Fourier Transform is calculated and normalized to a unit integral for both. In this way two unidimensional digital filters have been obtained and convolved with the high resolution image by means of the following:

$$L_3\left(\lambda_c(i), h, x, y\right)$$
$$= \int \left(\int L_2\left(\lambda_c(i), h, x', y'\right) \cdot ISR_x\left(x - x'\right) \cdot dx'\right) \quad (5)$$
$$\cdot ISR_y\left(y - y'\right) \cdot dy',$$

where:

(i) $ISR_x(x) = IFFT(MTF_x)$ the Instrument Spatial Response along x (e.g., along-track),

(ii) $ISR_y(y) = IFFT(MTF_y)$ the Instrument Spatial Response along y (e.g., across-track).

The along- and across-track $MTF_{x,y}$ are calculated taking into account the image degradation contributions reported in Table 1.

The image quality can be affected from many factors such as the size of detector (spatial aperture), the detector degradations (e.g., pixel cross talk or charge transfer & reading smearing in CCD), the integration time during image motion (temporal aperture) caused from satellite motion or the scanning system (resp. for a push broom or a whisk-broom system), the electronic filtering, the focal plane jitter (instrument micro-vibrations), the optics diffraction and aberrations [11].

These components can influence both the across-track and/or the along-track MTF depending on the direction of scanning and the disposition of detector. Some of these components are described in annex.

Examples of simulated MTF and SRF functions for airborne and spaceborne instruments that were generated with the simulator are reported in Section 3.1.

2.4. Radiometric Degradation. The fourth processing module accounts for radiometric degradation. A random noise term is added to the images to simulate the $Ne\Delta L$ (Noise Equivalent Difference Radiance in $W/m^2/sr$) of the optical instrument. The radiance $L_3(\lambda_c(i), x, y)$ of each pixel (x, y) and of the ith spectral band (with central wavelength $\lambda_c(i)$) is

TABLE 1: MTF subsystems contributions for a push broom system.

Components	Terms	Parameters
MTF Across-Track	Satellite vibration, optics diffraction, optics aberration; detector size, CCD charge transfer, and electronic filter	Jitter, detector pitch, central obscuration, focal length, pupil diameter, aberration coefficient, detector pitch, number of charge transfer, charge transfer efficiency, and filter order
MTF Along-Track	Satellite vibration, satellite motion, optics diffraction, optics aberration, and detector size	Jitter, detector pitch, Integration time, central obscuration, focal length, pupil diameter, aberration coefficient, and detector pitch

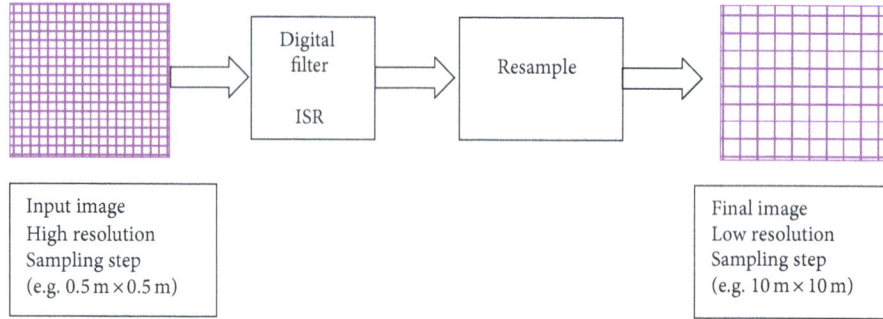

FIGURE 3: Spatial degradation module of the instrument simulator: convolution between the high spatial resolution image and the instrument spatial response (ISR) of the optical sensor that is to be simulated, followed by a resampling process (decimation).

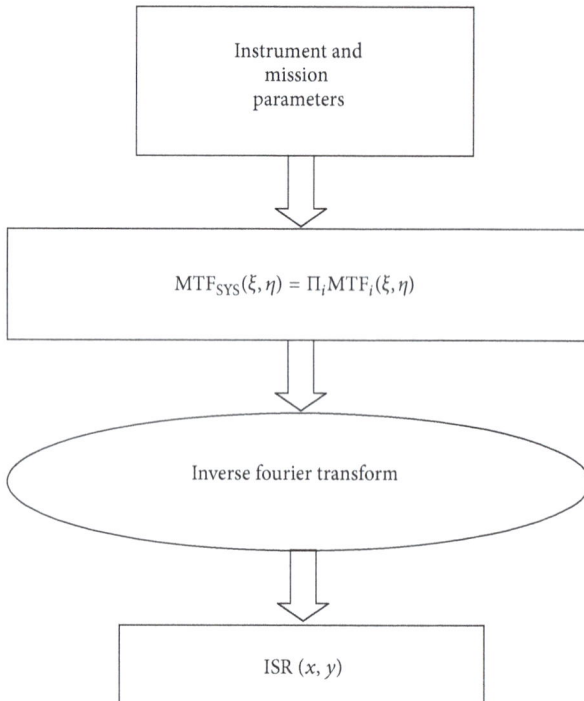

FIGURE 4: Flux diagram to calculate the instrument spatial response of the cascade model.

substituted with a random value taken from a Gaussian distribution, where $L_3(\lambda_c(i), x, y)$ represents the mean radiance value (W/m^2/sr) and $Ne\Delta L$ (W/m^2/sr) the noise equivalent radiance, which is the standard deviation of the instrument temporal noise. The relationship between $Ne\Delta L$ and the pixel radiance value L_3 is described in the following:

$$Ne\Delta L\left(\lambda_c(i), x, y\right) = \sqrt{A\left(\lambda_c(i)\right) + B\left(\lambda_c(i)\right) \cdot L_3\left(\lambda_c(i), x, y\right)}, \quad (6)$$

where $A(\lambda_c(i))$ is the noise variance of the detector (dark current, read-out and Johnson noises) plus FEE/ADC (Front End Electronics/Analog to Digital Converter) for the ith spectral band and the product $[B(\lambda_c(i)) \cdot L_3(\lambda_c(i), x, y)]$ is the photon noise variance, which is proportional to the input signal $[L_3(\lambda_c(i), x, y)]$. A Gaussian distribution function for the noise is a good approximation also for the photon noise, because the Poisson distribution approximates a Gaussian function for a high number of generated photocarries.

$Ne\Delta L$ is the minimum variation of the input radiance which can be measured and represents the radiometric resolution of the instrument. Another representation of the sensor noise can be derived from the signal-to-noise ratio (SNR) for each pixel and for each wavelength. The SNR can be obtained from the following:

$$SNR_i(\lambda) = \frac{L_3\left(\lambda_c(i), x, y\right)}{Ne\Delta L\left(\lambda_c(i), x, y\right)}$$
$$= \frac{L_3\left(\lambda_c(i), x, y\right)}{\sqrt{A\left(\lambda_c(i)\right) + B\left(\lambda_c(i)\right) \cdot L_3\left(\lambda_c(i), x, y\right)}}. \quad (7)$$

The photon noise formulation reported in (6) ($[B(\lambda_c(i)) \cdot L_3(\lambda_c(i), x, y)]$) is based on the relationship between the number of acquired electrons Ne and the integrated input radiance L_{IN} (W/m^2/sr) for each spectral channel as described in the following:

$$N_e (e^-) = \overline{L}_{IN} \left(\frac{W}{m^2 \, sr} \right) \cdot A_{pupil} (m^2) \cdot \Omega \, (sr) \cdot \overline{\tau}$$

$$\cdot \left(\frac{\overline{\lambda}}{h \cdot c} \right) (J^{-1}) \cdot t_{int} (sec) \cdot \eta \, (e^-/phot)$$

$$= K \cdot \overline{L}_{IN} \left(W/m^2 \, sr \right),$$

(8)

with

(i) $\overline{\tau}$ = Total mean in band (with ICR) instrument transmittance;

(ii) A_{pupil} = Input pupil area (m^2);

(iii) Ω = Scene pixel IFOV (sr);

(iv) t_{int} = Integration time (sec);

(v) η = Mean in band (with ICR) detector quantum efficiency (electrons/photon);

(vi) $hc/\lambda = (1{,}98E-19 \, Joule*\mu m)/\lambda$ = Energy of a photon of wavelength λ (μm);

(vii) L_{IN} (W/m^2/sr) = Spectrally integrated mean radiance in the ICR;

(viii) $K(e^-/(W/m^2 \, sr)) = A_{pupil}(m^2) \cdot \Omega(sr) \cdot \overline{\tau} \cdot (\overline{\lambda}/(h \cdot c)) \cdot t_{int}(sec) \cdot \overline{\eta}(e^-/phot))$ the coefficient of proportionality between the number of acquired electrons N_e and the input radiance L_{IN}.

The photon noise equivalent difference radiance $Ne\Delta L$ is related to the photon noise equivalent difference electron $Ne\Delta N_e$, which can be obtained from the standard deviation of the Poisson noise distribution. This standard deviation is equal to the square root of the number of electrons itself and is described from the following:

$$Ne\Delta N_e = \sqrt{N_e},$$

$$Ne\Delta L = \frac{\sqrt{N_e}}{K} = \frac{\sqrt{K \cdot \overline{L}_{IN}}}{K} = \sqrt{\frac{\overline{L}_{IN}}{K}} \equiv \sqrt{B \cdot \overline{L}_{IN}},$$

$$B \left(\frac{W/m^2 \, sr}{e^-} \right) = \frac{1}{K} = \frac{1}{A_{pupil} \cdot \Omega \cdot \overline{\tau} \cdot \left(\overline{\lambda}/(h \cdot c) \right) \cdot t_{int} \cdot \overline{\eta}}.$$

(9)

The A and B coefficients, which depend from the selected spectral channel and are fed as input to the simulator, can be derived from the radiometric model of the simulated optical sensor or they can be evaluated on the basis of acquired images of homogeneous targets acquired by the sensor [13–16].

Two additional procedures have been implemented to permit the analysis of the simulated images (Sections 2.5 and 2.6).

2.5. Atmospheric Correction. The first permits the retrieval of surface reflectance from airborne and spaceborne sensor radiances. Two standard methods can be used: one is based on Modtran code, by inverting (1), to obtain the surface reflectance from the instrument radiance and the second based on the standard ENVI-FLAASH [17] software, which allows aerosols to be estimated by means of the dark pixel method (water bodies, shadowed area, and dense vegetation) and the water vapour map to be estimated by means of the 820, 940, and 1135 nm absorption bands ratio method [18].

2.6. Synthetic Image Generation Module. The second procedure permits to quantitatively evaluate the impact of instrumental parameters on simulated image quality when a low noise airborne input image is not available.

In particular it allows the creation of black and white bar test images with different modulations (square or sinusoidal), periods and shading, to the scope to evaluate the impact on the image quality of instrument parameters such as MTF and noise as a function of the spatial sampling interval and the target reflectivity, and to analyse the minimum detectable albedo contrast as a function of spatial frequency and illumination conditions (Figure 5).

It is also possible to generate synthetic hyper-spectral surface reflectance images at the desired spatial and spectral resolution by using as input a thematic map of the zone under investigation (derived synthetically or from a classification) and a spectral library of the surface materials of interest. A statistical mixing of spectral signatures for each zone with the Dirichlet method permits to control the percentage of the statistical variability [19].

The following further statistical variability, devoted to a better representation of a real scenario, can be introduced [20]:

(i) a uniform or Gaussian variability for each spectral signature due to a possible spatial variation of the substance composition, such as contaminants, oxidation, and ageing, and so forth,

(ii) a beta function distributed statistical variation of illumination, which takes into account possible image errors due uncompensated observation and surface slope angles,

(iii) a Gaussian variability due to scenario noise, coming from uncompensated atmospheric and environment effects or uncompensated errors of sensors used to obtain the spectral library data.

Then the surface reflectance $\overline{R}$, represented from a column vector for each wavelength, is obtained by means of the following matrix mixing relationship:

$$\overline{R} = \overline{\overline{P}} \cdot \gamma \cdot \overline{\overline{\Psi}} \cdot \left[\overline{A} \cdot t + \overline{M} \cdot (1 - t) \right] + \overline{N},$$

(10)

with

(i) $\overline{\overline{P}}$ a matrix representing the end-members (pure elements) reflectance for each wavelength;

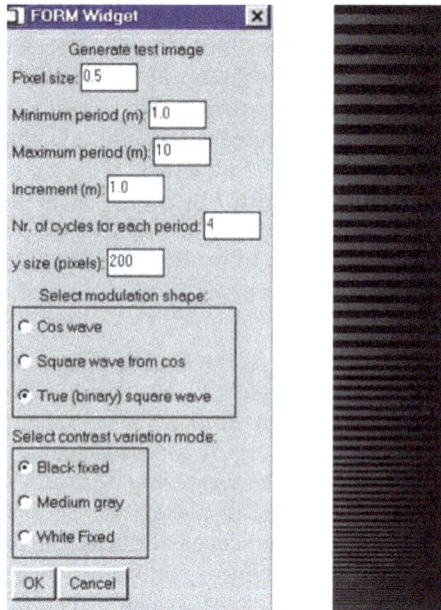

FIGURE 5: Input mask to generate the synthetic black and white images.

(ii) $\overline{A}$ a column vector representing the statistical variability of abundances, according to a Dirichlet distribution;

(iii) $\overline{M}$ a column vector representing the mean value of abundances for each wavelength;

(iv) t a scalar parameter which describes the degree of statistical mixing ($t = 0$ no mixing, $t = 1$ all random mixing);

(v) γ a statistical parameter obtained from a beta distribution, describing the illumination variation for each pixel;

(vi) $\overline{\overline{\Psi}}$ a diagonal matrix derived from a uniform or a Gaussian density, representing the spatial variation of end members;

(vii) $\overline{N}$ a column vector representing a Gaussian scenario noise, that is uncompensated atmospheric retrieval and/or sensors errors used to obtain the library data set.

3. Simulations Examples

Several simulation tests were performed to assess the potential of the tool for the instrument image quality and applications evaluation in the framework of the study and the testing phases of the Selex Galileo SIMGA airborne hyper spectral camera and the HypSeo (ASI-PRISMA precursor [21]) spaceborne hyper spectral and panchromatic cameras phase A study.

Such activities allowed also the validation of the simulator by means of real SIMGA data acquired on clay soil targets during an airborne campaign of 2009 in Mugello (Tuscany, I)

test site, where ground truth data were collected simultaneously at the same time of overflights [16].

Some examples of simulation of a 3D map representation of the ISR function have been produced for the purpose of evaluating the instrument image quality, which is generally given from the FWHM of the instrument spatial response or by the ratio between the integral of the spatial response within a delimited spatial domain (e.g., 1 Spatial Sampling Interval) and the integral in all spatial domain, which is generally called integrated energy (in percentage unit).

A more detailed analysis based on another image quality parameter (SNR*MTF) has been done to trade-off the image quality of a panchromatic camera as a function of some instrument parameters (e.g., pupil diameter and spatial sampling) for different atmospheric conditions (summer/winter and rural/urban aerosol) aiming to a better definition of the instrument requirements.

Finally VIS and SWIR radiance and reflectance simulated images have been generated for some specific targets related to civil (land use) and dual use applications for terrestrial and marine environments to the scope of understanding the instrument capabilities for targets' discrimination. These two dual use applications has been simulated during the testing phase of the airborne SG instrument (SIMGA) by means of targets of small green panels over vegetation cover and small grey panels under water, and then verified by means of an airborne campaign on a controlled area.

3.1. SRF 3D Maps for Integrated Energy Calculations. The simulator permits a 3D representation of the SRF map by using as input a delta function. As an example this representation has been done to evaluate the spatial resolution (defined in terms of percentage of integrated energy of SRF within a certain space domain) of the airborne SG SIMGA hyperspectral camera by taking into account both the laboratory measurements and the smearing effect introduced from the detector integration which occurs during platform motion. The along and the across-track MTF and SRF contributions are displayed in Figures 6(a) and 6(b), respectively, for the VIS and the SWIR channels. The instrument parameters used in the simulations are reported in Tables 2(a) and 2(b). From the tables it appears that the ratio between the FWHM of the SRF and the Spatial Sampling Distance (SSD) is much lower for SWIR channels (0.87 along scan*1.05 across scan) with respect to the VIS ones (2.70 along scan*1.49 across scan), showing that the SRF of VIS channels has a larger width (in $\pm 2/3$ pixels) with respect to that of the SWIR ones (± 1 pixel) (see also Figures 6(a) and 6(b)).

The integrated energy calculation performed within an area of 1 SSD*1 SSD of the VIS and SWIR 3D maps also confirms that the energy content within a pixel is much lower for VIS respect to SWIR channels (the same happens within the same ground size of 1.333 m*1.333 m):

(i) Integrated Energy in 0.706 m*0.706 m for VIS (1 SSD*1 SSD) = 19%,

(ii) Integrated Energy in 1.333 m*1.333 m for VIS (1.9 SSD*1.9 SSD) = 51%,

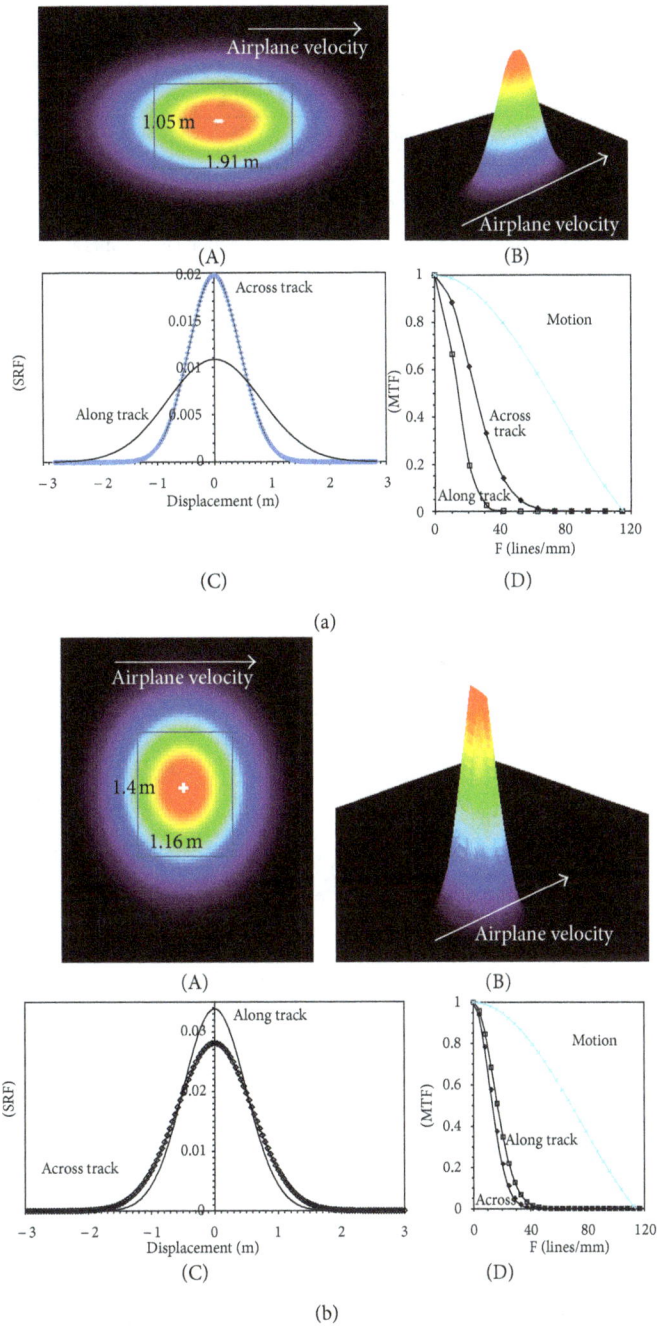

FIGURE 6: (a) Maps of Spatial Response Function (SRF) of SIMGA VIS channels ((a), (b)), plots of SRFs across and along-track (along and across-slit) for the central FOV pixel as a function of on-ground displacement (c). The simulation is obtained by using the static laboratory measurements after having introduced the smearing effect due to the along-track motion (Tables 2(a) and 2(b)). MTF motion and along/across-track components as a function of spatial frequency at detector level (d). (b) Maps of Spatial Response Function (SRF) of SIMGA SWIR channels (a), (b), plots of SRFs across and along-track (along and across-slit) for the central FOV pixel as a function of on-ground displacement (c). The simulation is obtained by using the static laboratory measurements after having introduced the smearing effect due to the along-track motion (Tables 2(a) and 2(b)). MTF motion and along/across-track components as a function of spatial frequency at detector level (d).

(iii) Integrated Energy in $1.333\,m * 1.333\,m$ for SWIR ($1\,SSD * 1\,SSD$) = 61%.

In conclusion the spatial resolution of VIS channels is coarser with respect to that of SWIR channels, also if the spatial sampling is better (0.706 m respect to 1.333 m).

A further exercise was done to simulate the 3D Hypseo SRF [16] by using the instrument parameter reported in Table 3 and the Hypseo MTF model [21]. The FWHM of the spatial response is $24.4\,m * 20.6\,m$ (along-scan * across-scan) while the integrated energy in $1\,SSD * 1\,SSD$ ($20\,m * 20\,m$) is

TABLE 2

(a) Full Width Half Maximum (FWHM) values of SIMGA Spatial Response (across and along slit) obtained by static measurements performed in laboratory (Gaussian fit to data, [9]) and smearing effect due to integration during motion

	FWHM (m) at 1 Km Across-SLIT (along-scan)	FWHM (m) at 1 Km Along-SLIT (across-scan)
VIS	1.91 (1.88 static)	1.05
SWIR	1.16 (1.13 static)	1.40

(b) SIMGA detector and optical parameters used for simulation (V = 30 m/s at 1 Km height)

	Detector pitch (micron)	Focal length (mm)	Pupil diameter (mm)	SSD (m) At H = 1 Km	FOV (deg)	Tdwell (msec)	Tint (msec)	Tint/Tdwell
VIS	12	17	7.08	0.706	±19	23.5	17	0.72
SWIR	30	22.5	11.25	1.333	±12	44.3	13	0.29

TABLE 3: Characteristics of HYPSEO Hyper-spectral and PAN cameras at 620 Km satellite altitude.

Parameters	Hyper-spectral camera VNIR band	Swir band	Pan camera Pan band			
Spectral Interval [nm]	400–1000	1000–2500	500–900			
Average Spectral sampling [nm]	10	10	500–900			
Swath [km]/FOV (°)	20/1.85		20/1.85	14/1.30	10/0.93	10/0.93
Spatial Sampling [m]	20	20	5	3.5	2.5	2.5
Useful Zone [km]	500		500			
Spectral channels (max)	≈60	≈150	1			
Aperture Diameter [mm]	150 (F/3.7)		150 (F/5.8)	150 (F/8.3)	150 (F/11.6)	300 (F/5.8)
Detector pixel dimension [μm]	18	18	7			
Array dimension	1000 × 256	1000 × 256	4000 × 1			
FPA type	Si on Hybrid CMOS	Cooled CMT on Hybrid CMOS	Si CCD			
SNR at ρ = 0.3 and SZA = 60°	≈200	≈50	144	71	32	92

53%, which is a value substantially equal to that estimated for the airborne SIMGA instrument at 1.33 m ∗ 1.33 m of pixel size.

3.2. Satellite Panchromatic Image Quality Requirements. The simulator permits to study the impact of system design parameters on the instrument image quality. To this scope a parametric analysis of the performance of the HypSEO-PAN Camera as a function of the pupil diameter dimension for different spatial sampling, atmospheric and illumination conditions was performed on the basis of simulated test images and instrument parameters (Table 4), to trade-off the instrument sizing with the image quality.

We have adopted as a first approximation of image quality criterion the Minimum Resolvable Contrast (MRC) at a certain spatial frequency f, which is equal to the inverse of the product [MTF$(f) \cdot$ SNR], where SNR is calculated for uniform scenes (spatial frequency f = 0) [11]. As a rule of thumb, the adopted value of MRC = 10% gives the following threshold relationship for target identification with spatial frequency f:

$$\text{MTF}(f) \cdot \text{SNR} > 10. \quad (11)$$

Two different kind of input images have been used for the simulations: a surface reflectance image, based on the IKONOS panchromatic camera at ~1 m spatial sampling and a bar synthetic image at a spatial sampling of 0.5 m.

In Figure 7 a comparison between simulations of the Hypseo panchromatic image, obtained from the IKONOS image, at different spatial sampling intervals and different Hypseo pupil diameters is shown for the low radiance case (case "B" in Table 4). The product MTF$(f) \cdot$ SNR has been calculated and the results are displayed in Table 5. The simulation with a high pupil diameter of 300 mm (case (b) in Table 5) is better from image quality point of view (lower GSD, high SNR∗MTF, and optimum targets discrimination in Figure 7(b)) but it has a large impact on instrument sizing. All other images are strongly affected from diffraction, due to the pupil size of 150 mm, but the case with 5 m of GSD (case (d) in Table 5) seems better for SNR∗MTF parameter and targets discrimination (in case (a) and (c) of Figure 7 the instrument noise overlays all other possible image features).

Another simulation with synthetic bars has been done to verify the previous results, by changing the sampling and the illumination conditions avoiding any possible effect coming from the degraded characteristics of the IKONOS image quality. In Figure 8 an HypSEO PAN simulation, from a synthetic bar image, at different spatial resolution (GSD) and pupil diameter (D) is shown for a high (case "A") and a low (case "B") radiance case, with parameters represented in

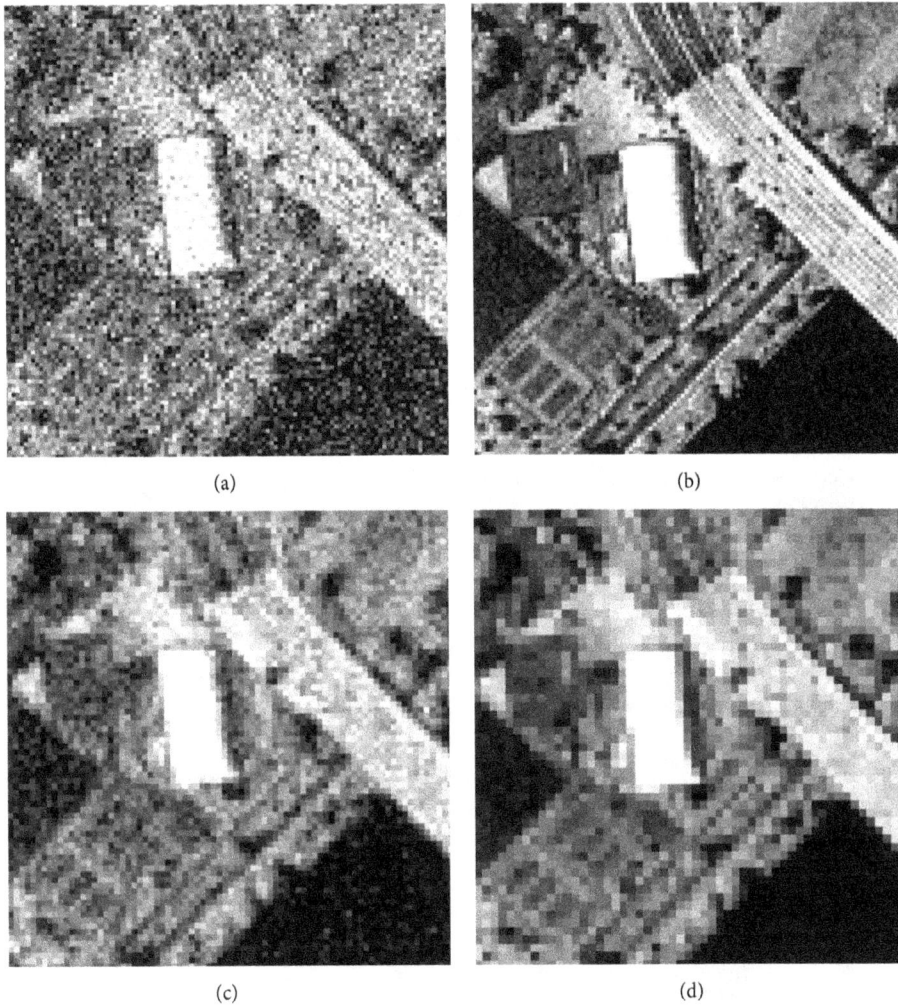

FIGURE 7: Simulated HypSEO Pan images (from IKONOS) for low TOA radiance (18 W$*$m$^{-2}*$sr$^{-1}*\mu$m^{-1} at albedo = 0.27, case "B" in Table 4), different spatial sampling (GSD) and pupil diameter (D): (a) GSD = 2.5 m, D = 150 mm (SNR = 11, MTF(f_{Nyquist}) = 0.09); (b) GSD = 2.5 m, D = 300 mm (SNR = 32, MTF(f_{Nyquist}) = 0.18); (c) GSD = 3.5 m, D = 150 mm (SNR = 27, MTF(f_{Nyquist}) = 0.17); (d) GSD = 5 m, D = 150 mm (SNR = 50, MTF(f_{Nyquist}) = 0.22).

TABLE 4: Illumination/atmospheric parameters used in the Hypseo PAN camera simulations.

	Atmosphere	Visibility	Day	Hour	Lat.	Long.	Sun Zenith Angle (SZA)	Altitude
Case A	Midlatitude summer	Rural VIS 23 Km	174	10:00 a.m.	45°	0°	32°	620 Km
Case B	Midlatitude winter	UrbanVIS 5 Km	355	10:00 a.m.	45°	0°	72°	620 Km

TABLE 5: Image quality parameters related to the simulations of Figure 7 with low illumination radiance (Case B in Table 4). The diffraction limit represents the Airy radius ($\lambda = 1\,\mu$m, H = 620 Km) of Rayleigh criterion. Case (b) is better respect to case (d) because SNR$*$MTF is related to a signal with lower Nyquist period.

Case	GSD	Pupil diameter	On ground diffraction limit	Nyquist period	SNR$*$MTF$_{\mathrm{Nyquist}}$	Image quality evaluation
(a)	2.5 m	150 mm	5.0 m	5 m	1	Low
(b)	2.5 m	300 mm	2.5 m	5 m	5.8	High
(c)	3.5 m	150 mm	5.0 m	7 m	4.6	Sufficient
(d)	5.0 m	150 mm	5.0 m	10 m	11	Sufficient

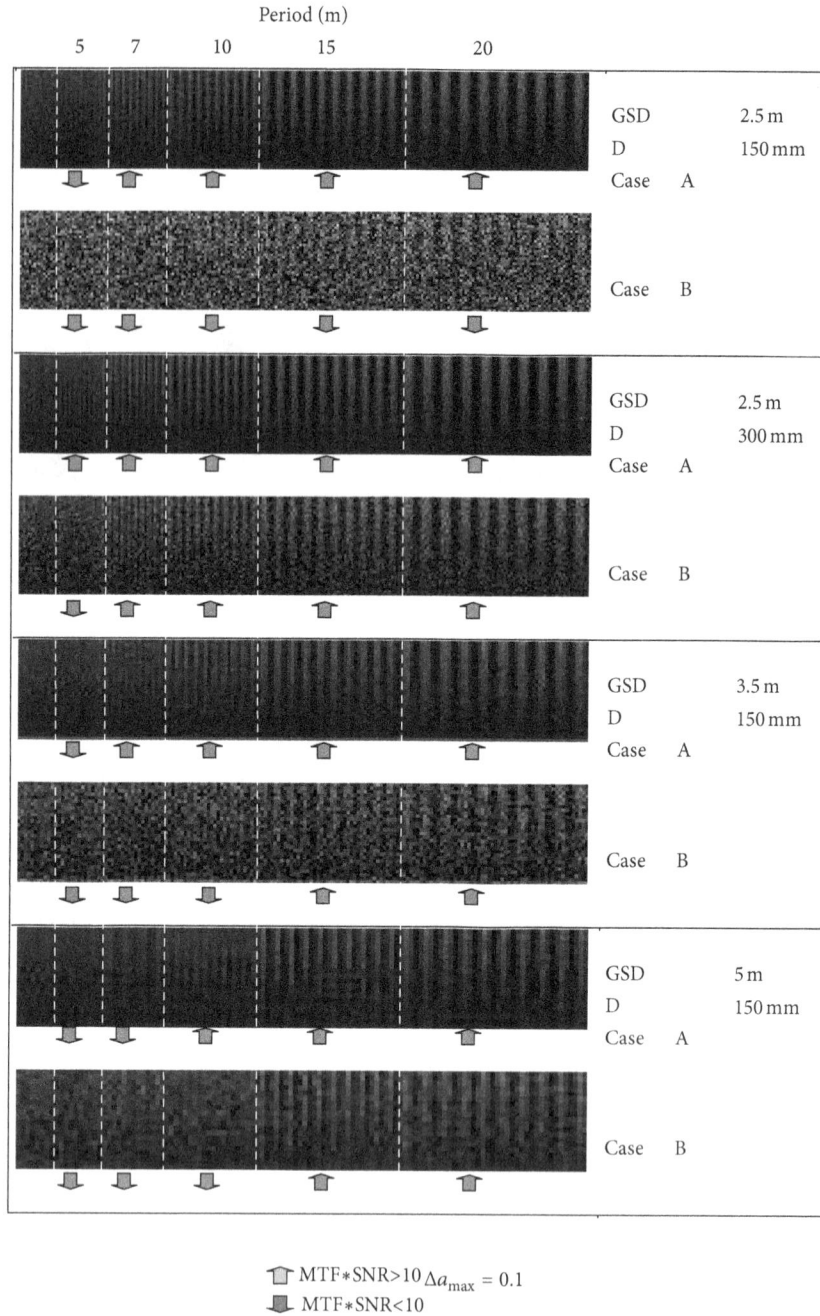

FIGURE 8: HypSEO PAN simulation (from synthetic bar image) at different spatial resolution (GSD), pupil diameter (D), and illumination conditions (high radiance case "A" and low radiance case "B" in Table 4); the contrast along the vertical axis changes between 10% and 20%; the up and down arrows indicate the periods for which the criterion (SNR∗MTF > 10) is satisfied or not.

Table 4. The input synthetic image in the horizontal direction is composed by 5 sequences of grey-black bars, each consisting of 10 cycles at fixed period (5, 7, 10, 15, and 20 m). In the vertical direction the albedo range of grey bars is between 10% and 20%, whereas the albedo of black bars is constant (10%). The up and down arrows in Figure 8 indicate the periods for which the criterion $MTF(f) \cdot SNR > 10$ is satisfied or not.

The radiance (L), the Signal to Noise Ratio (SNR), and the Modulation Transfer Function (MTF) values corresponding to the extreme simulated albedo values are represented in Table 6 while the product $MTF(f) \cdot SNR$ has been reported in Table 7.

The results confirm that image quality as defined from this kind of metric is improved by increasing pupil diameter (from 150 to 300 mm) at equal spatial sampling (GSD = 2.5 m), because of an increased SNR and a reduced diffraction effect on MTF. The image quality is also improved as spatial sampling decreases from 2.5 to 3.5 m (pupil diameter = 150 mm) because of an increased SNR.

TABLE 6: Radiance, SNR, and across-track MTF values corresponding to the extreme simulated albedo values, calculated for different ground spatial sampling (GSD), pupil diameter (D) and the two simulation conditions reported in Table 4.

Simulation scenario			SNR at L at a = 0.09		SNR at L at a = 0.20		MTF across-track at period (m)				
Case/SZA	GSD (m)	D (mm)	L (W/m²/sr/μm)	SNR	L (W/m²/sr/μm)	SNR	5	7	10	15	20
A/32°	2.5	150	40.0	22.5	73.4	37.7	0.09	0.26	0.46	0.65	0.74
B/72°			14.5	8.8	16.7	10.1					
A/32°	2.5	300	40.0	69.1	73.4	104.9	0.18	0.38	0.57	0.75	0.81
B/72°			14.5	31.0	16.7	34.9					
A/32°	3.5	150	40.0	52.0	71.5	79.8	0.04	0.17	0.37	0.58	0.68
B/72°			14.5	22.3	16.6	25.1					
A/32°	5	150	40.0	110.8	71.6	158.9	0.001	0.06	0.22	0.45	0.61
B/72°			14.5	54.0	16.6	60.0					

TABLE 7: Image quality parameter (SNR∗MTF) calculated at low and high illumination radiance (Table 4), 20% of on ground albedo and different spatial periods of Figure 8. The diffraction limit represents the Airy radius of Rayleigh criterion. Values with SNR∗MTF > 10 are bold.

Illumination	GSD	Pupil diameter	On ground diffraction limit	SNR∗MTF at r = 0.2 at spatial period					Quality evaluation
				5 m	7 m	10 m	15 m	20 m	
High	2.5 m	150 mm	5.0 m	3.4	9.8	**17.3**	**24.5**	**27.9**	Low
Low				1.0	2.6	4.6	6.6	7.5	
High	2.5 m	300 mm	2.5 m	**18.9**	**39.9**	**59.8**	**78.7**	**85.0**	High
Low				6.3	**13.3**	**19.9**	**26.2**	**28.3**	
High	3.5 m	150 mm	5.0 m	3.2	**13.6**	**29.5**	**46.3**	**54.3**	Suff.
Low				1.0	4.3	9.3	**14.6**	**17.1**	
High	5.0 m	150 mm	5.0 m	0.2	9.5	**34.9**	**71.5**	**96.9**	Suff.
Low				0.1	3.6	**13.2**	**27.9**	**36.6**	

TABLE 8: Main parameters for Modtran simulations.

Latitude	44°
Longitude	11.4°
Time	10.6
Day	23/9/09
Atmospheric model	Midlatitude summer
Aerosol model	Rural
Vis	23 Km
Water Vapour	0.4 (standard)
Ground elevation	0.25 Km
Scattering model	Scaled DISORT 4 streams
CO_2	390 ppm
Airplane altitude	1 Km

The HypSEO PAN nominal case Ground Spatial Sampling (GSD) of 5 m and pupil diameter (D) of 150 mm seems a good compromise in terms of image quality with respect to the others, because the simulation results are not so different with respect to the case with GSD = 3.5 m (D = 150 mm) (SNR∗MTF is higher for low radiance case) and from an instrument design point of view it appears more feasible with respect to the best case with GSD = 2.5 m (D = 300 mm).

For PAN nominal case the above MTF(f) · SNR criterion is satisfied only for periods larger than the Nyquist period of 10 m at high radiances, but some oscillations affected by aliasing can also be observed at low periods as 7 m.

Finally an example of simulated Hypseo PAN image (GSD = 5 m, D = 150 mm) obtained from airborne high resolution MIVIS data in a forest environment has been performed (Figure 9(c)) to the scope of testing image fusion methods based on the sharpening of hyperspectral image by means of panchromatic observations [22].

3.3. Satellite Hyperspectral Land Use Classification. Another important use of the simulator has regarded the demonstration of potential applications of the HYPSEO SG spaceborne hyperspectral camera.

A simulation of the HYPSEO SG space-borne hyperspectral camera was performed by using as input the airborne MIVIS reflectance images acquired on a Tuscany (I) test site (S. Rossore Park and Arno River mouth) at 2.5 m spatial resolution [23]. The instrumental parameters are reported in Table 3 [24].

A MIVIS reflectance image is transformed into the satellite HYPSEO radiance (Figure 9) (H = 620 Km) by using the atmospheric model parameters of Table 4. Then the HYPSEO radiance image is obtained by means of a spectral resampling of the MIVIS image to the 210 spectral

FIGURE 9: Airborne and simulated space-borne image on S. Rossore Park (I) test site: (a) MIVIS airborne reflectance (2.5 m), used as input; (b) simulated hyper-spectral HYPSEO radiance (20 m); (c) simulated PAN HYPSEO radiance (5 m) image.

bands of Hypseo, with a Gaussian Instrument Chromatic Response (with FWHM = 10 nm) and a spatial resampling to the HYPSEO spatial sampling interval of 20 m by using the simulated spatial response, and adding the noise by means of parameters coming from the HYPSEO radiometric model. Moreover an HYPSEO reflectance image has been obtained after removal of atmospheric effects introduced by MODTRAN code (Table 4).

A land use classification map based on the Spectral Angle Mapper (SAM) algorithm [25] from the HYPSEO simulated reflectance image is shown in Figure 10. The confusion matrix shows a good correlation between classified and ground truth data, compared with multispectral sensors [23], confirming the instrument capabilities for this kind of application.

3.4. Target "Camouflage" in Rural Background.

The dual use capability for targets discrimination with camouflage panels embedded in vegetation has been evaluated during the testing phase of the SIMGA airborne hyperspectral instrument. To this scope some simulations was done during the SIMGA project phase. The simulated instrument SIMGA reflectance images been obtained by using the MODTRAN code in

a standard atmospheric condition (Table 8), the measured instrument spatial response (Figures 6(a) and 6(b) and Tables 2(a) and 2(b)) and the instrument noise [16]. In Figure 11 a SIMGA reflectance image of simulated green panels over vegetation after FLAASH inversion algorithm is shown. The result of simulation showed that green panels were clearly distinguished respect to vegetation, because of their higher reflectance in the SWIR bands (1.2 and 1.6 micron), so validating the utility of the hyperspectral sensor for this kind of application. Moreover a validation of the simulation was obtained during an airborne campaign performed in S. Rossore park (Tuscany, I), where different green panels were placed over green grass. In Figure 12 the green panels are clearly distinguished in the SWIR bands while the contrast in VIS bands is negligible.

3.5. Underwater Submerged Targets.

Another dual use capability regarding the discrimination of underwater submerged targets was tested by means of SIMGA image simulations and verified with overflights in a controlled zone.

In order to test the detection capabilities of small grey panels under water a direct bathymetric model has been

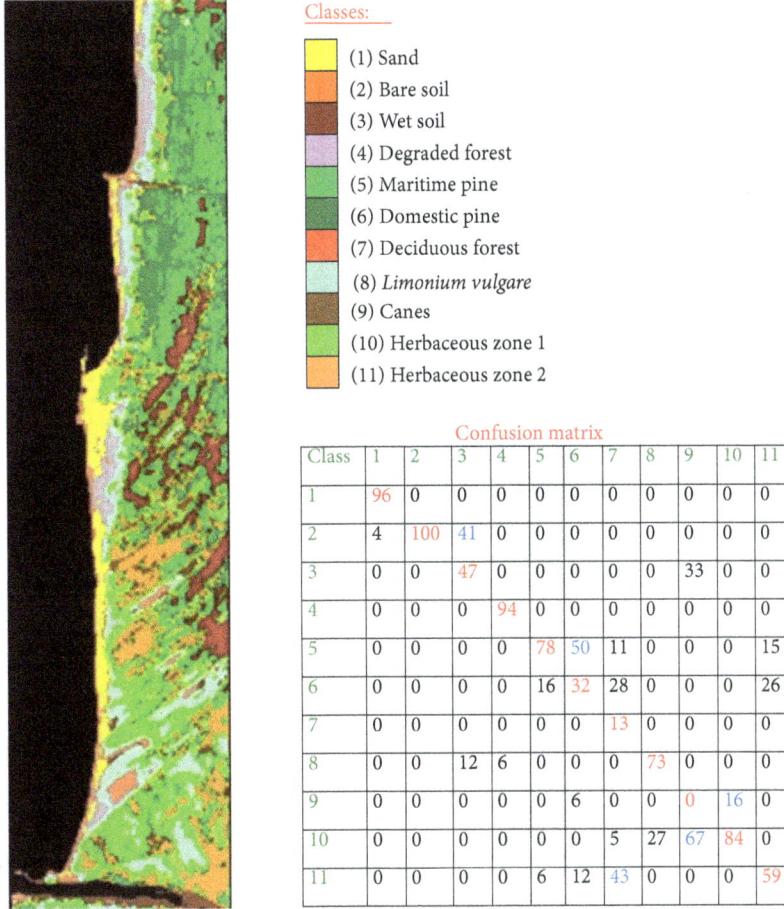

Classes:

- (1) Sand
- (2) Bare soil
- (3) Wet soil
- (4) Degraded forest
- (5) Maritime pine
- (6) Domestic pine
- (7) Deciduous forest
- (8) *Limonium vulgare*
- (9) Canes
- (10) Herbaceous zone 1
- (11) Herbaceous zone 2

Confusion matrix

Class	1	2	3	4	5	6	7	8	9	10	11
1	96	0	0	0	0	0	0	0	0	0	0
2	4	100	41	0	0	0	0	0	0	0	0
3	0	0	47	0	0	0	0	0	33	0	0
4	0	0	0	94	0	0	0	0	0	0	0
5	0	0	0	0	78	50	11	0	0	0	15
6	0	0	0	0	16	32	28	0	0	0	26
7	0	0	0	0	0	0	13	0	0	0	0
8	0	0	12	6	0	0	0	73	0	0	0
9	0	0	0	0	0	0	6	0	0	16	0
10	0	0	0	0	0	0	5	27	67	84	0
11	0	0	0	0	6	12	43	0	0	0	59

FIGURE 10: Example of forest SAM classification map of S. Rossore test site (Italy) based on simulated Hypseo hyper-spectral image (20 m). Confusion matrix shows a good correlation between classified and ground truth data.

developed to simulate the total reflectance of shallow waters on the basis of chlorophyll, sediment, and yellow substance content, the bottom and panel reflectance and the water depth height.

The total reflectance (Figure 13) $R(\lambda)$ has been calculated by means of the surface $R_{\text{surface}}(\lambda)$ and subsurface $R_{\text{sub-surface}}(\lambda)$ reflectance with the following relationship [26]:

$$R(\lambda) = R_{\text{surface}}(\lambda) + R_{\text{sub-surface}}(\lambda) * \left[\frac{1 - \rho_F\left(\vartheta'\right)}{n^2} \right], \quad (12)$$

With

(i) $\rho(\theta')$ the Fresnel reflectivity at the interface air-water which takes into account the reflection of the subsurface radiance into the water (~0.021),

(ii) θ' is the incident angle of the radiation coming from below the water which generates a refraction in air at the angle θ in the observation direction $[n * \sin(\theta') = \sin(\theta)$ with $n = 1.8368$ the water refraction index],

(iii) R_{surface} reflectance depends from surface roughness and foam, but in this analysis has been taken constant and equal in first approximation to 0.021.

The subsurface reflectance is obtained by means of a two-flux algorithm (Figure 13) which yields the following analytical relationship [27] for a water layer of uniform optical properties and thickness H (m), above a reflecting bottom with reflectance ρ_{bottom}:

$$R(\lambda) = R^0(\lambda) + \left[\frac{\left(\rho_{\text{bottom}}(\lambda) - R^0(\lambda)\right) \cdot \left(1 - R^0(\lambda)^2\right)}{R^0(\lambda) \cdot \left(\rho_{\text{bottom}}(\lambda) - R^0(\lambda)\right) + \left(1 - R^0(\lambda)\,\rho_{\text{bottom}}(\lambda)\right) \cdot e^{2H \cdot \sqrt{(A(\lambda)^2 + 2A(\lambda)B(\lambda))}}} \right], \quad (13)$$

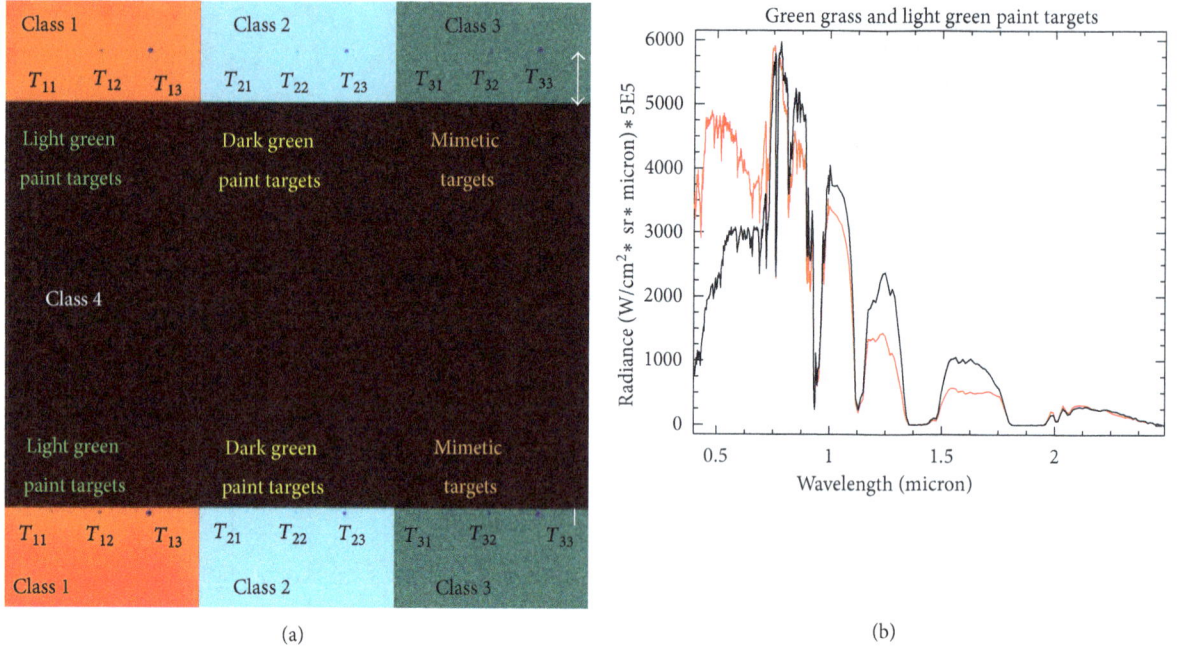

(a)   (b)

FIGURE 11: Simulated false color SIMGA radiance image (SWIR bands) of green panels over vegetation (a). The scenario has a dimension of $740*740\,\text{m}^2$ and it is composed of 4 classes of background scenario: class 1 = green grass, class 2 = bare soil, class 3 = linear mixing of green grass (1/3), dry grass (1/3), and bare soil (1/3); class 4 = pinewood [linear mixing of green (1/2) and dry (1/2) needles]. 18 targets are used grouping in three different sizes ($T\_1 = 1 \times 1\,\text{m}^2$, $T\_2 = 2 \times 2\,\text{m}^2$, and $T\_3 = 4 \times 4\,\text{m}^2$) and three different types (light green paint, dark green paint, and mimetic paint). Mimetic paint is composed of 1/3 light green paint, 1/3 dark green paint, and 1/3 light grey paint. The same type and size of targets are disposed in the upper and in the lower part of the figure at different distances (resp., 40 m and 4 m) respect to the class 4 scenario. Green panels are clearly distinguished as little dark points respect to vegetation, because of their lower radiance in the SWIR bands (at 1.25 and 1.6 micron). Comparison between green grass (upper black curve in the SWIR bands) and light green paint panel (lower red curve in the SWIR bands) (b).

where

(i) $R^0(\lambda) = B(\lambda)/(A(\lambda) + B(\lambda) + \sqrt{A(\lambda)^2 + 2A(\lambda)B(\lambda)})$,

(ii) $A(\lambda)$ and $B(\lambda)$ represent, respectively, the total absorption and scattering coefficients, including that of water, chlorophyll, and yellow substance,

(iii) the range of validity is $B/(A + B) \leq 0.6$.

The total absorption and backscattering coefficients are calculated from a three component water colour model ([28, 29]), which has been adapted for class 1 and class 2 waters [30].

In this model the total absorption and backscattering coefficients (m^{-1}) are obtained as a linear combination of that of water, chlorophyll, sediment, and yellow substance with the following relationships:

$$B(\lambda) = 0.5 \cdot B_w(\lambda) + 0.005 \cdot B_c(\lambda) + 0.015 \cdot B_s(\lambda),$$
$$A(\lambda) = A_w(\lambda) + a_c(\lambda) \cdot C + a_s(\lambda) \cdot S + A_Y(\lambda),$$
$$(14)$$

where

(i) the suffix w, c, s, Y means, respectively, water, chlorophyll, sediment, and yellow substance,

(ii) C, S, and Y represent the chlorophyll, sediment, and yellow substance content (C in mg/m^3, S, and Y in g/m^3),

(iii) a_c, a_s, a_Y, respectively, the chlorophyll, sediment and yellow substance specific absorption coefficients (m^2/mg and m^2/g, resp.) shown in Figure 14 [30]

(iv) $B_w(\lambda) = 0.002\,\text{m}^{-1}\,(\lambda(\text{nm})/550)^{-4.3}$,

(v) $B_s(\lambda) = 1\,(\text{m}^2/\text{mg})*S\,(\text{g/m}^3)*[(\lambda(\text{nm})/550)^{-1}]$,

(vi) $B_c(\lambda) = 0.12\,C^{0.63}\,[a_c(550)/a_c(\lambda)]\,(\text{m}^{-1})$,

(vii) $a_c(550) = 0.0189\,\text{m}^2/\text{mg}$ if $C < 1\,\text{mg/m}^3$,

(viii) $a_c(550) = 0.00486\,\text{m}^2/\text{mg}$ if $C > 1\,\text{mg/m}^3$,

(ix) $A_Y(\lambda) = A_Y(443\,\text{nm})*\exp[-0.014*(\lambda(\text{nm}) - 443)]\,(\text{m}^{-1})$,

(x) $A_Y(443\,\text{nm}) = (0.12\,\text{m}^{-1})*Y$,

(xi) $a_s(443\,\text{nm}) = 0.034\,\text{m}^2/\text{g}$,

(xii) $a_c(443\,\text{nm}) = 0.07\,\text{m}^2/\text{mg}$ if $C < 1\,\text{mg/m}^3$,

(xiii) $a_c(443\,\text{nm}) = 0.018\,\text{m}^2/\text{mg}$ if $C > 1\,\text{mg/m}^3$.

The concentration of the three water components can be divided in

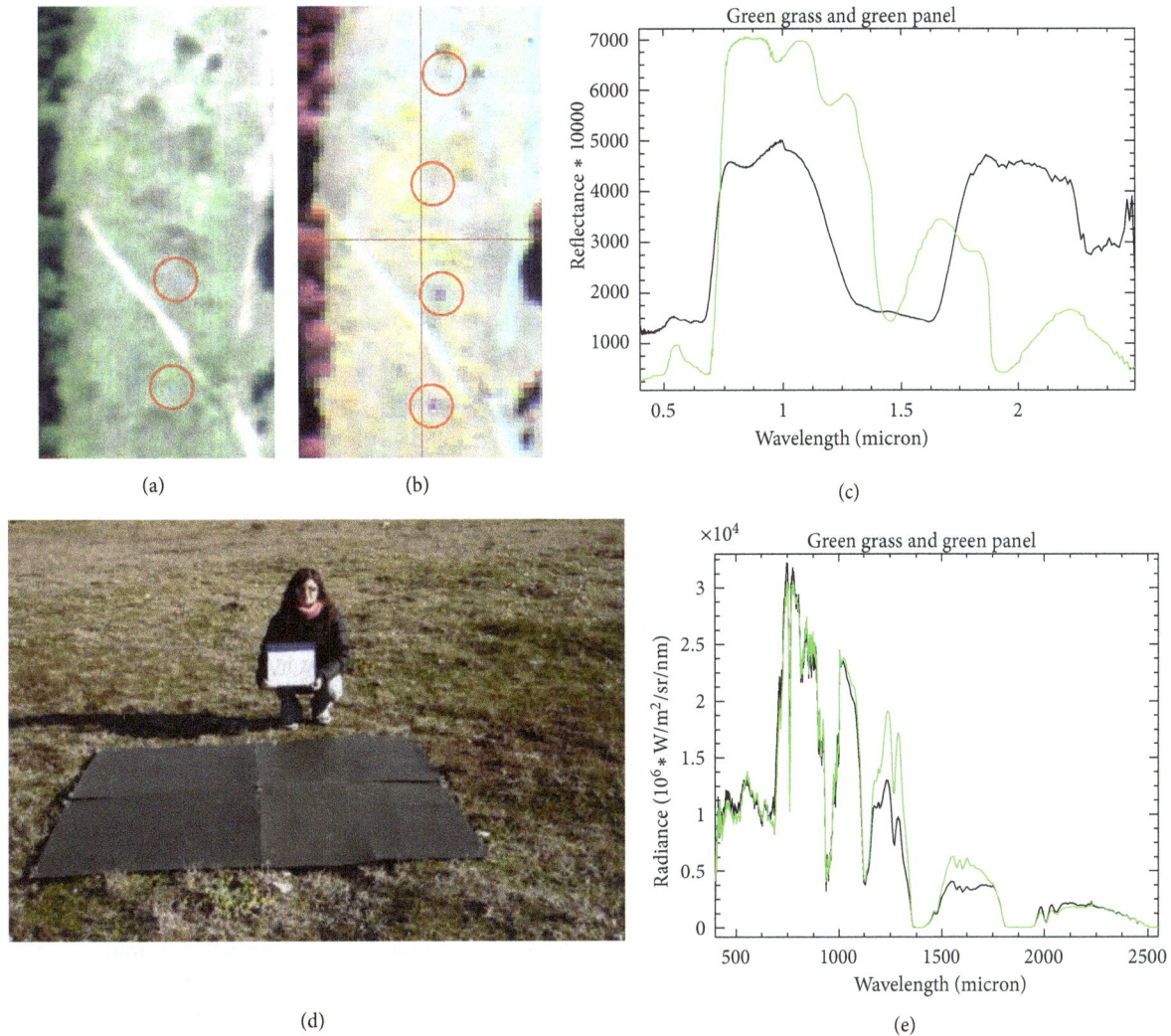

FIGURE 12: SIMGA false colour radiance images on panels over vegetation (S. Rossore park): VNIR false colour (a), SWIR false colour (b). Comparison between reflectance of green panel (lower black curve at 1.6 micron) and green grass (upper green curve at 1.6 micron) (c), photo of green panel over grass (d), and comparison between SIMGA radiance acquired on green grass (upper green curve at 1.6 micron) and green panel (lower black curve at 1.6 micron) (e). Panels are clearly distinguished in the SWIR bands, while the contrast respect to vegetation in VIS bands is negligible.

(i) completely correlated type 1 waters characterised by a rather stable correlation between optically active substances, with phytoplankton concentration as dominant, in this case the yellow substance backscattering and sediment absorption coefficients has been considered as related to the chlorophyll with the following relationships [12, 29]:

$$Bs \, (550 \, \text{nm}) = 0.2 \cdot C^{0.63} \cdot \text{m}^{-1},$$

$$A_Y \, (443 \, \text{nm})$$
$$= \frac{0.2}{0.8} \left[A_w \, (443 \, \text{nm}) + a_c \, (443 \, \text{nm}) \cdot C + a_s \, (443 \, \text{nm}) \cdot S \right],$$

$$S \left(\text{g/m}^3 \right) = \left[\frac{10^{-0.25}}{2} \right] \cdot C^{0.57} \left(\text{mg/m}^3 \right),$$

$$(15)$$

(ii) completely uncorrelated coastal type 2 waters, with no correlation between the three water components, when high concentration of sediments and yellow substances exist,

(iii) partially correlated coastal type 2 waters for which it is possible to retrieve a partial correlation between the three water components [30]. Examples are given by the following relationships:

(1) Gulf of Naples [31]

$$\text{Log} \, (S) = -0.25 + 0.57 \cdot \text{Log} \, (C)$$
$$\text{Log} \left(A_Y \, (440) \right) = -1.20 + 0.47 \cdot \text{Log} \, (C),$$

$$(16)$$

(2) Northern basin of the Adriatic Sea [32]

$$\text{Log} \, (S) = -0.026 + 0.59 \cdot \text{Log} \, (C)$$
$$\text{Log} \left(A_Y \, (440) \right) = -1.28 + 0.38 \cdot \text{Log} \, (C),$$

$$(17)$$

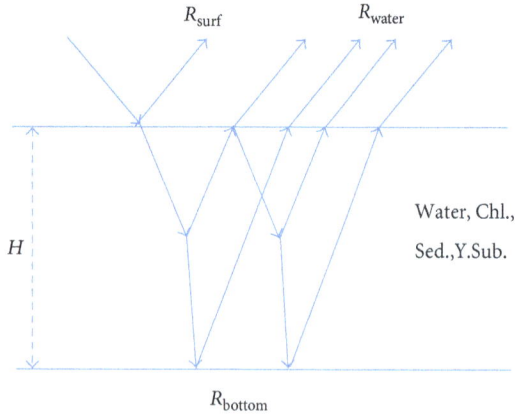

FIGURE 13: Two fluxes direct bathymetry reflectance model (R_{surf} = surface reflectance due to roughness and foam, R_{bulk} = Sub-surface water reflectance due to particulate concentration C = Chlorophyll, S = sediment, Y = yellow substance, and R_{bottom} = reflectance of bottom.

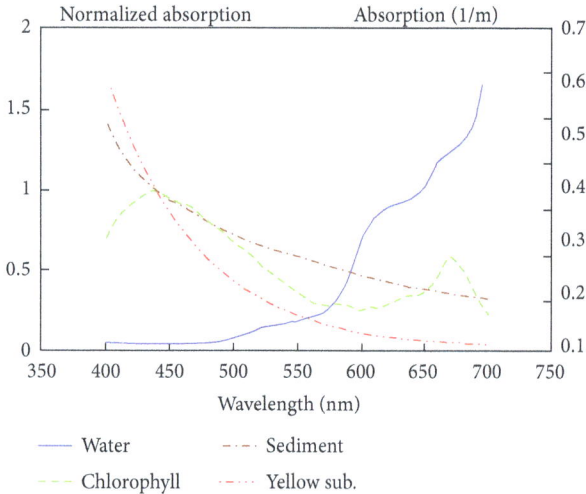

FIGURE 14: Specific absorption spectra of chlorophyll a, suspended sediment, and yellow substance normalized to the value at the 440 nm wavelength (left scale). Absorption spectrum of water is in m^{-1} (right scale) [12].

(3) Tirrenian Sea near Migliarino-S. Rossore (Tuscany) [33]

$$\text{Log}(C) = \text{Log}(0.0206) + 2.0615 \cdot \text{Log}(S), \qquad (18)$$

A reflectance simulation of S. Rossore waters at 1 m of bottom depth obtained by means of the two flux model is displayed in Figure 15. The Total Reflectance represents simulations of correlated waters type 2 model (C = 3 mg/m^3, S = 1.8 g/m^3, Y = 0.6 g/m^3), with the S. Rossore (I) bottom sand and grey panels reflectance measurements performed with a Field Spec portable spectrometer.

Finally simulated SIMGA reflectance and radiance images of a marine environment (sand, waters with sand bottom at 2 m and 8 m of depths, panels of 1 m∗1 m and 2 m∗2 m under water at 1 m and 0.2 m of depth) have been performed

(Figure 16). SIMGA radiance has been simulated at 1.5 km airborne altitude with MODTRAN code and at the SIMGA spatial (1 m for VIS and 2 m for SWIR bands) and spectral resolution (2.4 nm for VIS and 10.8 nm for SWIR). The simulated SIMGA reflectance image has been obtained through the inversion of MODTRAN parameters used for the direct simulation and results show that all grey panels (both at 0.2 m or 1 m depth) can be clearly distinguished both in low (2 m) and high (8 m) depth waters (Figure 16).

This result was validated (Figure 17) by means of SIMGA overflights on the Morto mouth river (S. Rossore park in Tuscany, I), where two different grey panels were submerged. The two panels are clearly detectable in the visible part of the spectrum, so demonstrating the capability of the SIMGA hyperspectral instrument for this kind of application.

4. Conclusions

An end-to-end software tool (SG_SIM) for the simulation of airborne/satellite optical sensors images has been implemented in ENVI-IDL environment. Input images can be either high resolution airborne or synthetic data. The simulator features three separate modules: the reflectance scenario, which generates a desired reflectance image with spectral mixtures, the atmosphere module, which converts the input reflectance map into the at-sensor radiance image, and the instrument module, which simulates the main degradations introduced by the instrument (ISR, MTF, ICR and noise). As other end-to-end simulators the SG_SIM Simulator integrates a complete atmospheric radiative transfer modelling which could easily refined through the implementation of most MODTRAN options and it includes all main functions and features necessary for a complete hyperspectral image simulation such as ISR&MTF, ICR and noise sources. Compared to the other simulators (e.g., SENSOR, [4]), SG_SIM allows also the control of spectral mixing and the generation of synthetic scenario, but is lacking of MWIR/LWIR spectral bands, 3D reflectance simulation, and DEM ray-tracing functions as included in CAMEO [5]. The implementation and further development of the SG_SIM approach was boosted significantly by the Selex Galileo S.p.A. airborne imaging system SIMGA and by other phase 0/A studies carried out for preliminary evaluations of image quality and product accuracy from new classes of space-borne optical sensors. The validation of the simulator is reported in [16], whereas in this paper the simulator's theoretical basis and some simulation examples have been described. For the simulated cases the following results can be outlined:

(i) the 3D representation of the SRF allows the visual inspection of the spatial pixel response for image quality analysis,

(ii) the potentials of the simulator for the HYPSEO Panchromatic camera trade-off analysis between project parameters (pupil diameter, optics degradations, detector noise, etc.) and system performances (SNR, spatial resolution, etc.) have been demonstrated by

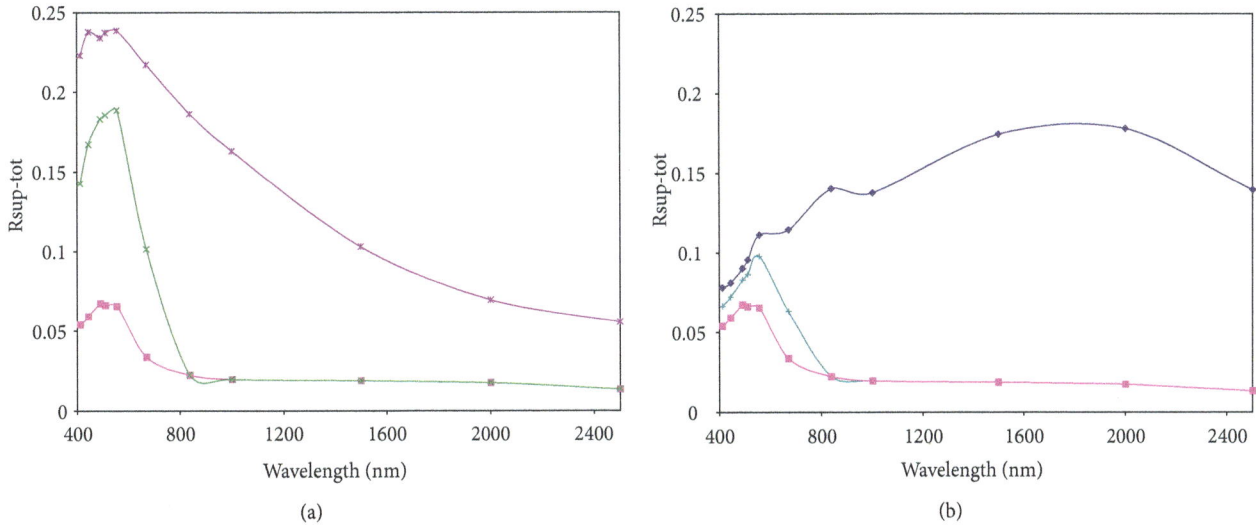

FIGURE 15: Total Reflectance curves (green curves in the middle part of Figures 15(a) and 15(b)) from the two fluxes direct bathymetry reflectance model obtained from simulations with different type of S. Rossore bottom targets: grey panel (upper maroon part of Figure 15(a)) (a) sand (upper blue part of Figure 15(b)) (b). The simulation is performed at a bottom depth (H) of 1 m with partially correlated waters type 2 model ($C = 3\,\mathrm{mg/m^3}$, $S = 1.8\,\mathrm{g/m^3}$, $Y = 0.6\,\mathrm{g/m^3}$) (lower pink reflectance curves in Figures 15(a) and 15(b)).

using as inputs synthetic bars and an IKONOS images with different radiance and surface albedo levels,

(iii) the potentials of the HYPSEO hyperspectral camera for vegetation mapping has been demonstrated on the basis of a MIVIS airborne scene rescaled at satellite level and ground truth data,

(iv) the potentials for detection of camuffled targets in a rural background has been demonstrated in the SWIR bands by means of a simulation of a synthetic scenario with green panels at different size,

(v) the potentials for the identification of submerged targets in the visible spectral range at airborne level (1 m of spatial resolution) have been demonstrated by means of the simulation of a synthetic scenario with submerged grey panels and the implementation of a direct bathymetric-water color model to generate surface reflectance as input to the scene simulator,

(vi) real airborne data on submerged and camuffled targets have confirmed the results from simulations performed before the flight campaign,

These results demonstrate the potentials of the proposed simplified end-to-end simulator as a preliminary aid tool (during phase 0/A) for the dimensioning of new optical instruments to trace the link between user and instrument requirements.

5. Annex

The formulation of the following MTF components implemented in the SG_SIM model are described in the following paragraphs:

(i) Detector pixel size

(ii) Detector cross talk

(iii) CCD detector charge transfer

(iv) Image motion during integration time

(v) Electronic filtering

(vi) Focal plane random jitter during integration time

(vii) Optics diffraction and aberrations.

5.1. Detector Pixel Size. The finite size of detector pixel permits the spatial integration of the signal coming from a finite spatial region on ground and this introduces a sort of degradation of the original high resolution image. This effect is analogous to a spatial filter windowing with a rect function which is 1 within a certain spatial rectangular domain and 0 outside:

$$\mathrm{PSF}_{\mathrm{pixel}}(x, y) = \mathrm{rect}(\Delta x, \Delta y) = \mathrm{rect}_x(\Delta x) \cdot \mathrm{rect}_y(\Delta y). \tag{19}$$

The transfer function of this function obtained by its Fourier Transform is represented with the following relationship:

$$\mathrm{MTF}_{\mathrm{pixel}}(k_x, k_y) = \mathrm{sin}\,c_x(\pi \cdot k_x \cdot \Delta x) \cdot \mathrm{sin}\,c_y(\pi \cdot k_y \cdot \Delta y), \tag{20}$$

where

(i) k_x, k_y are the spatial frequencies along the x and y direction,

(ii) $\Delta x, \Delta y$ are the detector size along x and y directions,

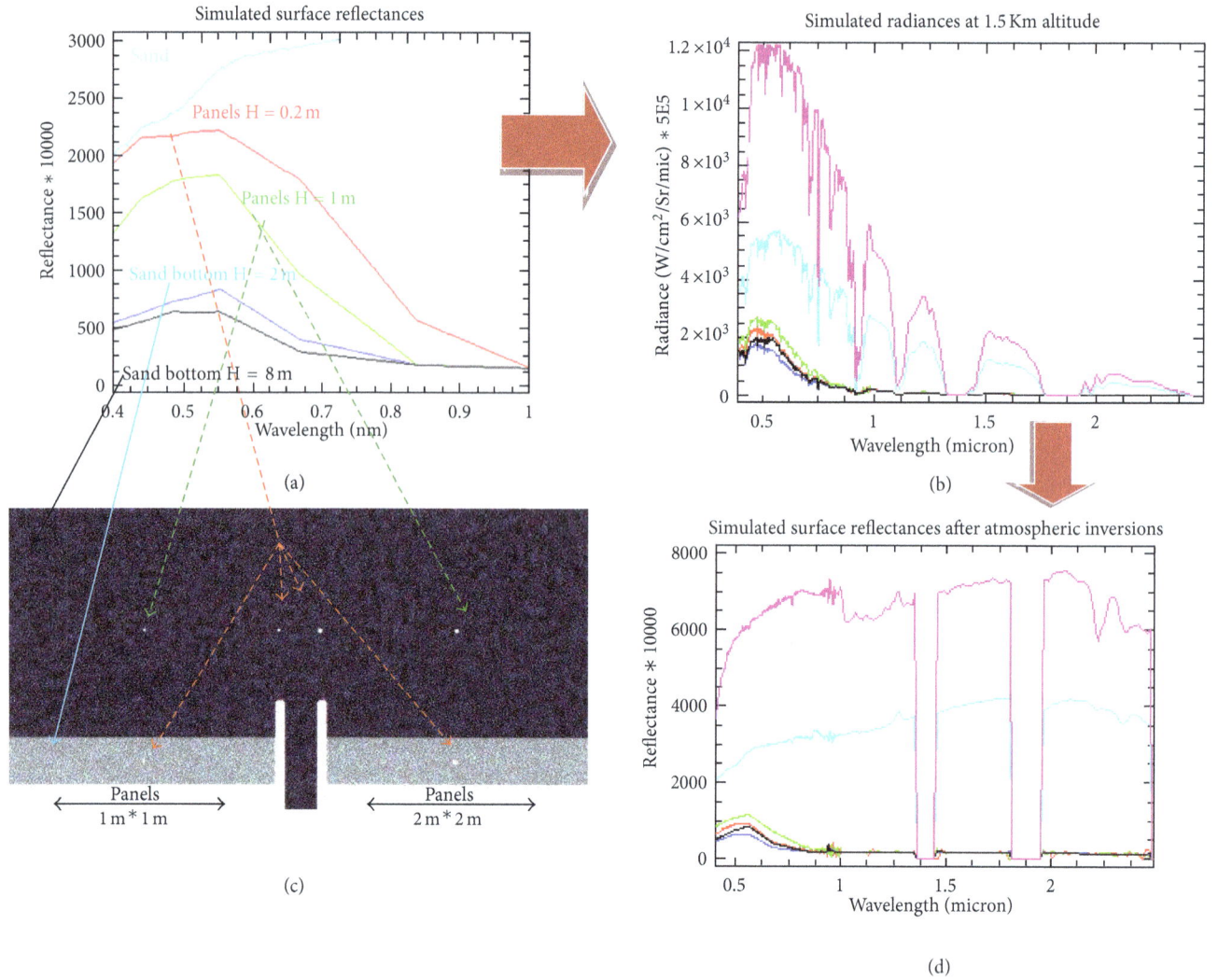

FIGURE 16: Simulated radiance/reflectance image of marine environment (sand, waters with sand bottom at 2 m & 8 m depths, panels of 1 m * 1 m & 2 m * 2 m under water at 2 m & 0.2 m depth, chlorophyll of 3 mg/m^3, sediment of 1.8 g/m^3, and yellow substance of 0.6 g/m^3) obtained with SIMGA model at 1.5 km altitude (spatial resolution 1 m VIS, 2 m SWIR, spectral resolution 2.4 nm VIS, 10.8 nm SWIR). Simulated synthetic reflectance (waters with sand bottom at 2 m or 8 m of depth, panels under water at 0.2 m and 1 m of depth, beach surface sand) from two-fluxes model (a), radiances at 1.5 Km altitude (b), surface reflectance (also beach sand and concrete pier are represented in the upper part of Figures 16(b) and 16(d)) after atmospheric inversion algorithm based on Modtran code with the same parameters as the direct simulation (d), and RGB synthetic radiance image representing the simulated scenario (c).

(iii) the sinc functions are expressed from the following relationships:

$$\sin c_x \left(\pi \cdot k_x \cdot \Delta x \right) = \frac{\sin \left(\pi \cdot k_x \cdot \Delta x \right)}{\left(\pi \cdot k_x \cdot \Delta x \right)},$$

$$\sin c_y \left(\pi \cdot k_y \cdot \Delta y \right) = \frac{\sin \left(\pi \cdot k_y \cdot \Delta y \right)}{\left(\pi \cdot k_y \cdot \Delta y \right)}. \quad (21)$$

5.2. Detector Cross Talk. The detector cross talk between two successive pixels is taken into account as a first approximation by assuming a trapezoidal spatial windowing filter, instead of a rectangular one, which can be obtained by means of a convolution between two rect functions, one representing the detector size $(\Delta x, \Delta y)$ and the other representing the

cross talk size $(\delta x, \delta y)$ between two successive pixels $(\delta x < \Delta x, \delta y < \Delta y)$:

$$\begin{aligned} \mathrm{PSF}_{\mathrm{pixel-2}} \left(x, y \right) &= \mathrm{rect} \left(\Delta x, \Delta y \right) \otimes \mathrm{rect} \left(\delta x, \delta y \right) \\ &= \left[\mathrm{rect}_x \left(\Delta x \right) \otimes \mathrm{rect}_x \left(\delta x \right) \right] \\ &\quad \cdot \left[\mathrm{rect}_y \left(\Delta y \right) \otimes \mathrm{rect}_y \left(\delta y \right) \right]. \end{aligned} \quad (22)$$

The Transfer function is obtained from the following relationship:

$$\begin{aligned} \mathrm{MTF}_{\mathrm{pixel-2}} &\left(k_x, k_y \right) \\ &= \left[\sin c_x \left(\pi \cdot k_x \cdot \Delta x \right) \cdot \sin c_x \left(\pi \cdot k_x \cdot \delta x \right) \right] \\ &\quad \cdot \left[\sin c_y \left(\pi \cdot k_y \cdot \Delta y \right) \cdot \sin c_y \left(\pi \cdot k_y \cdot \delta y \right) \right]. \end{aligned} \quad (23)$$

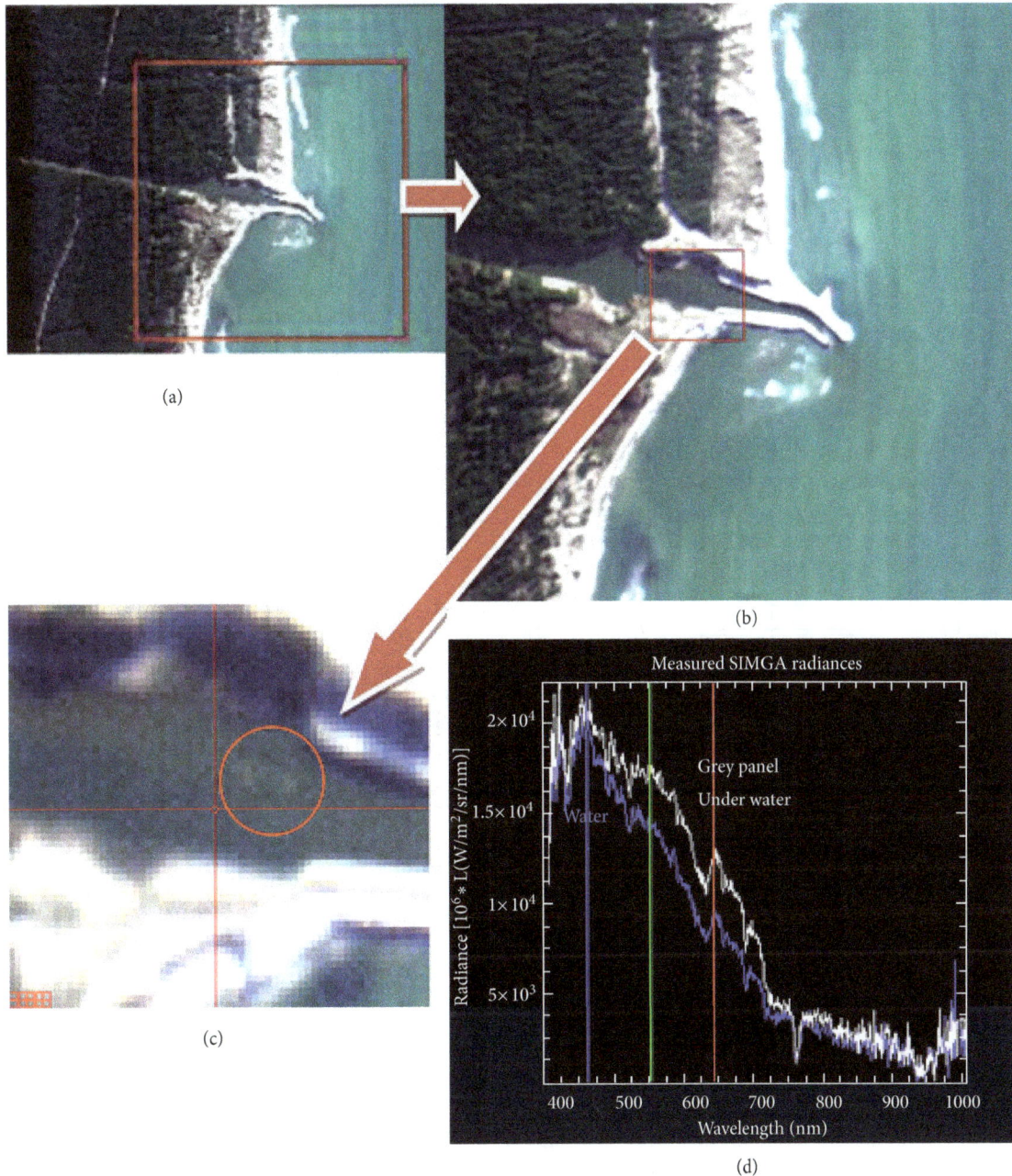

(a)

(b)

(c)

(d)

FIGURE 17: SIMGA RGB radiance images ((a), (b), (c)) on panels under water (S. Rossore park, Fiume Morto river), comparison between radiance measurements of grey panel under water and water (d).

5.3. CCD Detector Charge Transfer. For a CCD (Charge Capacitance Device) detector the reading of electrons acquired in each pixel of the matrix is performed by mean of a charge transfer from a pixel to the other. In this way the total transfer efficiency is related to the pixel-to-pixel Charge Transfer Efficiency (CTE) and the total number of transfers N_{trans}. As the CTE is not 1 some losses are present at the end of the reading time which implies a reduction of image contrast and then an MTF less than 1 in the direction of output register:

$$\text{MTF}_{\text{CCD-CTE}}\left(k, k_{\text{Nyq}}\right) = e^{-N_{\text{trans}} \cdot (1-\text{CTE}) \cdot [1-\cos(\pi \cdot k/k_{\text{Nyq}})]}.$$

(24)

With

(i) CTE the pixel Charge Transfer Efficiency (greater then 99,99%),

(ii) N_{trans} the total number of charge transfers,

(iii) k the spatial frequency and k_{Nyq} the Nyquist frequency of the system (half of the sampling frequency $k_{\text{Nyq}} = 1/(2p)$, with p the detector pitch).

5.4. Image Motion During Integration Time. The effect of temporal acquisition (integration time t_{int} greater than zero) during the image motion (with velocity v), which happens along the scan direction (the satellite velocity in a push broom system) introduces an image blur, which can be taken into account with a PSF similar to a rect function, which represents the temporal aperture along the motion:

$$\text{PSF}_{\text{temporal}}(x) = \text{rect}_x(v \cdot t_{\text{int}}). \quad (25)$$

With the following MTF:

$$\text{MTF}_{\text{along-motion}}(k_x, k_y) = \sin c_x(\pi \cdot k_x \cdot v \cdot t_{\text{int}}). \quad (26)$$

The worst case happens when the integration time is equal or larger than the dwell time, that is, the spatial displacement equivalent to a pixel size, while the best case (MTF $\cong$ 1) is for a short integration time.

5.5. Electronic Filtering. An electronic system can introduce a temporal smoothing due to a finite frequency bandwidth and thus a reduced space/temporal response. This effect has been simulated by using a general formulation based on the following Butterworth filter response:

$$\text{MTF}_{\text{filter}}(f) = \frac{1}{\sqrt{1 + \left(f/\left(N_f \cdot f_{\text{Nyq}}\right)\right)^{2 \cdot N_b}}}. \quad (27)$$

With

(i) f the frequency,

(ii) f_{Nyq} the Nyquist frequency,

(iii) N_b the order of Butterworth filter,

(iv) N_f the ratio between the 3 dB filter frequency and the Nyquist frequency $N_f = f_{3\text{dB}}/f_{\text{Nyq}}$; this ratio should be between 2.2 and 3 for a good reproduction of a square wave.

The above equation correctly reproduces the behaviour of the classical low-pass filter for $N_b = 1$.

5.6. Focal Plane Random Jitter during Integration Time. For high frequencies random vibration of the focal plane a Gaussian spatial response (PSF) can be taken into account. The Fourier Transform of PSF is still a Gaussian function, representing the MTF, with the following relationship:

$$\text{MTF}_{\text{jitter-random}}(f) = e^{-2 \cdot \pi^2 \cdot [\text{jitter} \cdot p \cdot f]^2}. \quad (28)$$

With

(i) p the detector pitch,

(ii) jitter $*$ p the fraction of pixel representing the rms values of random fluctuations,

(iii) f the spatial frequency at detector level.

5.7. Optics Diffraction and Aberrations. The MTF related to diffraction from optics has been evaluated by using the O'Neill formulas, valid for diffraction in presence of a telescope with central obscuration.

The following formulation for MTF diffraction term is used [34]:

$$\text{MTF}_{\text{diffraction}}(f) = \frac{(A + B + C)}{(1 - \eta^2)}, \quad (29a)$$

with η, f, and other parameters defined as follows:

(i) η = obscuration factor = ratio between the obscuration diameter and the pupil diameter,

(ii) f = spatial frequency at detector level (cm^{-1}),

(iii) $\omega = f / f$cut-off,

(iv) f cut-off = optics cut-off (cm^{-1}) at detector level = $1/(\lambda * f$-number),

(v) λ = wavelength;

(vi) f-number = F/D, ratio between the focal length F and the pupil diameter D.

The A, B, and C parameters are defined from the following relationships:

$$\omega \leq 1, \quad A = \frac{2}{\pi} \cdot \left[\arccos(\omega) - \omega\sqrt{1 - \omega^2}\right]$$
$$\omega > 1, \quad A = 0, \quad (29b)$$

$$\eta = 0, \quad t = 0, \quad \phi = 0, \quad B = 0;$$

$$\eta \neq 0, \quad t = \frac{\omega}{\eta}, \quad \phi = \arccos\left[\frac{1 + \eta^2 - 4\omega^2}{2\eta}\right]$$

$$t > 1, \quad B = 0 \quad (29c)$$

$$t \leq 1, \quad B = 2\eta^2 \frac{\left[\arccos(t) - t\sqrt{1 - t^2}\right]}{\pi} - 2\eta^2$$

$$\omega \leq \frac{1 - \eta}{2}, \quad C = -2\eta^2$$

$$\omega < \frac{1 + \eta}{2}, \quad C = \left\{2\eta\sin(\phi) + \left(1 + \eta^2\right)\phi\right.$$

$$\left. -2\left(1 - \eta^2\right)\arctan\left[\left(\frac{1 + \eta}{1 - \eta}\right)\tan\left(\frac{\phi}{2}\right)\right]\right\}$$

$$\cdot \frac{1}{\pi} - 2\eta^2,$$

$$\omega \geq \frac{1 + \eta}{2}, \quad C = 0. \quad (29d)$$

The above MTF formulation for optics diffraction can be simplified to the following well known diffraction relationship in absence of central obscuration, which is zero for $f \geq f_{\text{cut-off}}$:

$f \leq f_{\text{cut-off}}$,

MTF

$$= \frac{2}{\pi} \cdot \left[\arccos\left(\frac{f}{f_{\text{cut-off}}}\right) - \left(\frac{f}{f_{\text{cut-off}}}\right) \sqrt{1 - \left(\frac{f}{f_{\text{cut-off}}}\right)^2} \right],$$

$f > f_{\text{cut-off}}, \quad \text{MTF} = 0,$

$$f_{\text{cut-off}} = \frac{1}{\lambda \cdot f_{\text{number}}}.$$

(29e)

Regarding possible optics aberrations the model takes into account, as a first approximation, the following exponential fitting function:

$$\text{MTF}_{\text{aberration}}(f) = e^{-k \cdot (f/f_{\text{cut-off}})^x},$$ (30)

with k and x representing two empirical parameters used to approximate all optics degradation effects.

Acknowledgments

The authors wish to thank L. Tommasi of Selex Galileo, R. Bonsignori of Eumetsat (formerly with Selex Galileo) and F. Pecchioni of University of Florence for the useful technical discussions and contributions. The authors also wish to acknowledge the ASI HYPSEO Phase A/B study under ASI/CSM/vdc/299/00 Contract for instrument specification and modelling.

References

[1] J. S. Pearlman, P. S. Barry, C. C. Segal, J. Shepanski, D. Beiso, and S. L. Carman, "Hyperion, a space-based imaging spectrometer," *IEEE Transactions on Geoscience and Remote Sensing*, vol. 41, no. 6, pp. 1160–1173, 2003.

[2] ENVI software Exelis Visual Information Solutions, http://www.exelisvis.com/language/en-us/productsservices/envi.aspx.

[3] MODTRAN software, http://modtran5.com/.

[4] A. Börner, L. Wiest, P. Keller et al., "SENSOR: a tool for the simulation of hyperspectral remote sensing systems," *ISPRS Journal of Photogrammetry and Remote Sensing*, vol. 55, no. 5-6, pp. 299–312, 2001.

[5] I. R. Moorhead, M. A. Gilmore, A. W. Houlbrook et al., "CAMEO-SIM: a physics-based broadband scene simulation tool for assessment of camouflage, concealment, and deception methodologies," *Optical Engineering*, vol. 40, no. 9, pp. 1896–1905, 2001.

[6] S. A. Cota, C. J. Florio, D. J. Duvall, and M. A. Leon, "The use of the general image quality equation in the design and evaluation of imaging systems," in *Remote Sensing System Engineering II*, vol. 7458 of *Proceedings of SPIE*, 2009.

[7] S. A. Cota, J. T. Bell, R. H. Boucher et al., "PICASSO: an end-to-end image simulation tool for space and airborne imaging systems," *Journal of Applied Remote Sensing*, vol. 4, no. 1, Article ID 043535, 2010.

[8] S. A. Cota, T. S. Lomhein, C. J. Florio et al., "PICASSO: an end-to-end image simulation tool for space and airborne imaging systems: II. Extension to the Thermal Infrared—equation and methods," in *Imaging Spectrometry XVI*, vol. 8158 of *Proceedings of SPIE*, p. 81580, 2011.

[9] D. Labate, F. Butera, L. Chiarantini, and M. Dami, "SIMGA HYPER: Hyperspectral Avionic System Calibration Results," Technical note Galileo Avionica, 19 February 2007.

[10] J. W. Goodman, *Introduction to Fourier Optics*, McGraw-Hill, New York, NY, USA, 1968.

[11] G. C. Holst and T. S. Lomhein, *CMOS/CCD Sensors and Camera Systems*, JCD Publishing and SPIE Press, 2007.

[12] S. Tassan, "SeaWiFS potential for remote sensing of marine Trichodesmium at sub- bloom concentration," *International Journal of Remote Sensing*, vol. 16, no. 18, pp. 3619–3627, 1995.

[13] L. Alparone, M. Selva, B. Aiazzi, S. Baronti, F. Butera, and L. Chiarantini, "Signal-dependent noise modelling and estimation of new-generation imaging spectrometers," in *Proceedings of the 1st Workshop on Hyperspectral Image and Signal Processing: Evolution in Remote Sensing (WHISPERS '09)*, pp. 1–4, Grenoble, France, August 2009.

[14] L. Alparone, M. Selva, L. Capobianco, S. Moretti, L. Chiarantini, and F. Butera, "Quality assessment of data products from a new generation airborne imaging spectrometer," in *Proceedings of the IEEE International Geoscience and Remote Sensing Symposium (IGARSS '09)*, vol. 4, pp. 422–425, July 2009.

[15] B. Aiazzi, L. Alparone, S. Baronti, F. Butera, L. Chiarantini, and M. Selva, "Benefits of signal dependent noise reduction for spectral analysis of data from advanced imaging spectrometers," in *Proceedings of the Workshop on Hyperspectral Image and Signal Processing: Evolution in Remote Sensing (WHISPERS '11)*, Lisbon, France, June 2011.

[16] P. Coppo, L. Chiarantini, and L. Alparone, "Design and validation of an end-to-end simulator for imaging spectrometers," *Optical Engineering*, vol. 51, no. 11, Article ID 111721, 2012, Special session on Hyperspectral Imaging Systems.

[17] ENVI FLAASH atmospheric correction software and Exelis Visual Information Solutions, http://www.exelisvis.com/portals/0/pdfs/envi/Flaash_Module.pdf.

[18] S. Adler-Golden, A. Berk, L. S. Bernstein et al., "FLAASH, a MODTRAN4 atmospheric correction package for hyperspectral data retrievals and simulations," in *Proceedings of the 7th Jet Propulsion Laboratory (JPL) Airborne Earth Science Workshop*, JPL Publication 97-21, pp. 9–14, 1998.

[19] C. Ann Bateson, G. P. Asner, and C. A. Wessman, "Endmember bundles: a new approach to incorporating endmember variability into spectral mixture analysis," *IEEE Transactions on Geoscience and Remote Sensing*, vol. 38, no. 2, pp. 1083–1094, 2000.

[20] J. M. P. Nascimento and J. M. B. Dias, "Does independent component analysis play a role in unmixing hyperspectral data?" *IEEE Transactions on Geoscience and Remote Sensing*, vol. 43, no. 1, pp. 175–187, 2005.

[21] D. Labate, M. Ceccherini, A. Cisbani et al., "The PRISMA payload optomechanical design, a high performance instrument for a new hyperspectral mission," *Acta Astronautica*, vol. 65, no. 9-10, pp. 1429–1436, 2009.

[22] A. Garzelli, B. Aiazzi, S. Baronti, M. Selva, and L. Alparone, "Hyperspectral image fusion," in *Proceedings of the Hyperspectral 2010 Workshop*, pp. 17–19, Frascati, Italy, March 2010, (ESA SP-683, May 2010).

[23] P. Coppo, L. Chiarantini, F. Maselli, S. Migliorini, I. Pippi, and P. Marcoionni, "Application test of the OG-HYC hyperspectral camera," in *Sensors, Systems, and Next-Generation Satellites V*, vol. 4540 of *Proceedings of SPIE*, pp. 147–158, September 2001.

[24] A. Bini, D. Labate, A. Romoli et al., "Hyperspectral earth observer (HYPSEO) program," in *Proceedings of the 52nd International Astronautical Federation (IAF '01)*, Toulouse, France, October 2001.

[25] ENVI Spectral Angle Mapper (SAM) algorithm and Exelis Visual Information Solutions, http://www.exelisvis.com/portals/0/tutorials/envi/SAM_SID_Classification.pdf.

[26] I. S. Robinson, *Satellite Oceanography*, John Wiley & Sons, New York, NY, USA, 1985, Ellis Orwood Lim.

[27] S. Tassan, "An algorithm for the identification of benthic algae in the Venice Lagoon from thematic mapper data," *International Journal of Remote Sensing*, vol. 13, no. 15, pp. 2887–2909, 1992.

[28] A. Morel and Prieur, "Analysis of variations in ocean colour," *Limnology and Oceanography*, vol. 22, pp. 709–722, 1977.

[29] S. Sathyendranath, L. Prieur, and A. Morel, "A three-component model of ocean colour and its application to remote sensing of phytoplankton pigments in coastal waters," *International Journal of Remote Sensing*, vol. 10, no. 8, pp. 1373–1394, 1989.

[30] S. Tassan, "Local algorithms using SeaWiFS data for the retrieval of phytoplankton, pigments, suspended sediment, and yellow substance in coastal waters," *Applied Optics*, vol. 33, no. 12, pp. 2369–2378, 1994.

[31] S. Tassan and M. Ribera d'Alcalá, "Water quality monitoring by thematic mapper in coastal environments. A performance analysis of local biooptical algorithms and atmospheric correction procedures," *Remote Sensing of Environment*, vol. 45, no. 2, pp. 177–191, 1993.

[32] B. Sturm, "Ocean colour remote sensing: a status report," in *Satellite Remote Sensing for Hydrology and Water Management*, E. C. Barret, Ed., pp. 243–277, Gordon and Breach Science Publishers, New York, NY, USA, 1990.

[33] P. Coppo, L. Chiarantini, F. Maselli et al., "Test Applicativi della camera Iperspettrale Cosmo-Skymed," Contratto ALS-US-SBC-0058/99, doc. N. SKC-GAL-TN-008, Dicembre 2000.

[34] E. L. O'Neill, "Transfer function for an annular aperture," *The Journal of the Optical Society of America*, vol. 46, pp. 285–288, 1956.

10

In Search of Early Time: An Original Approach in the Thermographic Identification of Thermophysical Properties and Defects

Daniel L. Balageas[1, 2]

[1] Composite Materials and Systems Department, ONERA, BP 72, 92322 Châtillon Cedex, France
[2] TREFLE Department, ENSAM, Institute of Mechanics and Engineering of Bordeaux (I2M), Esplanade des Arts et Métiers, 33405 Talence Cedex, France

Correspondence should be addressed to Daniel L. Balageas; daniel.balageas@u-bordeaux1.fr

Academic Editor: Carlo Corsi

Active thermography gives the possibility to characterize thermophysical properties and defects in complex structures presenting heterogeneities. The produced thermal fields can be rapidly 3D. On the other hand, due to the size of modern thermographic images, pixel-wise data processing based on 1D models is the only reasonable approach for a rapid image processing. The only way to conciliate these two constraints when dealing with time-resolved experiments lies in the earlier possible detection/characterization. This approach is illustrated by several different applications and compared to more classical methods, demonstrating that simplicity of models and calculations is compatible with efficient and accurate identifications.

1. Introduction

The evolution of thermophysical properties metrology and nondestructive evaluation (NDE) is characterized by the increased use of refined inverse techniques [1] requiring the establishment of models taking into account many parameters, although among these parameters often only one parameter is of interest for the experimenter. This complexity is particularly important when the analyzed thermal fields are 3D, a situation characteristic of the experiments realized with thermographic systems producing sequences of large thermal images of complex structures presenting important heterogeneities in their thermal properties, internal geometries, and boundary conditions. This approach leads to time-consuming calculations that may be prohibitive for thermographic data processing. Furthermore, it happens that in many situations the reality remains more complex than the sophisticated model used.

How to conciliate the existence 3D thermal situations involving numerous parameters and the necessity to have rapid calculations compatible with the very high number of information to process in a thermographic image sequence?

A solution lies in the use of 1D thermal models for pixel-wise data processing and the choice of a limited early time domain (for time-resolved techniques) or high frequency domain (for modulated techniques) for which the measured temperatures are essentially depending on the sole parameter to be identified and weakly affected by the 3D heat transfer. To achieve that, the proposed approach consists in performing the identification at the earlier possible time (or at the higher possible frequency) after the thermal stimulation. Here, we will mainly consider time-resolved techniques and early time detection (emerging signal).

The present work wants to show the following.

(i) A detailed procedure can be defined for early detection and characterization (see Section 2).

(ii) That this approach is not new. In a first step, it has been applied in the field of thermophysical properties measurements (period 1970–1990). This review is the subject of Section 3.

(iii) The application to NDE, started at the beginning of the 90's, continues to give rise to new developments (Section 4).

The goal of the present work is to put into perspective results spread over three decades, showing the unity of the approaches up to now hidden by the diversity of the applications and to emphasize that in thermal methods precociousness is as important as signal-to-noise ratio.

2. Presentation of the Early Detection Approach

The early detection approach can be considered as the sequence of the following six operations:

(i) choice of a model simpler than the actual configuration, generally 1D,

(ii) choice of an early time window for the analysis of the thermograms, in such a way that very few parameters (one if possible) be influent,

(iii) inverse problem solving in these conditions,

(iv) Analysis of the time evolution of the so-identified parameter for the assessment of its accuracy,

(v) choice of a fitting function,

(vi) extrapolation to zero time (thermophysics application) or zero contrast (NDE) for obtaining the most precise parameter estimate.

Most of the examples that will be given consider models with only one parameter to be identified. Nevertheless, more complex situations are possible. The second example of Section 3 presents a model in which three parameters are to be identified. Consequently the procedure is more complex, involving two time window analysis and time extrapolations followed by two space extrapolations thanks to the use of several samples of different thicknesses.

3. Application in the Field of Thermophysical Properties Measurements

The early detection approach defined in Section 2 is illustrated in Section 3.1 in the case of a simple configuration (rear face flash diffusivity experiment), universally known under the name of Parker's method [2], applied to the diffusivity measurement of an homogeneous slab.

In Section 3.2, a more complex procedure is presented dealing with the same type of experiment (rear face flash diffusivity), identifying the *in situ* diffusivities of the components (matrix and reinforcement) of a 3D C/C composite and the thermal contact resistance characterizing the interface between them.

3.1. Rear Face Flash Diffusivity Measurement. In flash thermal diffusivity measurements on the face opposite to the pulsed heat deposition the sample heat losses distort the rear face temperature time history. These thermal losses occurring necessarily after the flash, the temperature is all the less disturbed as time is nearer the origin. Consequently, the diffusivity identification by extrapolating towards the initial time the apparent diffusivity-versus-time history resulting from the use of an adiabatic solution was proposed in 1982

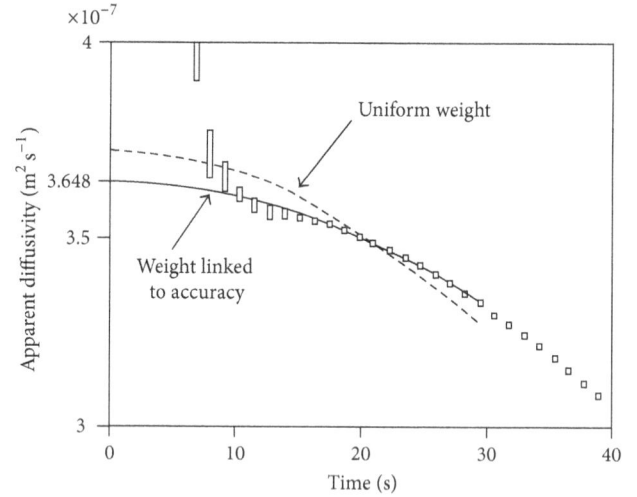

FIGURE 1: Test no. 2 on plaster identified apparent diffusivity law and regression parabola with a uniform weight and with a nonuniform weight linked to the accuracy of the measurement, taken from [4].

[3, 4]. It was demonstrated that the early phase of the thermogram (for instance for a defect Fourier number $Fo_d = \kappa t / z_d^2 < 0.1$, where κ is the diffusivity of the material and z_d the defect depth) may be used for this early detection if an estimate of the final adiabatic temperature increase, ΔT_{lim}, is deduced without any assumption concerning the heat losses from the maximum temperature reached with losses, ΔT_{max}. A relation between these two parameters was proposed

$$\text{Log}_{10}\left[\text{Log}_{10}\left(\frac{\Delta T_{lim}}{\Delta T_{max}}\right)\right] = -\frac{0.667 t_{max}}{t_{1/2}} + 1.113, \quad (1)$$

where t_{max} and $t_{1/2}$ are, respectively, the measured occurrence times of the maximum and half maximum temperature increase at the center of the sample. This relation allows to normalize the thermogram: $\Delta T(t)/\Delta T_{lim} = (\Delta T(t)/\Delta T_{max}) \times (\Delta T_{max}/\Delta T_{lim})$, and using a look-up table to build an apparent diffusivity curve, $\kappa(t)$, from the analytical solution for the 1D rear face adiabatic sample submitted to a pulsed stimulation. Whatever is the accuracy of the estimated ΔT_{lim}, it was verified that the law $\kappa(t)$ starts at $t = 0$ from the actual diffusivity value, and that if the estimation is sufficiently accurate, the apparent diffusivity evolution in the early times ($Fo_d < 0.1$) can be approximated by a parabolic law whose peak is obtained for $t = 0$. To improve the accuracy of the parabolic fitting using least mean squares, the experimental points were weighted by the accuracy of the diffusivity measurement deduced from the a sensitivity analysis. It was imposed to the parabola to have a zero slope at $t = 0$.

Figure 1 presents, taken from [4], the extrapolation in the case of a measurement on a 10 mm thick disk-shaped plaster sample, with a radius over thickness ratio $R/L = 1.206$. An extrapolated value of diffusivity is found $3.65 \cdot 10^{-7}$ m^{2}s^{-1}.

Table 1 summarizes the diffusivity results obtained by this method and compares them to results obtained with exactly the same samples by three methods (round robin tests) as follows.

TABLE 1: Present method result and comparison to Parker's and Degiovanni's methods.

Material	Test number	Sample			Identified diffusivity (m^2 s^{-1}) $\times 10^7$				
		Thickness (mm)	R/L	Parker's method	Degiovanni [5]			Degiovanni [6]	Balageas [4]
					$\kappa_{2/3}$	$\kappa_{1/2}$	$\kappa_{1/3}$		
Plaster	#1	9.96	1.206	4.759	3.675	3.703	3.689	3.69	**3.703**
Plaster	#2	9.96	1.206	4.734	3.618	3.644	3.645	3.64	**3.648**
Chalk	#3	6.16	1.950	3.548	2.861	2.881	2.886	2.89	**2.905**

(i) The Parker's formula, $\kappa = 0.139 L^2/t_{1/2}$, with $t_{1/2}$ the half rise time, which considers the sample adiabatic and leads to considerable errors due to the low conductivities of the three materials chosen for the tests.

(ii) The more precise method at the time of the publication of the present method: the partial times method of Degiovanni [5]. The method of partial times considers for the identification several pairs of points of the thermogram corresponding to the following normalized temperature increases $\Delta T(t)/\Delta T_{max}$: 1/3 and 5/6, 1/2 and 5/6, 2/3 and 5/6. It is interesting to compare the Degiovanni's method with the present early detection approach because, as shown in Figure 2, the extrapolation method results are coherent with the evolution with time of the identified diffusivities of the points considered in the data processing of the Degiovanni's partial times method. We see that the present method corrects the bias of this method, which increases with identification times.

(iii) The method of partial moments, proposed by Degiovanni [6] that is still considered as a reference. There is a perfect agreement with this method. The results of the partial moments are not plotted in Figure 2 because they would not be distinguishable from those of the present method.

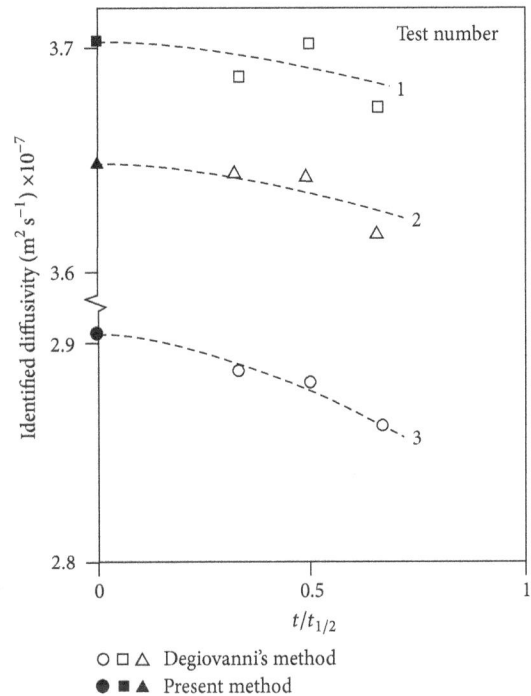

FIGURE 2: Compared results of Degiovanni's partial times method and present method, taken from [4].

3.2. Rear Face Flash Measurement of the In Situ Diffusivities of the Components of Periodic Directional-Reinforced Composites

3.2.1. Periodic Directional Reinforced Composites (PDRCs).
The types of materials here considered are PDRCs (periodic directional reinforced composites) in which the reinforcement is arranged following preferential directions. For instance, unidirectional carbon/epoxy composites and 3D carbon/carbon composites belong to this family of materials and the identification method here presented can be applied to them. These materials are difficult to homogenize when there is a large difference between the thermal conductivities of the composite components (matrix and reinforcement) and when the main heat transfer is parallel to a reinforcement direction [7].

Considering the same type of measurements as in the previous section (rear face flash diffusivity), a more complex procedure is presented, which identifies the *in situ* diffusivities of the matrix and reinforcement and the thermal contact resistance characterizing the interface between them in the case of a 3D C/C composite [8–10]. The method supposes that the composite sample is quasi adiabatic (no heat losses).

The simplest PDRCs that can be imagined are presented in Figure 3. They are unidirectional PDRCs. The two components are arranged in alternated slabs parallel to the imposed heat flux (Figure 3(a)) or following a chessboard pattern (Figure 3(b)). The dimensional geometric parameters are the space period of the pattern, ω, the thickness of the medium, L, and the volume ratio of the components (τ_i). For these first two composites $\tau_1 = \tau_2 = 0.5$. The more thermally conductive component is supposed to be the reinforcement, since it is usually the case (i.e., in C/epoxy or C/C composites).

A slightly more general arrangement for 1D PRDC is given in Figure 4(a). Figure 4(b) corresponds to a more realistic arrangement representative of carbon/epoxy composite and Figure 4(c) to a 3D C/C material with orthogonal x, y, z reinforcements.

Based on numerous numerical simulations [9], it has been demonstrated that the transient thermal behavior of 1D

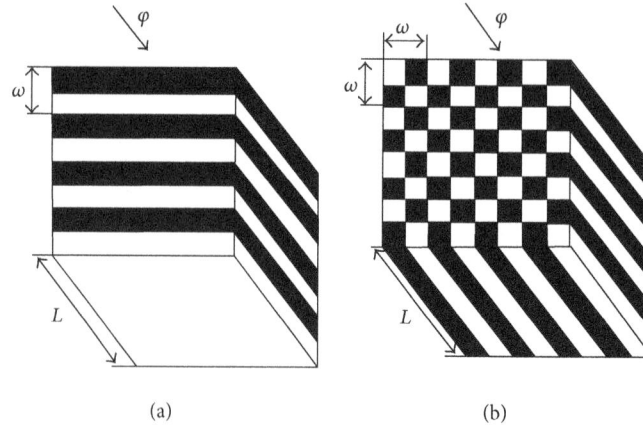

FIGURE 3: Simple configurations of unidirectional PDRCs: (a) stack of parallel layers, (b) chessboard pattern (from [9]).

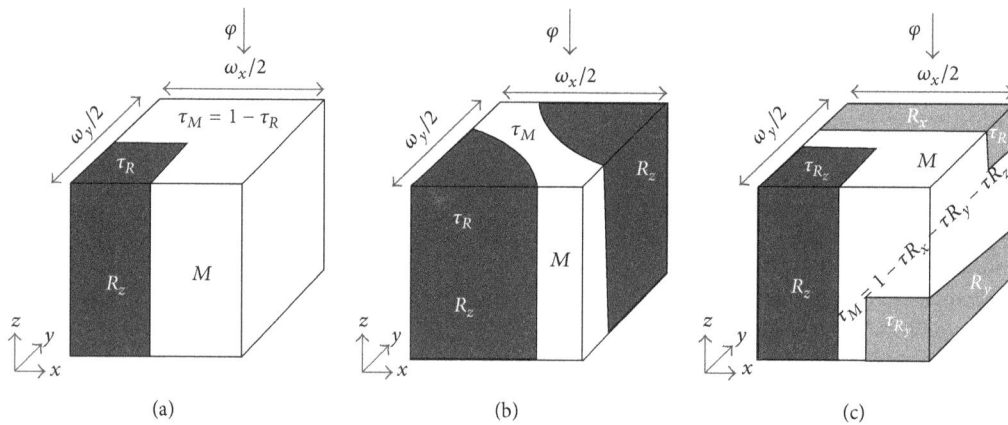

FIGURE 4: (a) Model of the elementary volume (1/4 of the repetitive cell) of a 1D PDRC, (b) idem for a 1D C/epoxy composite, (c) idem for a 3D C/C with x, y, z orthogonal reinforcements, from [10, 11].

PDRCs, when considering the mean temperature evolution, $\overline{T}(t)$ depends only on five nondimensional parameters:

(i) volume content of the reinforcement, τ_R;

(ii) the reinforcement-to-matrix ratio of thermal conductivities, $k_{R/M} = k_R/k_M$;

(iii) the reinforcement-to-matrix ratio of volume specific heats, $C_{R/M} = C_R/C_M$;

(iv) the specific contact surface between reinforcement and matrix per unit surface of composite: $\sigma = L\Sigma$, where L is the sample thickness, and Σ the specific contact surface between reinforcement and matrix per unit volume, parameter which depends on the shape of the section of the reinforcement;

(v) the specific contact thermal resistance between reinforcement and matrix: $\rho = Rk_M\Sigma$, R being the contact thermal resistance of the interface.

This mean temperature $\overline{T}(t)$ is easily measured by using IR radiometer or thermography if the emissivity of the surfaces are homogenized by a black coating insuring uniform emissivity.

3.2.2. Principle of the Identification of the Reinforcement Diffusivity and Homogenized Diffusivity from the Time Evolution of the Rear Face Mean Temperature $\overline{T}(t)$ in the Case of 1D DRC. Let us consider a 1D PDRC with a periodic pattern such as those of Figures 4(a) and 4(b). A typical mean temperature evolution of the rear face of the sample in flash diffusivity measurement is given in Figure 5(a) and compared to the theoretical response of a homogeneous material.

At the beginning, the energy that arrives at the rear face of a sample of thickness L has travelled through the more conductive component of the PDRC, here the reinforcement. If the identification is performed for each and every points of the curve, the identified diffusivity varies with time and the earlier the considered point, the nearer of the reinforcement diffusivity it is. Thus, following the early detection approach,

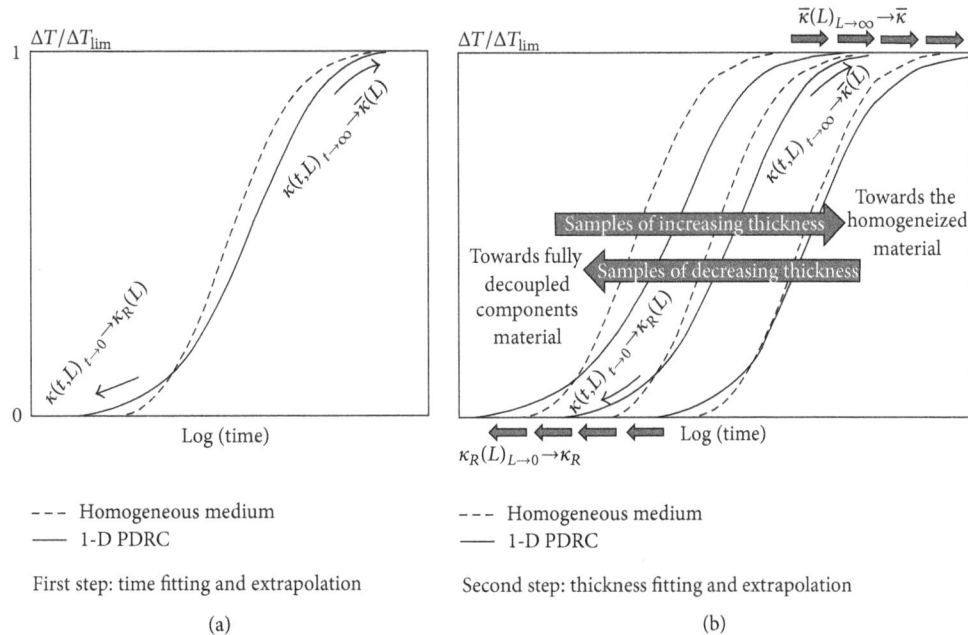

FIGURE 5: *In situ* characterization of a 1D PDRC. Principle of the early time and later time identifications (a), followed by spatial extrapolations to zero and infinite sample thicknesses (b).

by fitting the early part of the apparent diffusivity law and extrapolating it to zero time, an estimate of the reinforcement diffusivity is reached. This diffusivity is depending on L

$$\kappa(t, L)_{t \to 0} \longrightarrow \kappa_R(L). \qquad (2)$$

Using several samples of different thicknesses the law $\kappa_R(L)$ can be extrapolated to zero thickness and the so-extrapolated diffusivity can be considered as the best possible estimate of the reinforcement diffusivity

$$\kappa_R(L)_{L \to 0} \longrightarrow \kappa_R. \qquad (3)$$

On the contrary, the energy arriving at the end of the experiment can be considered as representative of a partially homogenized material. In effect, for long times, 3D heat transfers have enough time to become important and homogenize the temperature of the two components (reinforcement and matrix). These 3D effects, which are generally considered as a disturbing phenomenon, have in the present case a beneficial role, becoming a source of information about the material. The identified diffusivity law presents a final asymptote that can be considered as an approximate estimate of the fully homogenized material

$$\kappa(t, L)_{t \to \infty} \longrightarrow \overline{\kappa}(L). \qquad (4)$$

Thanks to the series of measurements performed with the different samples of various thicknesses, the law of partially homogenized diffusivity versus thickness can be extrapolated to an infinite thickness and the so-extrapolated value can be considered as the best estimate of the diffusivity of the fully homogenized material equivalent to the composite

$$\overline{\kappa}(L)_{L \to \infty} \longrightarrow \overline{\kappa}. \qquad (5)$$

In a third step, with a few assumptions which depends on the considered composite, from κ_R and $\overline{\kappa}$, it is possible to deduce the value of the diffusivity of the less conductive component (generally the matrix), κ_M, and the thermal resistance, R, characterizing the interface between the two components (see in particular [10, 12, 13]).

3.2.3. Application to 3D C/C Composites. The method has been applied to two 3D C/C composites in which the reinforcement is constituted of bundles of carbon fibers highly anisotropic and highly axially conductive, oriented in the three Cartesian directions x, y, z.

The first step consists to consider the 1D PDRC model of Figure 4(a) as an approximation equivalent to the 3D PDRC of Figure 4(c) in the case of the 3D C/C composite, considering that, due to the anisotropy of the reinforcement and the same chemical nature of the matrix and the reinforcement, it is possible to replace the actual matrix and the transverse reinforcement (following x and y directions) by an equivalent matrix. The identification procedure of Section 3.2.2 is then applied to this 1D PDRC. The operation can be repeated with the two other orientations of the samples to obtain the diffusivity of the three reinforcements if the composite is not equilibrated.

Figure 6 presents the experimental thermograms obtained for the 3D C/C no. 3 the flux being parallel to the z-reinforcement. They are compared to the thermograms of the equivalent homogenized material with the identified homogenized diffusivity, $\overline{\kappa}$, and the thermograms calculated using the equivalent 1D PDRC model with the identified parameters. The identified parameters are presented in Table 2 for two materials (3D C/C no. 1 and 3D C/C no. 3 for two orientations). These results show that even for a coupon

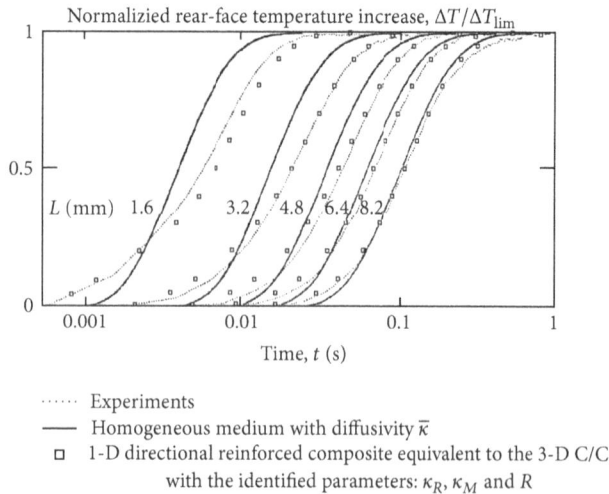

FIGURE 6: Normalized experimental pulse diffusivity rear face thermograms (mean temperature) obtained with a series of 3D C/C coupons of different thicknesses. Comparison to the homogeneous medium solution with the identified homogenized diffusivity, $\overline{\kappa}$, and to the 1D directional reinforcement composite equivalent to the 3D C/C using the identified parameters: axial thermal diffusivity of the reinforcement, κ_R, diffusivity of the matrix, κ_M, and thermal contact resistance of the reinforcement/matrix interface, R. Taken from [10, 13].

thickness of 8.2 mm, the material transient behavior is not exactly that of a homogeneous material. This thickness, knowing that the z-space period is 0.8 mm, represents ten times the space period.

These results were obtained with an IR radiometer viewing an area of the rear face much larger than a mesh of material to obtain a mean temperature measurement. It could be easily performed with an IR camera.

The main merit of the method is the fact that the measured properties are relative to *in situ* materials, which is of prime importance since the material process has a strong influence on the resulting final thermal properties of the product, in particular the axial diffusivity of the reinforcement.

3.2.4. Complementarity of Front-Face and Rear Face Pulse Measurements. The front-face and rear face pulse experiments are complementary [14]. In effect, let us apply the zero-time extrapolation to the apparent diffusivity identified from the rear face mean-temperature time evolution (see Figure 7(a)) and to the apparent effusivity identified from the front-face mean-temperature time evolution (see Figure 7(b)). These operations permit, respectively, the identification of characteristic thermal properties of the reinforcement parallel to the flux (longitudinal diffusivity) and of the equivalent matrix resulting from the homogenization of the matrix (isotropic effusivity) and of the two reinforcements perpendicular to the heat flux (radial effusivity). In other words, at the beginning of the pulse experiments, on the front-face the heating is driven by the less conductive component of

the composite and on the rear face by the more conductive component.

4. Application to Nondestructive Evaluation (NDE)

4.1. Notion and Interest of the Emerging Contrast. For NDE, the experimental data from which defect parameters can be identified are not thermograms (temperature increase, a function of time: $\Delta T(x, y, t) = T(x, y, t) - T(x, y, t = 0)$) but the time evolution of the contrasts between the temperature rise of a defective zone and the one of a sound zone taken as a reference, $Cr(x, y, t) = (\Delta T(x, y, t)_d - \Delta T(x, y, t)_s)/\Delta T(x, y, t)_s$). Consequently, the early detection approach in NDE consists in exploiting the contrast when it is just emerging from the noise, reason why we proposed to call it "the emerging contrast." Figure 8 presents the principle of the detection of the emerging contrast.

This attitude is contrary to common practice. In effect, traditionally, pulse thermography users favored the maximum contrast to identify the depth and thermal resistance of defects, starting from the *a priori* justified reason that it corresponds to the best SNR. This choice, which could be understood when the performances of cameras were limited (NETD ≥ 100 mK), is unfortunate and inappropriate. Nevertheless, it is still used by most of the users of the pulse technique. The early detection approach by the use of the emerging contrast is an efficient alternative that deserves to be promoted, as it will be demonstrated here.

4.1.1. "Universality" of the Emerging Contrast (Pulse Thermography). Let us consider the simple 1D configuration of a thermally imperfect interface (thermal resistance) inside two layers of solid materials. The interface is characterized by an extrinsic parameter, its location depth, z_d, and by an intrinsic thermal property, its thermal resistance, R_d. If $R_d = 0$, the structure is considered as sound and if $R_d > 0$, it is considered as being defective, with a resistive defect which has to be detected and characterized.

The problem is that the time-evolution of Cr depends on a large number of factors: the depth z_d, the thermal resistance, R_d, but also the total thickness of the structure, L, the thermal properties of the two layers effusivities and diffusivities, and the boundary conditions characterized by heat transfer coefficients h.

Figure 9 presents the wide variety of contrast evolutions that can be encountered. These data are the synthesis of simulation results taken from [15]. All parameters are non-dimensional and the graph is in log-log scales to conveniently show the various time scales. The possible domain in which the contrast may evolve, here lightly coloured in grey, is enormous, and it is obvious that to solve the inverse problem it is necessary to rely on a model taking into consideration all the aforementioned parameters.

A simple explicit expression relating the time of occurrence and the amplitude of the maximum relative contrast, a universally known and used approach, cannot be *a priori*

TABLE 2: *In situ* identified thermal properties of two 3D C/C composites, taken from [10].

Identified thermal parameters	Materials		
	3-D C/C no. 1	3-D C/C no. 3 //	3-D C/C no. 3 ⊥
Mean volume heat capacity[a]			
$\bar{c}$ (MJ m^{-3} K^{-1})	1.4	1.4	1.4
Homogenized properties			
$\bar{\kappa}$ (cm^2 s^{-1})	1.09	0.90	1.06
$\bar{k}$ (W m^{-1} K^{-1})	154	127	149
Reinforcement//heat flux			
κ_R (cm^2 s^{-1})	3.0	2.6	3.1
k_R (W m^{-1} K^{-1})	423	366	437
Equivalent matrix			
κ_M (cm^2 s^{-1})	0.27	0.33	0.48
k_M (W m^{-1} K^{-1})	38	46	68
Reinforcement/matrix interface			
ρ	0.01	1.6	6.6
R (m^2 K W^{-1})	$2.0 \cdot 10^{-7}$	$2.8 \cdot 10^{-5}$	$5.7 \cdot 10^{-5}$

[a] Estimated value.

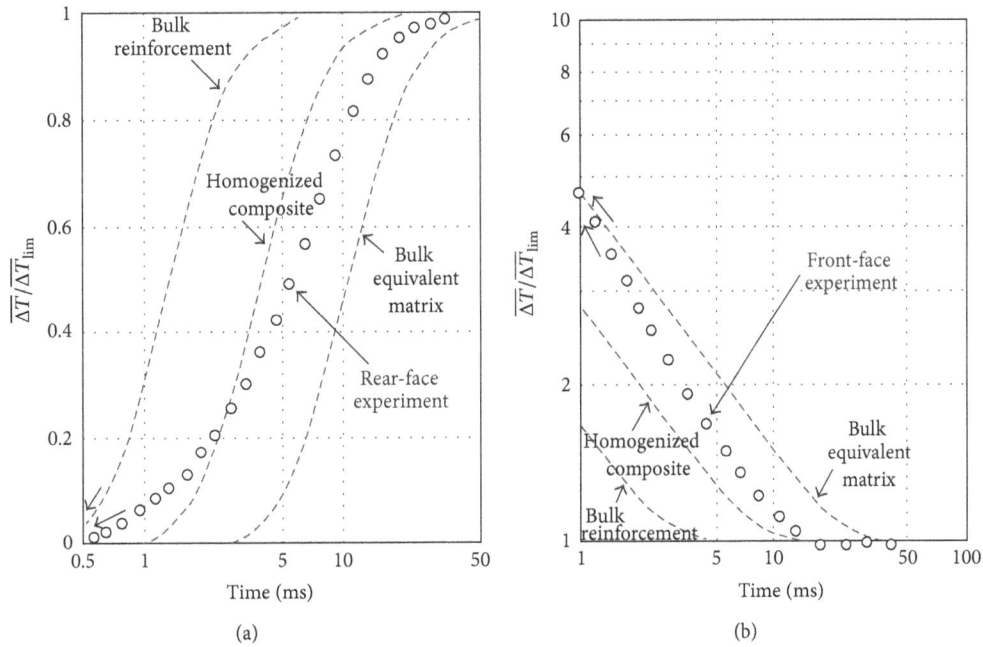

FIGURE 7: Compared mean temperature time histories of the front- and rear faces in the case of the 3D C/C #3// and comparison to the calculated thermograms corresponding to the thermal properties of the bulk reinforcement (longitudinal properties), bulk equivalent matrix, and fully homogenised composite, taken from [14].

satisfying when one considers the wide scatter existing on these two parameters.

Nevertheless, the main lesson which can be drawn from this graph is the existence, at the origin, of a narrow tail-shaped domain in which all curves merge, thus where a unique correlation between the relative contrast Cr and the defect Fourier number Fo_d can be used for the identification of the defect depth whatever is the thermal resistance and the other already mentioned parameters.

If we suppose that one can evaluate the occurrence time of a contrast of 1%, the maximum error on the identified depth is ±7%. Of course an earlier identification, based on a still lower contrast, leads to a more precise identified depth. The aim to be reached is clear; the problem is in the way to practically achieve the required experimental accuracy.

This simple example illustrates the main virtue of the emerging contrast: the "universality" of its applicability.

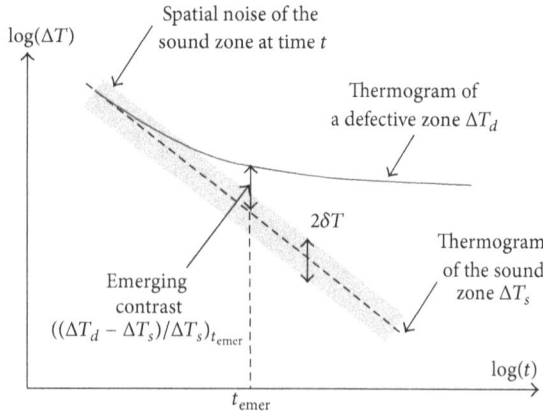

FIGURE 8: Principle of the detection of the emerging contrast.

Another illustration of this universality is given by the following study of the characterization of a defect embedded at different depths in a slab of a homogeneous sample. Let us consider a slab of thickness L, with a 1D defect located at the depth z_d and characterized by a thermal resistance R_d. For such a defect embedded in a semi-infinite medium, the thermal contrast associated with the 1D defect is depending on two non dimensional parameters: the Fourier number related to the defect depth, $Fo_d = \kappa t/z_d^2$, and the non dimensional thermal resistance $R^* = R_d/(z_d/k)$, the ratio of the defect thermal resistance to the one of the material layer in between the defect and the front-surface (see for instance [16, 17]. In these expressions, κ and k are, respectively, the diffusivity and the thermal conductivity of the material in between the monitored surface and the defect. In the case of a slab, the situation is more complex as seen in Figure 10(a). This Figure presents the Fourier-evolution of the relative contrast for various values of the normalized defect depth, z_d/L, and for $R^* = 1$. Similar (but different) nets would be obtained for other values of this last parameter. The shape of the contrast curves presents a maximum highly dependent on R^* and z_d/L parameters. The influence of the rear face "echo" gains in importance with increasing z_d/L and drastically decreases the amplitude of the relative contrast when the defect is approaching from this rear face, making the detection more and more problematic. Like in Figure 9, the large spreading of the net is converging to a narrow "tail" in which all curves merge for early times when the contrast is emerging form the noise. Thus, considering the emerging contrast domain ($Fo_d < 0.5$ and Cr of a few percents), for identification purpose, we can use the unique curve $Cr (Fo_d, R^* = 1)$ corresponding to the case of the defect embedded in a semi-infinite medium, for which there is no rear face reflection (see Figure 10(b)).

4.1.2. Identification Based on Emerging Contrast Needs Simpler Model than That Based on Maximum Contrast. Figure 10(b) presents the emerging contrast domain taken from Figure 10(a) and compares the contrast curves net to the two grey dotted curves of relative contrast related to the defects

of infinite thermal resistance and $R^* = 1$ in a semi-infinite medium, curves taken from [19, 20]. We see that this second curve remains near of the curves of the net for contrast of a few percents when z_d/L is smaller than 0.8. This means that we can use the semi-infinite medium solution $Cr (Fo_d, R^*)$ for identifying the depth and thermal resistance of a defect in this early time domain, instead of using a $Cr (Fo_d, R^*, z_d/L)$ function.

4.1.3. Early Detection and Characterization Lead to Less Blurred Images and More Accurate Identified Defect Parameters. Both Figures 9 and 10 show that the gain in precociousness of the identification when going from maximum contrasts to emerging contrasts can be very important (up to one order of magnitude). This gain is important since less time is given to 3D internal heat diffusion effects. This means from a qualitative point of view that the defect images will be less blurred, an important quality that increases the detectivity of the method, and quantitatively this produces a gain in accuracy for the identified parameters. This has been demonstrated in [19, 20].

4.2. Rapid Survey of the Early Detection and Characterization Approach in the NDE Literature. At the beginning of the 90's, the attention of several authors was drawn to the fact that it would be better to consider the contrast at its beginning to achieve an early detection.

The idea of an early identification of the defect depth, z_d, from the emerging contrast time was first formulated by Bontaz [21, 22] who proposed the following empirical relation:

$$t_0 = \frac{z_d^2}{2\pi\kappa}, \tag{6}$$

where t_0 is the origin time of the contrast, called by the authors the "divergence" time. This relation can be formulated using the defect Fourier number

$$Fo_d = \frac{\kappa t_0}{z_d^2} = \frac{1}{2\pi} = 0.159. \tag{7}$$

The weakness of this approach was double: (i) the relation between defect depth and divergence time was totally empirical and no precise procedure was proposed since the emergence had to be localized by the simple examination of the contrast curve without any indication given concerning the relative contrast value to consider for this determination of t_0; (ii) for the thermal resistance, R_d, the identification was similar to the temporal moment approach already proposed for diffusivity identification by Degiovanni [6], and for defect characterization by Balageas et al. [23], and was based on the use of the time integral of the full contrast from its origin to its extinction, then comprising data related to high Fourier numbers and consequently corrupted by 3D conduction effects.

Finally, Krapez et al. [18, 24, 25] kept the idea of performing the identification as early as possible with the emerging contrast, but developed a precise methodology

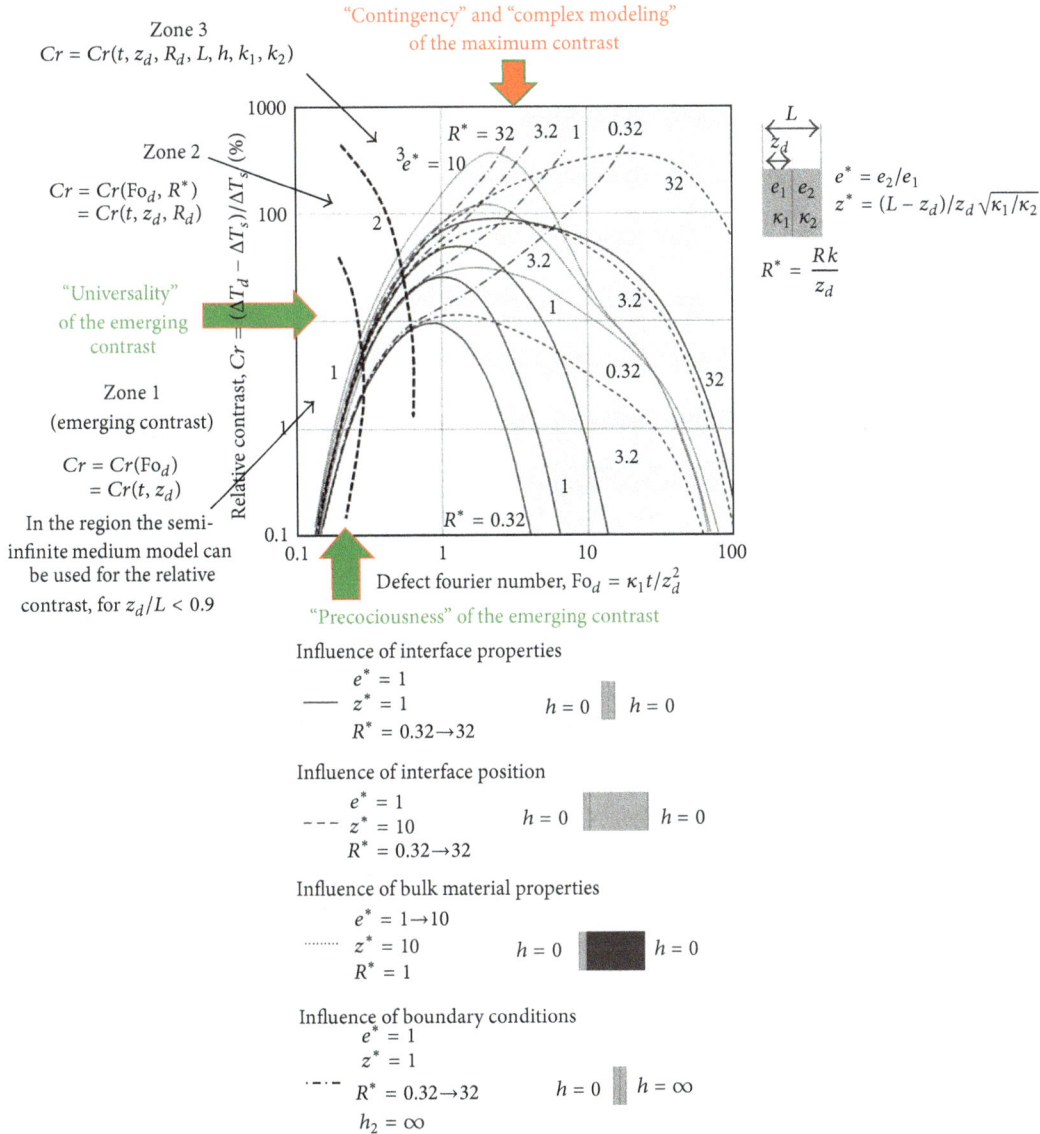

FIGURE 9: The virtues of the emerging contrast: universality, precociousness, and simple modeling. Comparison to the maximum contrast.

for the identification leading to propose explicit analytical formulas for identifying both depth and thermal resistance of a defect

$$z_d = \sqrt{\kappa_z t \text{Ln}\left[\frac{2}{Cr(t)}\right]},$$

$$R_d = \frac{Cr(t)\,\kappa_z t / k_z z_d}{\exp\left(-z_d^2/\kappa_z t\right) - Cr(t)/2},$$ (8)

expressions in which $Cr(t)$ is the relative contrast measured at time t, and k_z and κ_z are, respectively, the thermal conductivity and thermal diffusivity of the material between the monitored surface and the defect. The identification was achieved in two steps. The first relation was used for relative

contrasts near of 1 to 3% and once the defect depth identified, the second formula was used for identifying the thermal resistance for slightly larger contrast (3 to 6%).

In parallel to this research, another way to perform early detection was explored. Almond et al. [26, 27] found from numerical simulations and from experiments on a sample with artificial defects of same thermal resistance, located at the same depth, but with different sizes, that the "thermal contrast slope at the beginning" was independent of the size of the defect and just related to the defect depth. He suggested that the "short-time slope" of the contrast-time curve can be used to assess the defect depth rather than the peak time (time of the maximum contrast). Thomas, Favro and coll [28–30], and Ringermacher et al. [31] proposed to use the time of occurrence of the "peak slope" of the contrast to identify the

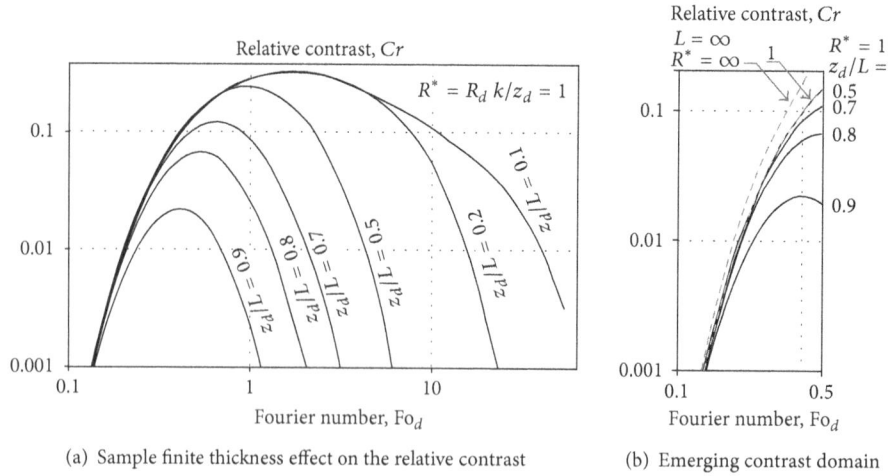

(a) Sample finite thickness effect on the relative contrast

(b) Emerging contrast domain

FIGURE 10: On the left: Influence of the normalized defect depth, z_d/L, on the relative contrast generated by a defect of normalized thermal resistance $R^* = 1$ in a slab of thickness L. On the right: Comparison in the emerging contrast domain between the contrasts generated by the same defect in a semi-infinite medium (solution used for the depth and thermal resistance identification by the early detection technique) and in a slab of thickness L.

defect depth. This proposal remained qualitative, no explicit relation relating the defect depth to this time being proposed by these authors.

Recently, Sun [32] studied quantitatively the relation between the contrast slope peak time, t_{ps}, and the defect depth using a normalized time $\omega_{def} = \pi^2 Fo_{def} = \pi^2 \kappa t_{ps}/z^2$ def. His study, limited to the case of an infinite defect thermal resistance (case of a hole machined in a slab), shows that the maximum slope is reached for a $Fo_{def} = 0.368$, which allows to identify the defect depth

$$z_d = 1.65\sqrt{\kappa t_{ps}}. \tag{9}$$

Nevertheless, this relation is only valid when the relative depth (ratio of the defect depth to the total sample thickness) is lower than 0.5. For higher values, the characteristic Fourier number is a strong function of the normalized defect depth, decreasing continuously down to 0.25 for a defect approaching the sample rear face.

This approach, although constituting a progress compared to the use of the occurrence time of the maximum contrast, is not as interesting as the early detection from the emerging contrast because the contrast slope peak occurs later than the emergence of the contrast ($Fo_{def} = 0.368$ instead of $Fo_{def} = 0.159$ for the Bontaz method, see (7)).

4.3. Recent Developments in Early Detection and Characterization of Defects

4.3.1. Improvement of the Emerging Contrast Technique by Linear Extrapolation to Nill Contrast. Balageas [19, 20] recently reworked the method developed in the 90's [18, 24, 25], following rigorously the early detection procedure described in Section 2 and using the identification formulas (8). In particular, he proposed and experimentally validated the extrapolation to zero contrast of the law of the identified

defect depth as a function of the contrast. By this way, the most accurate estimate of the defect depth is obtained. A relation between the error on the so-identified defect depth and the value of the defect thermal resistance is found, allowing to correct the first estimate of the defect depth once estimated the thermal resistance, which improves the accuracy of the method.

The early detection approach using the described procedure allows to reach the optimum accuracy on both the depth and the thermal resistance of defects: between 0.1% and 10% for the depth, and less than 30% for the thermal resistance (see Figure 11). The values of accuracy here given are intrinsic to the method and do not include the influence of the experimental noise and the 3D heat transfer effect linked to the limited extent of defects. The experimental noise can be reduced with the modern thermographic cameras characterized by an NETD of 20 mK or less, and by the use of a preprocessing technique of the thermograms such as the TSR method (see following Section).

4.3.2. Improvement of the Emerging Contrast Technique by the Combined Use of the Thermographic Signal Reconstruction Technique (TSR). The TSR method [33–37], well known and largely used in pulse thermographic NDE, consists in the fitting of the experimental log-log plot thermogram by a logarithmic polynomial

$$\ln(\Delta T) = a_0 + a_1 \ln(t) + a_2[\ln(t)]^2 \cdots + a_n[\ln(t)]^n, \tag{10}$$

and the use for NDE purpose of the 1st and 2nd logarithmic derivatives of the thermogram, the derivation being achieved directly on the polynomial, then with a limited increase of the temporal noise.

The advantages of the fitting are as follows: (i) a noticeable noise reduction; (ii) the replacement of the sequence of temperature rise images, $\Delta T(i,j,t)$, by the series of $(n + 1)$ images of the polynomial coefficients,

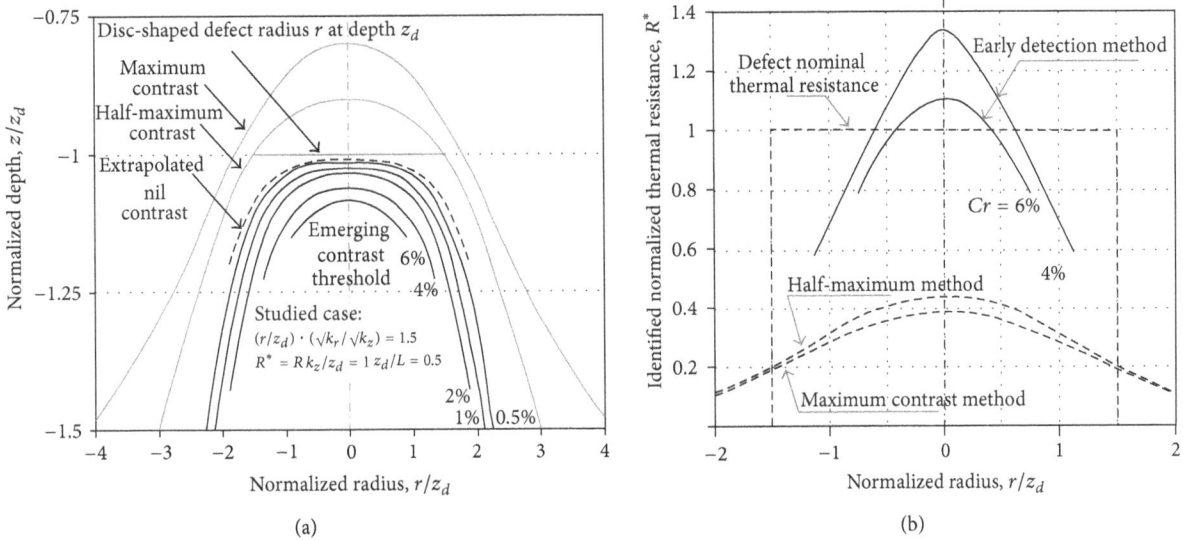

FIGURE 11: Simulation results: (a) identification of the depth profile of a circular defect ($r_d/z_d = 1.5$ and $R_d^* = R_d k_z/z_d = 1$) using the early detection/characterization with extrapolation to zero contrast and comparison to the results obtained with the maximum and half maximum contrast techniques (simulation results taken from Krapez and Balageas [18]), revisited using a linear extrapolation [19, 20]; (b) identification of the thermal resistance profile, taken from [20].

$a_0(i, j), \ldots, a_n(i, j)$, which allows a drastic reduction of the data amount; (iii) the possibility of reconstructing a full thermographic sequence (from which the name of the method follows).

The logarithmic derivations are interesting too, because they produce a remarkable increase of detectivity and a "precession" of the detection [19, 20], which improve the precociousness of the defect detection.

So, it is easy to understand that both methods (early detection by emerging contrast and TSR) pursue identical purposes, and that consequently coupling them is beneficial. The increase of signal-to-noise ratio (SNR) given by the logarithmic fitting is particularly welcome since experimental crude thermal contrasts may have weak SNR, especially when deep defects are considered. In this case, the TSR technique is used as a preprocessing tool before application of the early detection process. An example of such a coupling is shown in [20].

A deeper coupling could be used, consisting in considering the emerging contrast of the 1st or 2nd logarithmic derivatives instead of the emerging contrast of the thermogram itself for the characterisation of the defect. Such an early detection method, which remains until now to be established regarding the quantitative identification of the defect parameters, is better than the one based on the half-rise time of the first derivative or the time of the maximum of the second derivative. Nevertheless, if the sole qualitative aspect is considered (detection of defects from thermographic images), the use of early images of the first and second logarithmic derivatives produces sharper defect images, a noticeable improvement compared to traditional maximum contrast images. Figure 12, taken from [19], demonstrates the ability of this approach to produce images with good SNR

and high sharpness, making the pulse thermographic NDE technique comparable to the better-established techniques, in particular ultrasonics.

4.3.3. Application of the Early Detection and Characterization of Defects to Step-Heating Thermography. The early detection and characterization approach based on the emerging contrast has been extended to step-heating thermography, leading to comparable improvements. The theory is given in [38], leading to identification formulas for defect depth and thermal resistance analogous to the ones found for pulse heating experiments (8).

5. Conclusion and Perspectives

A review of the literature of the early time detection approach in the field of thermophysical properties measurements and NDE has been made.

This approach has been built progressively and never presented as a well-defined general procedure. This is due to the fact that the applications of the method were spread in time (3 decades) and pertaining to different fields. Putting these works in perspective, it has been possible to describe and formalize the general procedure here called "the early detection and characterization."

In the field of thermophysical properties measurements, two examples of flash diffusivity identification using rear face thermograms have been presented. They illustrate the philosophy of the approach and are well suited to thermography.

This paper demonstrates that in NDE by pulse-stimulated thermography, the generally followed attitude that consists in taking into account the sole signal-to-noise ratio when optimizing an identification process is an error. The optimization

(a)   (b)

Ultrasonic absorption image

The artificial defect   Unexpected defects

(c)   (d)

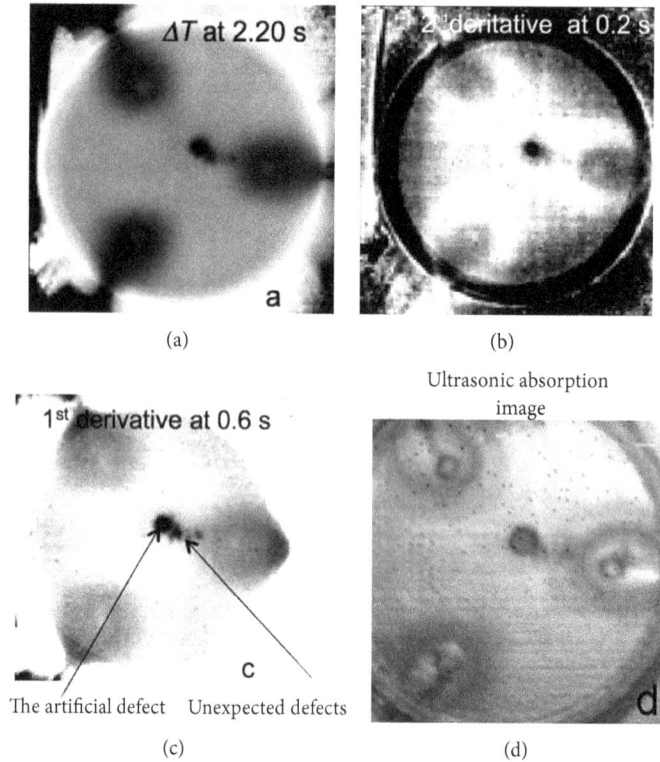

FIGURE 12: Comparison between the best images obtained by the TSR method coupled with an early detection approach and an ultrasonic D-scan: (a) Thermograms, ΔT at 2.20 s; (b) 2nd derivative at 0.20 s; (c) 1st derivative at 0.6 s; (d) ultrasonic absorption image. The artificial defect to detect is located at the center of the sample. Unexpected defects were discovered in the near vicinity of the artificial defect. The chosen times of observation have not been optimized for imaging the 3 peripheral inner inserts that are sound regions.

must consider with the same weight the signal-to-noise ratio and the precociousness for both qualitative and quantitative purposes.

A way to conciliate both signal-to-noise ratio and precociousness is now more easily feasible by combining the early detection/characterization and the thermographic signal reconstruction (TSR) technique. Presently, this combination has given rise to recent developments in the field of NDE, increasing the attractiveness of time-resolved thermography.

References

[1] R. B. Orlande, O. Fudym, D. Maillet, and R. M. Cotta, Eds., *Thermal Measurements and Inverse Techniques*, CRC Press, 2011.

[2] W. J. Parker, R. J. Jenkins, C. P. Butler, and G. L. et Abbot, "Flash method of determining thermal diffusivity, heat capacity and thermal conductivity," *Journal of Applied Physics*, vol. 32, no. 9, pp. 1679–1684, 1961.

[3] D. Balageas, "Flash Thermal diffusivity measurements using a novel temperature-time history analysis," in *Proceedings of the 1st International Joint Conferences on Thermophysical Properties*, no. TP 1981-62, ONERA, Gaithersburg, Md, USA, June 1981.

[4] D. L. Balageas, "Nouvelle méthode d'interprétation des thermogrammes pour la détermination de la diffusivité thermique par la méthode impulsionnelle (méthode flash)," *Revue de Physique Appliquée*, vol. 17, pp. 227–237, 1982 (French).

[5] A. Degiovanni, "Diffusivité et méthode flash," *Revue Générale de Thermique*, vol. 16, no. 185, pp. 420–442, 1977 (French).

[6] A. Degiovanni, "Identification de la diffusivité thermique par l'utilisation des moments temporels partiels," *High Temperatures-High Pressures*, vol. 17, pp. 683–689, 1985 (French).

[7] D. L. Balageas and A. M. Luc, "Transient thermal behavior of directional reinforced composites: applicability limits of homogeneous property model," *AIAA Journal*, vol. 24, no. 1, pp. 109–114, 1986.

[8] A. M. Luc and D. L. Balageas, "Non-stationary thermal behavior of reinforced composites—a better evaluation of wall energy balance for convective conditions," in *Proceedings of the 1st International Joint Conferences on Thermophysical Properties*, no. TP 1981-61, ONERA, Gaithersburg, Md, USA, June 1981.

[9] A. M. Luc-Bouhali, R. M. Pujolà, and D. L. Balageas, "Thermal diffusivity in situ measurements of carbon/carbon composite reinforcements," in *Thermal Conductivity*, T. Ashworth and D. R. Smith, Eds., vol. 18, pp. 613–624, Plenum, London, UK, 1984.

[10] R. M. Pujolà and D. L. Balageas, "Derniers développements de la méthode flash adaptée aux matériaux composites à renforcement orienté," *High Temperature-High Pressure*, vol. 17, pp. 623–632, 1985 (French).

[11] D. L. Balageas, "Détermination par méthode flash des propriétés thermiques des constituants d'un composite à renforcement orienté," *High Temperatures-High Pressures*, vol. 16, pp. 199–208, 1984 (French).

[12] D. L. Balageas, "Détermination de la diffusivité thermique du milieu homogène équivalent à un matériau composite à renforcement orienté," *Comptes-Rendus des Séances de l'Académie des Sciences*, vol. 299, no. 4, pp. 143–148, 1984 (French).

[13] J.-P. Bardon, D. Balageas, A. Degiovanni, and J. Vuiiliermes, "Thermics of composites and interfaces: current status and perspectives," *La Recherche Aérospatiale*, no. 1989-6, pp. 37–45, 1989.

[14] D. Balageas, A. Déom, and D. Boscher, "Composite thermal properties measurements by pulse photothermal radiometry," in *Proceedings of the Eurotherm IV Conference Thermal Transfer in Composite Materials and Solid-Solid Interface*, pp. 92–95, Nancy, France, June-July 1988.

[15] J. C. Krapez, *Contribution à la caractérisation des défauts de type délaminage ou cavité par thermographie stimulée [Ph.D. thesis]*, Ecole Centrale de Paris, Châtenay-Malabry, France, 1991.

[16] D. L. Balageas, J. C. Krapez, and P. Cielo, "Pulsed photothermal modeling of layered materials," *Journal of Applied Physics*, vol. 59, no. 2, pp. 348–357, 1986.

[17] D. L. Balageas, A. A. Deom, and D. M. Boscher, "Characterization and nondestructive testing of carbon-epoxy composites by a pulsed photothermal method," *Materials Evaluation*, vol. 45, no. 4, pp. 461–465, 1987.

[18] J. C. Krapez and D. Balageas, "Early detection of thermal contrast in pulsed stimulated infrared thermography," in *Proceedings of the Quantitative Infrared Thermography, Editions Europ Thermosonde et Induction (QIRT '94)*, pp. 260–266, 1994, http://qirt.gel.ulaval.ca/dynamique/index.php?idD=56, paper # QIRT 1994-039.

[19] D. L. Balageas, "Defense and illustration of time-resolved thermography for NDE," in *Proceedings of the SPIE Thermosense III*, vol. 8013, pp. 8013V-1–88013V20, 2011, http://publications.onera.fr/exl-php/cadcgp.php.

[20] D. Balageas, "Defense and illustration of time-resolved thermography for NDE," *Quantitative InfraRed Thermography Journal*, vol. 9, no. 1, pp. 5–38, 2012.

[21] J. Bontaz, Ch. Fort, and B. Horbette, "Identification de la profondeur et de la valeur de la résistance thermique de contact dans des matériaux stratifiés par la méthode photothermique impulsionnelle," in *Proceedings of the Annual Conference of the Société Française des Thermiciens (SFT '90)*, pp. 221–224, Nantes, France, May 1990.

[22] J. Bontaz, *Une méthode photothermique impulsionnelle appliquée au contrôle de matériaux composites [Ph.D. thesis]*, University of Bordeaux, 1991.

[23] D. L. Balageas, D. M. Boscher, and A. A. Déom, *Temporal Moment Method in Pulsed Photothermal Radiometry. Application to Carbon Epoxy N.D.T.*, vol. 58 of *Springer Series in Optical Sciences*, Springer, 1987.

[24] J. C. Krapez, D. Balageas, A. Deom, and F. Lepoutre, "Early detection by stimulated infrared thermo-graphy. Comparison with ultrasonics and holo/shearo-graphy," in *Advances in Signal Processing for Non Destructive Evaluation of Materials*, X. P. V. Maldague, Ed., vol. 262 of *NATO ASI Series E*, pp. 303–321, Kluwer Academic, 1994.

[25] J.-C Krapez, F. Lepoutre, and D. Balageas, "Early detection of thermal contrast in pulsed stimulated thermography," *Journal de Physique IV*, vol. 4, no. C7, pp. 47–50, 1994.

[26] S. K. Lau, D. P. Almond, and J. M. Milne, "A quantitative analysis of pulsed video thermography," *NDT and E International*, vol. 24, no. 4, pp. 195–202, 1991.

[27] D. P. Almond and S. K. Lau, "A quantitative analysis of pulsed video thermography," in *Proceedings of the Quantitative Infrared Thermography, Editions Europ Thermosonde et Induction (QIRT '92)*, pp. 207–211, QIRT, Paris, France, 1992, http://qirt.gel.ulaval.ca/dynamique/index.php?idD=55, paper # QIRT, 1992-031.

[28] L. D. Favro, X. Han, P. K. Kuo, and R. L. Thomas, "Imaging the early time behavior of reflected thermal wave pulses," in *Proceedings of the Thermosense XVII: An International Conference on Thermal Sensing and Imaging Diagnostic Applications*, pp. 162–166, April 1995.

[29] X. Han, L. D. Favro, P. K. Kuo, and R. L. Thomas, "Early-time pulse-echo thermal wave imaging," *Review of Progress in Quantitative Nondestructive Evaluation*, vol. 15, pp. 519–524, 1996.

[30] R. L. Thomas, L. D. Favro, and P. K. Kuo, "Thermal wave imaging of hidden corrosion in aircraft components," Report AFOSR-TR-96, 1996.

[31] H. I. Ringermacher et al., "Towards a flat-bottom hole standard for thermal imaging," in *Review of Progress in Quantitative Nondestructive Evaluation*, D. O. Thompson and D. E. Chimenti, Eds., vol. 17, pp. 425–429, Plenum Press, New York, NY, USA, 1998.

[32] J. G. Sun, "Analysis of pulsed thermography methods for detect depth prediction," *Journal of Heat Transfer*, vol. 128, no. 4, pp. 329–338, 2006.

[33] S. M. Shepard, T. Ahmed, B. A. Rubadeux, D. Wang, and J. R. Lhota, "Synthetic processing of pulsed thermographic data for inspection of turbine components," *Insight*, vol. 43, no. 9, pp. 587–589, 2001.

[34] S. M. Shepard, J. R. Lhota, B. A. Rubadeux, D. Wang, and T. Ahmed, "Reconstruction and enhancement of active thermographic image sequences," *Optical Engineering*, vol. 42, no. 5, pp. 1337–1342, 2003.

[35] S. M. Shepard, Y. L. Hou, T. Ahmed, and J. R. Lhota, "Reference-free interpretation of flash thermography data," *Insight*, vol. 48, no. 5, pp. 298–307, 2006.

[36] S. M. Shepard, J. Hou, J. R. Lhota, and J. M. Golden, "Automated processing of thermographic derivatives for quality assurance," *Optical Engineering*, vol. 46, no. 5, Article ID 051008, 2007.

[37] S. M. Shepard, "Flash thermography of aerospace composites," in *Proceedings of the 4th Pan American Conference for NDT*, Buenos Aires, Argentina, October 2007, http://www.ndt.net/article/panndt2007/papers/132.pdf.

[38] D. Balageas and J. M. Roche, "Détection précoce et caractérisation de défauts par thermographie stimulée par échelon de flux et comparaison à la méthode impulsionnelle," in *Congrès Annuel de la Société Française de Thermique*, Bordeaux, France, May-June 2012, http://publications.onera.fr/exl-doc/DOC401721_s1.pdf.

11

Carrier Formation Dynamics of Organic Photovoltaics as Investigated by Time-Resolved Spectroscopy

Kouhei Yonezawa,[1] Minato Ito,[1] Hayato Kamioka,[1, 2] Takeshi Yasuda,[3] Liyuan Han,[3] and Yutaka Moritomo[1, 2]

[1] Graduate School of Pure and Applied Science, University of Tsukuba, Tsukuba 305-8571, Japan
[2] Tsukuba Research Center for Interdisciplinary Materials Science (TIMS), University of Tsukuba, Tsukuba 305-8571, Japan
[3] Photovoltaic Materials Unit, National Institute for Materials Science (NIMS), Tsukuba 305-0047, Japan

Correspondence should be addressed to Yutaka Moritomo, moritomo@sakura.cc.tsukuba.ac.jp

Academic Editor: Saulius Juodkazis

Bulk heterojunction (BHJ) based on a donor (D) polymer and an acceptor (A) fullerene derivative is a promising organic photovoltaics (OPV). In order to improve the incident photon-to-current efficiency (IPCE) of the BHJ solar cell, a comprehensive understanding of the ultrafast dynamics of excited species, such as singlet exciton (D*), interfacial charge-transfer (CT) state, and carrier (D⁺), is indispensable. Here, we performed femtosecond time-resolved spectroscopy of two prototypical BHJ blend films: poly(3-hexylthiophene) (P3HT)/[6,6]-phenyl C_{61}-butyric acid methyl ester (PCBM) blend film and poly(9,9'-dioctylfluorene-co-bithiophene) (F8T2)/[6,6]-phenyl C_{71}-butyric acid methyl ester ($PC_{70}BM$) blend film. We decomposed differential absorption spectra into fast, slow, and constant components via two-exponential fitting at respective probe photon energies. The decomposition procedure clearly distinguished photoinduced absorptions (PIAs) due to D*, CT, and D⁺. Based on these assignments, we will compare the charge dynamics between the $F8T2/PC_{70}BM$ and P3HT/PCBM blend films.

1. Introduction

Organic photovoltaics (OPV) is an environmentally friendly and low-cost technology, which converts the solar energy into electric one. The incident photon-to-current efficiency (IPCE) of the bulk heterojunction (BHJ) solar cell [1, 2] is governed by the three processes: (1) charge formation process at the donor (D)-acceptor (A) interface, (2) charge transport process within the organic semiconductor, and (3) charge collecting process on the Al and indium tin oxide (ITO) electrodes. The femtosecond time-resolved spectroscopy is one of the powerful tool to reveal the (1) charge formation dynamics, because we can trace the ultrafast dynamics of excited species, such as singlet exciton (D*), interfacial charge-transfer (CT) state, and carrier (D⁺) [3–5]. The photoirradiation of the D polymer (A molecule) excites an electron from the highest occupied molecular orbital (HOMO) to the lowest unoccupied molecular orbital (LUMO). We call such a photoexcited D (A) state as excitons

D* (A*). The D* (A*) state migrates within the D domain (or A domain) to reach the D-A interface. At the interface, the charge transfer between D and A produces an intermediate state (CT state). The CT state consists of electrostatically bound charge pairs, where the hole is primarily localized on the D HOMO and the electron on the A LUMO. Finally, the charge separation takes place to produce free carriers D⁺ (A⁻).

Historically, extensive spectroscopic investigations [3–12] have been carried out on the charge dynamics in poly(3-hexylthiophene) (P3HT)/[6,6]-phenyl C_{61}-butyric acid methyl ester (PCBM) blend film, due to its reproducible power conservation efficiency (PCE > 5% [13, 14]). In particular, the regioregularity of P3HT, as well as the annealing process, has a significant effect on the charge dynamics of the P3HT/PCBM blend film [4, 7, 8]. Among them, Hwang et al. [4] investigated the charge dynamics in a regioregular-P3HT (RR-P3HT)/PCBM blend film and proposed a two-step process for charge generation, that is, formation of the interfacial CT states (<250 fs) followed by the charge separation (=4 ps). We emphasize that their interpretation is based on the assignment of the photoinduced absorption (PIA) signals. They assigned the PIA at 1.7 eV to the donor polaron [15, 16], while they did that at 1.1–1.6 eV to the CT state. Unfortunately, the assignment of the PIAs of the RR-P3HT/PCBM blend film is still controversial. Another candidate of OPV is poly(9,9'-dioctylfluorene-co-bithiophene) (F8T2)/[6,6]-phenyl C_{71}-butyric acid methyl ester (PC$_{70}$BM), because the fluorene-based copolymers, for example, F8T2, are more stable than the thiophene-based polymers, for example, P3HT. The solar cell using the blend film shows a PCE of 2.2–2.3% [17, 18].

In this study, we investigated the charge dynamics of two prototypical blend films, that is, RR-P3HT/PCBM and F8T2/PC$_{70}$BM blend films. In order to assign the PIA signals, we decomposed differential absorption spectra into fast, slow, and constant components via two-exponential fittings at respective probe photon energies. The decomposition procedure clearly distinguished photoinduced absorptions (PIAs) due to D*, CT, and D$^+$. We observed exciton conversion into the CT state (D* → CT) in both the blend films. The conversion speed (=0.7 ps) in the F8T2/PC$_{70}$BM blend film is nearly the same as that (=1.2 ps) in the P3HT/PCBM blend film.

2. Experiment

2.1. Film Preparation for Optical Measurements. All materials (see Figure 1) were purchased from commercially available sources and used as received. F8T2 was purchased from American Dye Source. The weight average molecular weight (M_w), number average molecular weight (M_n), and polydispersity (M_w/M_n) were estimated to be 45000, 13000, and 3.4, respectively. RR-P3HT was purchased from Luminescence Technology Corp. The weight average molecular weight (M_w), number average molecular weight (M_n), and polydispersity (M_w/M_n) were estimated to be 44000, 22000, and 2.0, respectively. The fullerene derivatives, that is, PCBM and PC$_{70}$BM, were purchased from Solenne.

P3HT and P3HT/PCBM blend films were spin-coated on quartz substrates and annealed for 10 min at 110°C. Solutions of P3HT and that of blend with 50% PCBM by weight was prepared by dissolving the compounds in o-dichlorobenzene (20 mg polymer in 1 mL solution). The thickness of the P3HT and P3HT/PCBM blend films was 129 and 234 nm, respectively. F8T2 and F8T2/PC$_{70}$BM blend films were spin-coated on quartz substrates and annealed for 10 min at 80°C. Solutions of F8T2 and that of blend with 66% PC$_{70}$BM by weight were prepared by dissolving the compounds in o-dichlorobenzene (16 mg polymer in 1 mL solution). The thickness of the F8T2 and F8T2/PC$_{70}$BM blend films was 87 and 89 nm, respectively. All the film preparation and post-treatment were performed in an inert N_2 atmosphere. The F8T2/PC$_{70}$BM blend film shows a definite D/A interface [17, 19]. The average size of the PC$_{70}$BM domains is 230 nm in diameter.

2.2. Preparation of OPV and Characterization. The F8T2/PC$_{70}$BM OPV was fabricated in the following configuration: ITO/PEDOT: PSS (40 nm)/active layer/LiF (1.2 nm)/Al (80 nm). The patterned ITO (conductivity: 10 Ω/square) glass was precleaned in an ultrasonic bath of acetone and ethanol and then treated in an ultraviolet-ozone chamber. A thin layer (40 nm) of PEDOT: PSS was spin-coated onto the ITO and dried at 110°C for 10 min on a hot plate in air. The substrate was then transferred to an N_2 glove box and dried again at 110°C for 10 min on a hot plate. An o-dichlorobenzene solution of F8T2:PC$_{70}$BM (1:2 by weight) was subsequently spin-coated onto the PEDOT:PSS surface to form the active layer. The resultant substrates were then annealed at 80°C for 10 min in an N_2 glove box. Finally, LiF (1.2 nm) and Al (80 nm) were deposited onto the active layer by conventional thermal evaporation at a chamber pressure lower than 5×10^{-4} Pa, which provided the devices with an active area of 2×2 mm^2. For comparison, we fabricated RR-P3HT/PCBM OPV in a similar procedure. An o-dichlorobenzene solution of RR-P3HT: PCBM (1:1 by weight) was subsequently spin-coated onto the PEDOT: PSS surface to form the active layer. The resultant substrates were then annealed at 110°C for 10 min in an N_2 glove box.

The current density-voltage (J-V) curves were measured using an ADCMT 6244 DC voltage current Source/Monitor under AM 1.5 solar-simulated light irradiation of 100 mWcm^{-2} (Wacom Electric Co., Ltd.). The incident photon-to-current conversion efficiency (IPCE) was measured using a CEP-2000 system (Bunkoh-Keiki Co., Ltd.).

2.3. Time-Resolved Spectroscopy. Ultrafast time-resolved spectroscopy was carried out in a pump-probe configuration at room temperature (Figure 2). We employed a regenerative amplified Ti: sapphire laser with a pulse width of 100 fs and a repetition rate of 1000 Hz as the light source. The pump pulse wavelength was 400 nm, which was generated as second harmonics in a β-BaB$_2$O$_4$ (BBO) crystal. The excitation intensity was 27–54 μJ/cm^2. The frequency of the pump pulse was decreased by half (500 Hz) to provide "pump-on" and "pump-off" condition. A white probe pulse (450–1600 nm), generated by self-phase modulation in a sapphire plate, was focused on the sample with the pump pulse. Spot sizes of the pump and probe pulses were 2.5 and 1.3 mm in diameter, respectively. The transmitted probe spectra were detected using a 72 ch Si photodiode array (450–900 nm) and/or a 256 ch InGaAs photodiode array (800–1600 nm) attached to a 30 cm imaging spectrometer. The spectral data were accumulated for 10000 pulses to improve the signal/noise ratio. The differential absorption spectra (ΔOD) are expressed as ΔOD = $-\ln(I_{on}/I_{off})$, where I_{on} and I_{off} are the transmitted light intensity with and without pump excitation, respectively.

2.4. Optical Modulation Spectroscopy. Optical modulation spectroscopy was carried out in a pump-probe configuration at room temperature (Figure 3). A continuous wave Yttrium Aluminum Garnet (CW-YAG) laser (532 nm) was used as the excitation light source. The excitation intensity was

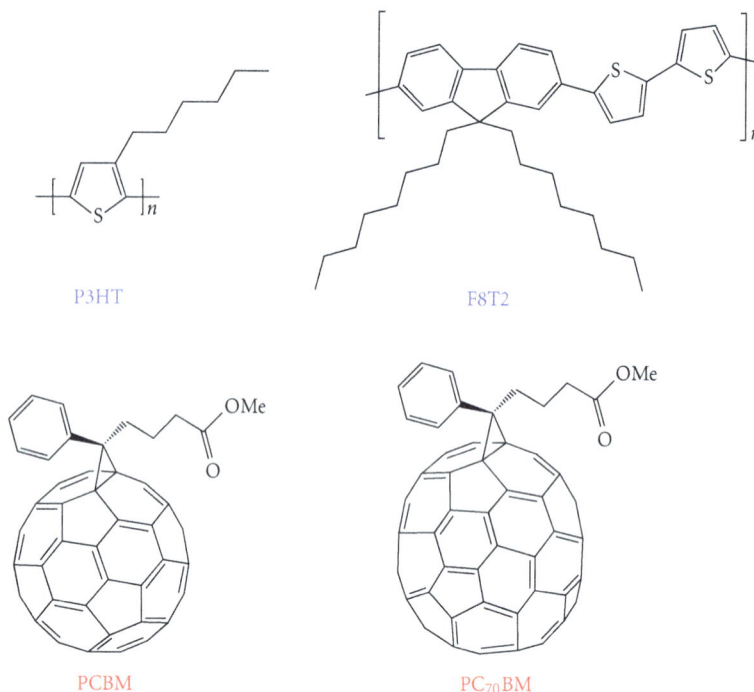

FIGURE 1: Molecular structures of poly(3-hexylthiophene) (P3HT), poly(9,9′-dioctylfluorene-co-bithiophene) (F8T2), [6,6]-phenyl C_{61}-butyric acid methyl ester (PCBM), and [6,6]-phenyl C_{71}-butyric acid methyl ester (PC$_{70}$BM).

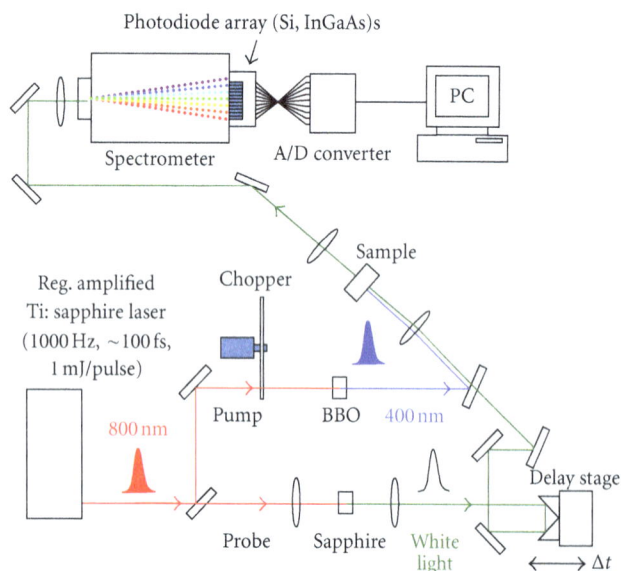

FIGURE 2: Schematic illustration of the experimental setup for time-resolved spectroscopy.

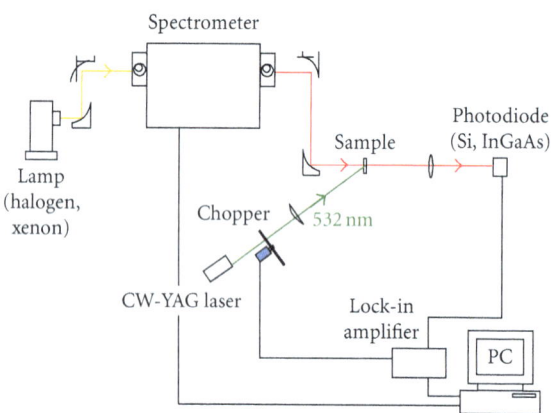

FIGURE 3: Schematic illustration of the experimental setup for optical modulation spectroscopy.

1.4 mW/cm^2, which was modulated with an optical chopper. The white light from the Halogen and/or Xe lamp was monochromatized with a 30 cm imaging spectrometer. The monochromatic probe light was focused on the sample with the pump light. The transmitted probe light was detected using a Si and/or InGaAs photodiode. A lock-in detection was adopted to extract the modulation signal. The optical modulation spectroscopy can clarify the PIA due to the long-lived D$^+$.

3. Results

3.1. OPV Properties of RR-P3HT/PCBM and F8T2/PC$_{70}$BM Blend Films. Figure 4(a) shows current density-voltage (J-V) curves of OPVs based on the RR-P3HT/PCBM and F8T2/PC$_{70}$BM blend films. The OPV based on F8T2 (HOMO of -5.46 eV) exhibits a high open-circuit voltage (V_{oc}) of 1.00 V, as compared with the value (=0.6 V) in typical OPVs based on RR-P3HT (HOMO of -4.70 eV). This is because V_{oc} is basically determined by the energy difference between D HOMO level and A LUMO (~-3.7 eV). The OPV based on the F8T2/PC$_{70}$BM blend film exhibits a short circuit current (J_{sc}) of 4.28 mAcm^{-2}, a V_{oc} of 1.00 V, a fill factor (FF)

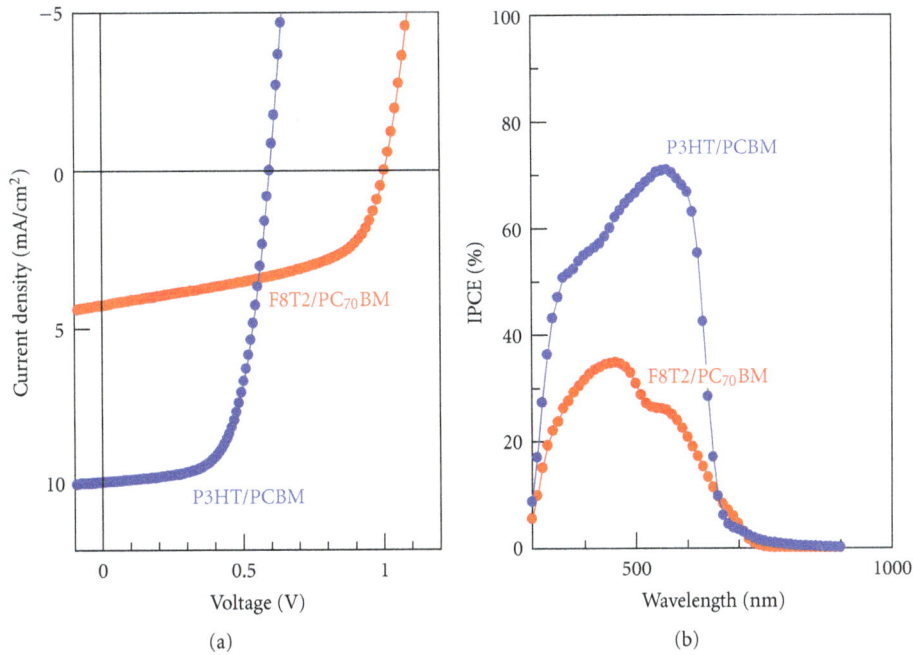

FIGURE 4: (a) *J-V* curves of OPVs based on the RR-P3HT/PCBM and F8T2/PC$_{70}$BM blend films. (b) IPCE spectra for OPVs based on the RR-P3HT/PCBM and F8T2/PC$_{70}$BM blend films.

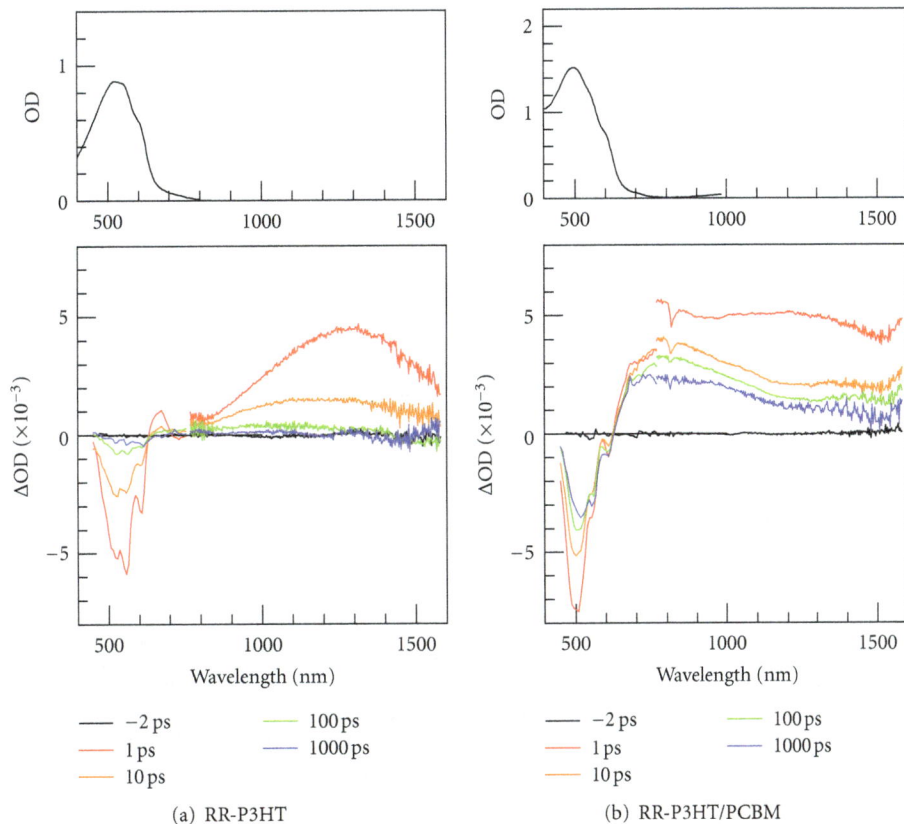

FIGURE 5: Absorption (OD) spectra and differential absorption (ΔOD) spectra of (a) neat RR-P3HT film and (b) RR-P3HT/PCBM blend films at 300 K.

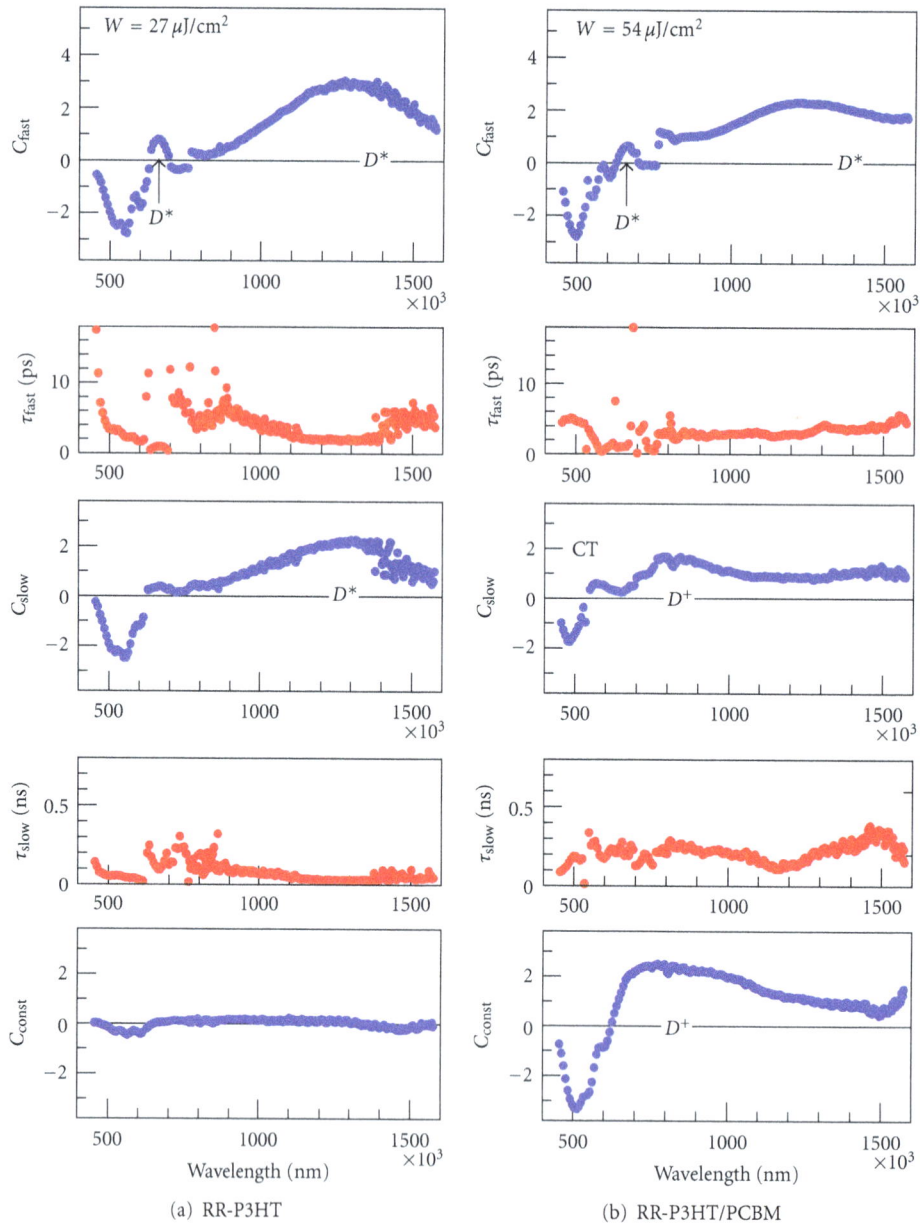

FIGURE 6: Decomposition of the time-resolved spectra of (a) neat RR-P3HT film and (b) RR-P3HT/PCBM blend films. The upper two panels are magnitudes (C_{fast}) and lifetimes (τ_{fast}) of the fast component. The middle two panels are magnitudes (C_{slow}) and lifetimes (τ_{slow}) of the slow component. The bottom panels are magnitudes (C_{const}) of the constant component. D*, CT, and D+ represent PIAs due to singlet exciton, CT state, and carrier, respectively.

of 0.53, and a PCE of 2.28%. The OPV based on the RR-P3HT/PCBM blend film exhibits a J_{sc} of 9.9 mAcm^{-2}, a V_{oc} of 0.60 V, a, FF of 0.64, and a PCE of 3.8%.

Figure 4(b) shows wavelength dependence of the photovoltaic response for the OPVs based on the RR-P3HT/PCBM and F8T2/PCBM blend films. In all the wavelength region, the IPCE values for the F8T2/PCBM blend film are smaller than those for the RR-P3HT/PCBM blend film. The IPCE values at 400 nm are 54% and 30% for the RR-P3HT/PCBM and F8T2/PCBM blend films, respectively.

3.2. Time-Resolved Spectra of RR-P3HT and RR-P3HT/PCBM Films. Figure 5 shows ΔOD spectra (lower panels) of (a) neat

RR-P3HT film and (b) RR-P3HT/PCBM blend films, together with their linear absorption spectra (OD: upper panels). As seen in the OD spectra, the pump pulse (at 400 nm) efficiently excites the D polymer. In all the films, the ΔOD spectra consist of negative signals in the short-wavelength region and positive signals in the long-wavelength region. The negative signal is ascribed to the ground state bleach (GSB) as well as the stimulated emission (SE) of singlet exciton luminescence. On the other hand, the positive signal is ascribed to PIAs due to the photogenerated species, such as D*, CT, and D+.

Generally speaking, the lifetime is a good physical quantity that distinguishes the PIAs due to different excited

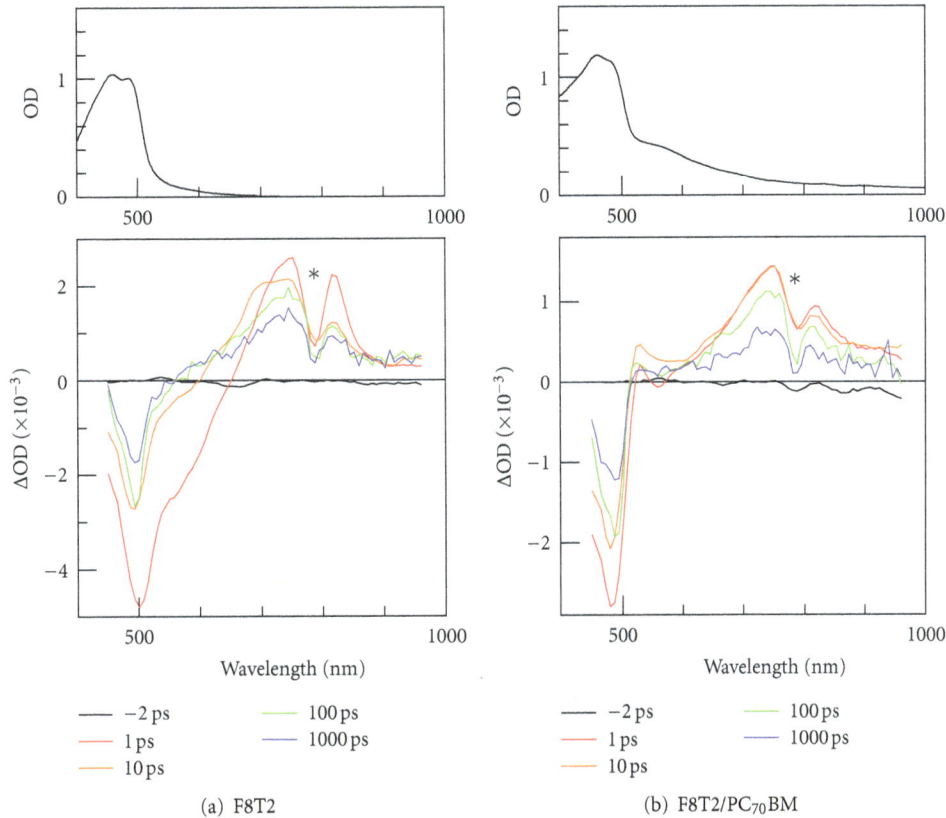

FIGURE 7: Absorption (OD) spectra and differential absorption (ΔOD) spectra of (a) neat F8T2 film and (b) F8T2/PC$_{70}$BM blend films at 300 K. Star symbols indicate an artificial structure due to the notch filter at 800 nm.

species: for example, lifetime of D* is much faster than that of D$^+$. In order to decompose the time-resolved spectra into fast, slow and constant components, we analyzed the decay curve at respective probe photon energies with two exponential functions:

$$\Delta OD = C_{fast} \times \exp\left(-\frac{t}{\tau_{fast}}\right) + C_{slow} \times \exp\left(-\frac{t}{\tau_{slow}}\right) + C_{const},$$

(1)

where C_{fast} (τ_{fast}), C_{slow}(τ_{slow}), and C_{const} are the magnitude (lifetime) of the respective components [20]. Thus obtained parameters, C_{fast}, τ_{fast}, C_{slow}, τ_{slow}, and C_{const} are plotted in Figure 6 against wavelength.

Figure 6(a) shows the spectral components of the neat P3HT film. In the C_{fast} component, a negative peak around 550 nm is ascribed to the GSB and SE. On the other hand, a positive sharp signal at 650 nm and broad signal at ~1300 nm are ascribed to the PIA due to D*. In the C_{slow} component, the broad PIA due to D* is discernible at ~1300 nm. No signal is observed in the C_{slow} component.

Figure 6(b) shows the spectral components of the RR-P3HT/PCBM blend film. In the C_{fast} component, the spectral profile is almost the same as the neat RR-P3HT film: negative signal due to GSB and SE at 500 nm and PIAs due to D* at 650 nm and ~1300 nm. In the C_{slow} component, the PIA due to D* completely disappears and new PIAs appears at 570 nm and ~800 nm. In the C_{const} component, broad PIA

at ~700 nm is observed. Jiang et al. [16] performed optical modulation spectroscopy in the neat RR-P3HT film and observed polaron signals at ~670 nm and ~1000 nm. They ascribed the former and latter signals to the free and localized carriers, respectively. According to their assignments, we ascribed the PIAs at ~700 nm (C_{const}) and at ~800 nm (C_{slow}) to the free and localized carriers, respectively. Based on the assignments of PIAs in the F8T2/PC$_{70}$BM blend film, we ascribed the PIA at 570 nm (C_{slow}) to the CT state (*vide infra*).

3.3. Time-Resolved Spectra of F8T2 and F8T2/PC$_{70}$BM Films. Figure 7 shows ΔOD spectra (lower panels) of (a) neat F8T2 film and (b) F8T2/PC$_{70}$BM blend films, together with the OD spectra (upper panels). As seen in the OD spectra, the pump pulse (at 400 nm) efficiently excites the D polymer. In all the films, the ΔOD spectra consist of negative signals in the short-wavelength region and positive signals in the long-wavelength region. The negative signal is ascribed to GSB and/or SE. On the other hand, the positive signal is ascribed to PIAs due to the photo-generated species, such as D*, CT, and D$^+$. We analyzed the decay curve at respective probe photon energies with two exponential functions (1). Obtained parameters C_{fast}, τ_{fast}, C_{slow}, τ_{slow}, and C_{const} are plotted in Figure 8 against wavelength.

Figure 8(a) shows the spectral components of the neat F8T2 film. In the C_{fast} component, a negative peak at 506 nm are ascribed to the GSB, whereas a negative peak at 553 nm

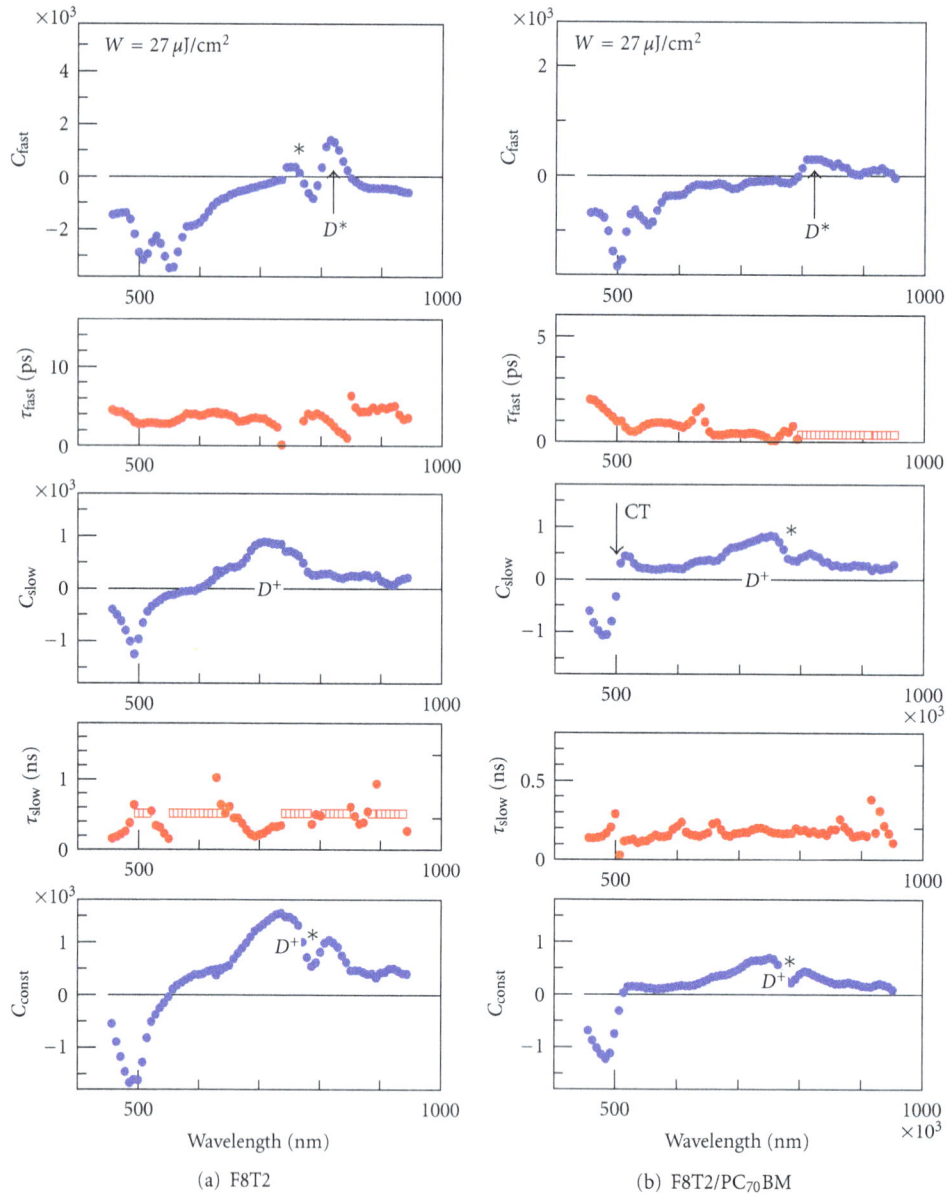

FIGURE 8: Decomposition of the time-resolved spectra of (a) neat F8T2 film and (b) F8T2/PC$_{70}$BM blend films. The upper two panels are magnitudes (C_{fast}) and lifetimes (τ_{fast}) of the fast component. The middle two panels are magnitudes (C_{slow}) and lifetimes (τ_{slow}) of the slow component. The bottom panels are magnitudes (C_{const}) of the constant component. Open symbols indicate that the parameter is fixed in the fitting procedure. Star symbols indicate artificial structures due to the notch filter at 800 nm. D*, CT, and D$^+$ represent PIAs due to singlet exciton, CT state, and carrier, respectively.

and the broad background extending from 500 to 700 nm is ascribed to the SE. Actually, the luminescence spectra of the F8T2 film extend from 500 to 700 nm with a peak at 550 nm. [21] On the other hand, a positive sharp signal around 800 nm is ascribed to the PIA due to D*. Actually, the exciton lifetime (=2.7 ps) is comparable to the decay time (~3 ps) of the SE signal. In the C_{slow} component, a positive broad signal is observed at ~700 nm. We ascribed the signal to the PIA due to D$^+$, because a long-lived polaron signal was observed at ~800 nm in the optical modulation spectrum of the F8T2/PC$_{70}$BM blend film (see Figure 9).

Figure 8(b) shows the spectral component of the F8T2/PC$_{70}$BM blend film. In the C_{fast} component, the GSB signal is observed at 501 nm, whereas the SE signal (500–700 nm) is rather suppressed. The suppression of the SE signal is probably due to the doping-induced luminescence quenching [21]. The PIA due to D* is discernible at 800 nm. In the C_{slow} component, the PIA due to D* completely disappears and new PIAs appears at 520 nm and ~700 nm. The PIA at ~700 nm is due to D$^+$, because a long-lived carrier signal was observed at ~800 nm in the optical modulation spectrum of the F8T2/PC$_{70}$BM blend film (see Figure 9). For the following reason, we ascribed the 520 nm signal to the PIA

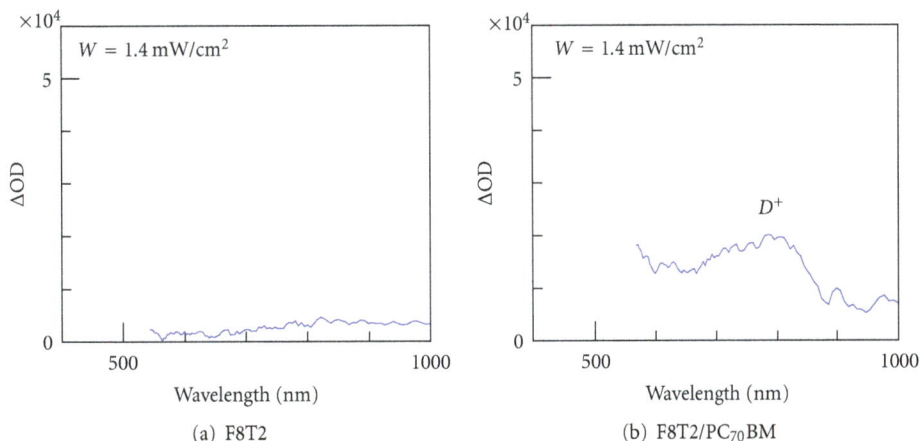

FIGURE 9: Optical modulation spectra of (a) neat F8T2 film and (b) F8T2/PC$_{70}$BM blend films.

due to the CT state. We compared the C_{slow} component in the 1 : 3 blend film (not shown) with that of the 1 : 2 film [19]. The average size of the PC$_{70}$BM domains is 300 nm (230 nm) in diameter in the 1:3 (1 : 2) blend film: the interface region is much reduced in the 1 : 3 film. We found that the 520 nm signal is suppressed in the 1 : 3 film, suggesting that the 520 nm signal relates to the interface. In addition, Lim et al. [21] reported an extra 570 nm absorption in the F8T2 film, in which 8% 2,3,5,6-tetrafluoro-7,7,8,8-tetracyanoquinodimethane (F$_4$TCNQ) is doped as an oxidant. They ascribed the 570 nm absorption to a CT complex, that is, F8T2$^+$-F$_4$TCNQ$^-$. Analogously, the CT state, that is, F8T2$^+$-PC$_{70}$BM$^-$, in our blend film is responsible for the PIA at 520 nm. In the C_{slow} component, PIA due to the CT state disappears, and only the PIA due to D$^+$ is observed.

Here, we note that the energy position and temporal behavior of the PIA at 570 nm in the RR-P3HT/PCBM blend film are analogues to the PIA at 570 nm in the F8T2/PC$_{70}$BM blend film. This strongly suggests that the PIA at 570 nm in the F8T2/PC$_{70}$BM blend film is also ascribed to the CT state.

3.4. Optical Modulation Spectra of F8T2 and F8T2/PC$_{70}$BM Films. Figure 9 shows optical modulation spectra of (a) F8T2 and (b) F8T2/PC$_{70}$BM blend films. No signal is observed in the neat F8T2 film. In the F8T2/PC$_{70}$BM blend film, characteristic positive signal is observed at ∼800 nm due to the long-lived carriers. Consistently, Ravirajan et al. [22] re-ported positive polaron signal at 720 nm in the chemically oxidized F8T2.

4. Discussions

Figure 10 shows temporal evolution of the PIAs of RR-P3HT/PCBM blend film due to D*, CT state, and D$^+$. Solid curves are the results of least-squares fitting with (1). The exciton lifetime (τ_{fast} = 0.9 ps) is close to the formation time (τ_{fast} = 1.2 ps) of the CT state, indicating exciton conversion into the CT state (D* → CT). The finite rise time of the CT state suggests that the state is created by exciton conversion at the interface, rather than by direct photo-generation. On

the other hand, the PIA due to D$^+$ exhibits a monotonic decrease with time, implying that D$^+$ is created mainly by direct photogeneration, not by conversion from the CT state. The number of D$^+$ gradually decreases and becomes 65% of the initial state at ∼1 ns. Note that the apparent constant component of 660 nm and at 560 nm should be ascribed to the broad PIA due to D$^+$ and GSB, respectively.

Figure 11 shows temporal evolution of the PIAs of F8T2/PC$_{70}$BM blend film due to D*, CT state, and D$^+$. Solid curves are the results of least-squares fitting with (1). The exciton lifetime (τ_{fast} = 2.0 ps) is close to the formation time (τ_{fast} = 0.7 ps) of the CT state, indicating exciton conversion into the CT state (D* → CT). The finite rise time of the CT state suggests that the state is created by exciton conversion at the interface, rather than by direct photo-generation. On the other hand, the PIA due to D$^+$ exhibits a monotonic decrease with time, implying that D$^+$ is created mainly by direct photo-generation, not by conversion from the CT state. The number of D$^+$ gradually decreases and becomes 40% of the initial state at ∼1 ns. Note that the apparent constant component of 800 nm should be ascribed to the broad PIA due to D$^+$.

Finally, let us comment on the plasmonic enhancement effect, which is a key technique to design the highly efficient OPV. For example, Poh et al. [23] clearly demonstrated that gold nanoparticles on the PEDOT-PSS enhance the PCE of the bilayer P3HT-C$_{60}$ device. They ascribed the enhancement to the enhanced absorption based on a finite-difference time-domain method (FDTD) simulation. We suspect that existence of the metal nanoparticle significantly influences the charge formation dynamics itself via the strong local field around the metal nanoparticles, especially when they locate near the D-A interface. We emphasize that the ultrafast spectroscopy is a powerful tool to clarify the plasmonic effects on the charge formation dynamics.

5. Summary

We investigated the charge dynamics of two prototypical blend films, that is, RR-P3HT/PCBM and F8T2/PC$_{70}$BM

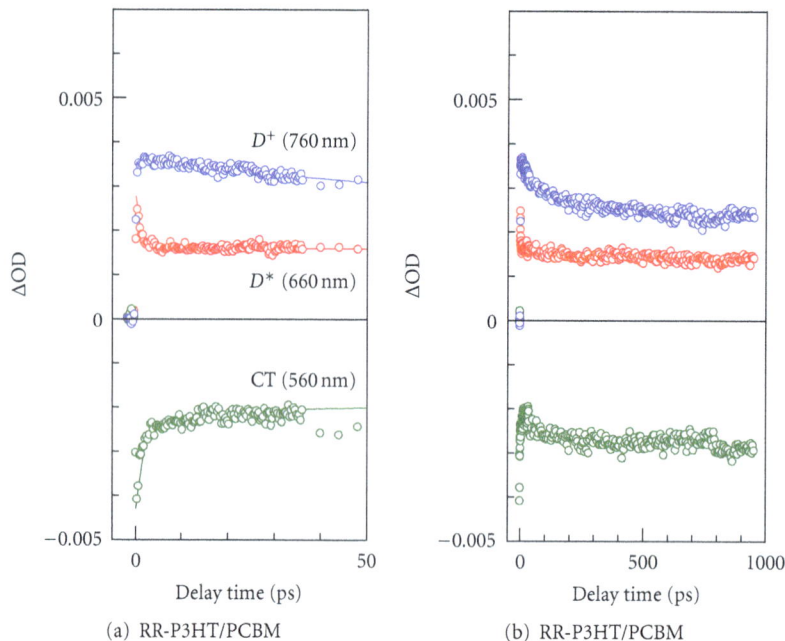

FIGURE 10: Temporal evolutions of PIAs of RR-P3HT/PCBM blend film due to D^*, CT state, and D^+: (a) fast region and (b) slow region. Solid curves in (a) are results of least-squares fitting with exponential functions: $\Delta OD = C_{fast} \times \exp(-t/\tau_{fast}) + C_{slow} \times \exp(-t/\tau_{slow}) + C_{const}$, C_{fast} is fixed at 0 in the PIA due to D^+.

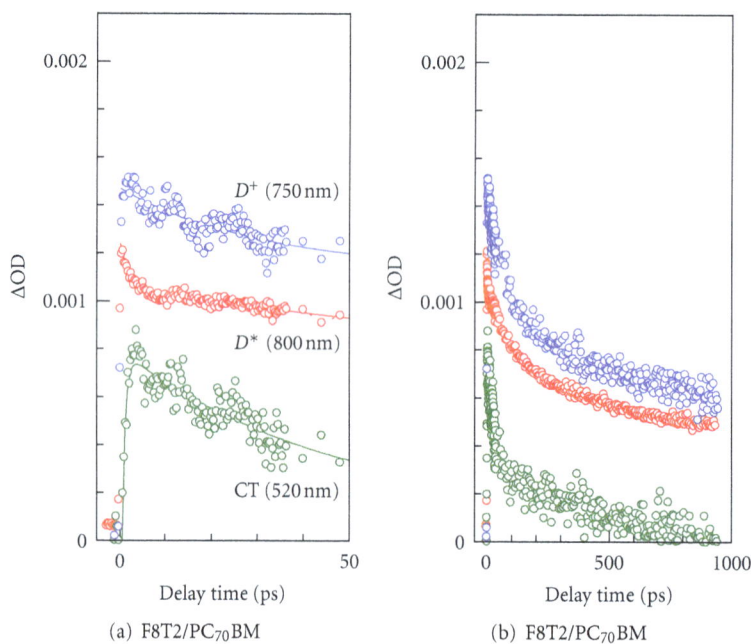

FIGURE 11: Temporal evolutions of PIAs of F8T2/PC$_{70}$BM blend film due to D^*, CT state, and D^+: (a) fast region and (b) slow region. Solid curves in (a) are results of least-squares fitting with exponential functions: $\Delta OD = C_{fast} \times \exp(-t/\tau_{fast}) + C_{slow} \times \exp(-t/\tau_{slow}) + C_{const}$, C_{fast} is fixed at 0 in the PIA due to D^+.

blend films. We observed exciton conversion into the CT state ($D^* \to CT$) in both the blend films. The conversion speed (=0.7 ps) in the F8T2/PC$_{70}$BM blend film is nearly the same as that (=1.2 ps) in the P3HT/PCBM blend film. Most of the carriers, however, are created by direct photogeneration, not by conversion from the CT state. The number of D^+ gradually decreases to 65% (40%) of the initial state at ~1 ns in the RR-P3HT/PCBM (F8T2/PC$_{70}$BM) blend film. The faster recombination ratio perhaps causes the lower PCE in the F8T2/PC$_{70}$BM blend film.

Acknowledgment

This work was supported by a Grant-in-Aid for Young Scientists (B) (22750176) for Scientific Research from the Ministry of Education, Culture, Sports, Science and Technology, Japan.

References

[1] M. Hiramoto, H. Fujiwara, and M. Yokoyama, "Three-layered organic solar cell with a photoactive interlayer of codeposited pigments," *Applied Physics Letters*, vol. 58, no. 10, pp. 1062–1064, 1991.

[2] N. S. Sariciftci, L. Smilowitz, A. J. Heeger, and F. Wudl, "Photo-induced electron transfer from a conducting polymer to buckminsterfullerene," *Science*, vol. 258, no. 5087, pp. 1474–1476, 1992.

[3] G. Grancini, D. Polli, D. Fazzi, J. Cabanillas-Gonzalez, G. Cerullo, and G. Lanzani, "Transient absorption imaging of P3HT:PCBM photovoltaic blend: evidence for interfacial charge transfer state," *Journal of Physical Chemistry Letters*, vol. 2, no. 9, pp. 1099–1105, 2011.

[4] I.-W. Hwang, D. Moses, and A. J. Heeger, "Photoinduced carrier generation in P3HT/PCBM bulk heterojunction materials," *Journal of Physical Chemistry C*, vol. 112, no. 11, pp. 4350–4354, 2008.

[5] S. Trotzky, T. Hoyer, W. Tuszynski, C. Lienau, and J. Parisi, "Femtosecond up-conversion technique for probing the charge transfer in a P3HT : PPCBM blend via photoluminescence quenching," *Journal of Physics D*, vol. 42, no. 5, Article ID 055105, 2009.

[6] J. Guo, H. Ohkita, H. Benten, and S. Ito, "Near-IR femtosecond transient absorption spectroscopy of ultrafast polaron and triplet exciton formation in polythiophene films with different regioregularities," *Journal of the American Chemical Society*, vol. 131, no. 46, pp. 16869–16880, 2009.

[7] J. Guo, H. Ohkita, H. Benten, and S. Ito, "Charge generation and recombination dynamics in poly(3-hexylthiophene)/fullerene blend films with different regioregularities and morphologies," *Journal of the American Chemical Society*, vol. 132, no. 17, pp. 6154–6164, 2010.

[8] R. Alex Marsh, J. M. Hodgkiss, S. Albert-Seifried, and R. H. Friend, "Effect of annealing on P3HT:PCBM charge transfer and nanoscale morphology probed by ultrafast spectroscopy," *Nano Letters*, vol. 10, no. 3, pp. 923–930, 2010.

[9] J. Piris, T. E. Dykstra, A. A. Bakulin et al., "Photogeneration and ultrafast dynamics of excitons and charges in P3HT/PCBM blends," *Journal of Physical Chemistry C*, vol. 113, no. 32, pp. 14500–14506, 2009.

[10] I. A. Howard, R. Mauer, M. Meister, and F. Laquai, "Effect of morphology on ultrafast free carrier generation in polythiophene: fullerene organic solar cells," *Journal of the American Chemical Society*, vol. 132, no. 42, pp. 14866–14876, 2010.

[11] S. Cook, R. Katoh, and A. Furube, "Ultrafast studies of charge generation in PCBM: P3HT blend films following excitation of the fullerene PCBM," *Journal of Physical Chemistry C*, vol. 113, no. 6, pp. 2547–2552, 2009.

[12] S. Cook, A. Furube, and R. Katoh, "Analysis of the excited states of regioregular polythiophene P3HT," *Energy and Environmental Science*, vol. 1, no. 2, pp. 294–299, 2008.

[13] W. Ma, C. Yang, X. Gong, K. Lee, and A. J. Heeger, "Thermally stable, efficient polymer solar cells with nanoscale control of the interpenetrating network morphology," *Advanced Functional Materials*, vol. 15, no. 10, pp. 1617–1622, 2005.

[14] Y. Kim, S. Cook, S. M. Tuladhar et al., "A strong regioregularity effect in self-organizing conjugated polymer films and high-efficiency polythiophene:fullerene solar cells," *Nature Materials*, vol. 5, no. 3, pp. 197–203, 2006.

[15] O. J. Korovyanko, R. Osterbacka, X. M. Jiang, and Z. V. Vardeny, "Theory of the electronic structure of the alloys of the actinides," *Physical Review B*, vol. 64, no. 23, Article ID 235122, 10 pages, 2001.

[16] X. M. Jiang, R. Osterbacka, O. Korovyanko et al., "Spectroscopic studies of photoexcitations in regioregular and regiorandom polythiophene films," *Advanced Functional Materials*, vol. 12, no. 9, pp. 587–597, 2002.

[17] T. Yasuda, K. Yonezawa, M. Ito, H. Kamioka, L. Han, and Y. Moritomo, "Photovoltaic properties and charge dynamics in nanophase-separated F8T2/PCBM blend films," *Journal of Photopolymer Science and Technology*. In press.

[18] J.-H. Huang, C.-P. Lee, Z.-Y. Ho, D. Kekuda, C.-W. Chu, and K.-C. Ho, "Enhanced spectral response in polymer bulk heterojunction solar cells by using active materials with complementary spectra," *Solar Energy Materials and Solar Cells*, vol. 94, no. 1, pp. 22–28, 2010.

[19] K. Yonezawa, H. Kamioka, T. Yasuda, L. Han, and Y. Moritomo, "Charge-transfer state and charge dynamics in poly(9, 9-dioctylfluorene-co-bithiophene) and [6, 6]-phenyl C_{70}-butyric acid methyl ester blend film," *Applied Physics Express*, vol. 4, no. 12, Article ID 122601, 2011.

[20] H. Kamioka, Y. Moritomo, W. Kosaka, and S. Ohkoshi, "Charge-transfer dynamics in cyano-bridged M_A–Fe system (M_A = Mn, Fe, and Co)," *Journal of the Physical Society of Japan*, vol. 77, no. 9, Article ID 093710, 2008.

[21] R. Lim, B.-J. Jung, M. Chikamatsu et al., "Doping effect of solution-processed thin-film transistors based on polyfluorene," *Journal of Materials Chemistry*, vol. 17, no. 14, pp. 1416–1420, 2007.

[22] P. Ravirajan, S. A. Haque, D. Poplavskyy, J. R. Durrant, D. D. C. Bradley, and J. Nelson, "Nanoporous TiO_2 solar cells sensitised with a fluorene-thiophene copolymer," *Thin Solid Films*, vol. 451-452, pp. 624–629, 2004.

[23] C. H. Poh, L. Rosa, S. Juodkazis, and P. Dastoor, "FDTD modeling to enhance the performance of an organic solar cell embedded with gold nanoparticle," *Optical Materials Express*, vol. 1, pp. 1326–1331, 2011.

12

Far-field Diffraction Properties of Annular Walsh Filters

Pubali Mukherjee[1] and Lakshminarayan Hazra[2]

[1] *MCKV Institute of Engineering, 243 G.T. Road, Liluah, Howrah 711204, India*
[2] *Department of Applied Optics and Photonics, University of Calcutta, 92 A.P.C. Road, Kolkata 700 009, India*

Correspondence should be addressed to Lakshminarayan Hazra; lnhaphy@caluniv.ac.in

Academic Editor: Augusto Belendez

Annular Walsh filters are derived from the rotationally symmetric annular Walsh functions which form a complete set of orthogonal functions that take on values either +1 or −1 over the domain specified by the inner and outer radii of the annulus. The value of any annular Walsh function is taken as zero from the centre of the circular aperture to the inner radius of the annulus. The three values 0, +1, and −1 in an annular Walsh function can be realized in a corresponding annular Walsh filter by using transmission values of zero amplitude (i.e., an obscuration), unity amplitude and zero phase, and unity amplitude and π phase, respectively. Not only the order of the Walsh filter but also the size of the inner radius of the annulus provides an additional degree of freedom in tailoring of point spread function by using these filters for pupil plane filtering in imaging systems. In this report, we present the far-field amplitude characteristics of some of these filters to underscore their potential for effective use in several demanding applications like high-resolution microscopy, optical data storage, microlithography, optical encryption, and optical micromanipulation.

1. Introduction

Annular apertures and different types of ring-shaped apertures continue to be investigated for catering to the growing exigencies in diverse applications, for example, high resolution microscopy, optical data storage, microlithography, optical encryption, and optical micromanipulation [1–5]. Not only for obvious energy considerations but also for their higher inherent potential in delivering complex far-field amplitude distributions, annular phase filters are being investigated in different contexts [6–9].

A systematic study on the use of phase filters on annular pupils can be conveniently carried out with the help of annular Walsh filters derived from the annular Walsh functions. Walsh functions form a closed set of normal orthogonal functions over a given finite interval and take on values +1 or −1, except at a finite number of points of discontinuity, where they take the value zero [10, 11]. The order of a Walsh function is directly related to the number of its zero crossings or phase transitions within the specified domain, and they constitute a closed set of normal orthogonal functions over the specified interval. They have the interesting property that an approximation of a continuous function over a finite interval by a

finite set of Walsh functions leads to a piecewise constant approximation to the function. Walsh filters of various orders may be obtained from corresponding Walsh functions, by realizing transmission values of +1 and −1 by 0 and π phase, respectively. Incidentally it may be noted that binary phase filters are being explored for many interesting light distributions [12].

Walsh functions have been used in the field of signal coding and transmission and in allied problems of information processing [13]. Two-dimensional Walsh functions in the usual rectangular coordinates have been used in digital image processing applications [14]. For treatment of problems of optical imaging, Walsh functions in polar co-ordinates have been utilised [15]. For systems with rotational symmetry about the axis, radial Walsh functions [16] have been developed as a special case of Walsh functions in polar co-ordinates and they were proved useful in the treatment of apodization problems [17, 18]. It has also been observed that not only the transverse amplitude distribution on the far-field plane but also the axial distribution of amplitude/intensity in the far-field is significantly modified in presence of radial Walsh filters on unobscured apertures [19].

Annular Walsh functions are a generalization of radial Walsh functions. For a specific central obscuration ratio, they are a complete set of orthogonal functions over the annulus. Annular Walsh filters may be considered to have ternary transmission values. The three values of transmission are zero amplitude over the central obscuration, unity amplitude and zero phase for value +1, and unity amplitude and π phase for value −1 in corresponding annular Walsh functions.

In this paper, we present some results of our investigations on the far-field amplitude characteristics of annular Walsh filters of orders 0, 1, 2, and 3 with different central obscuration ratios. After a brief description of annular Walsh functions in the next section, Section 3 presents the mathematical expression for the amplitude distribution in the far-field of an exit pupil with annular Walsh filters on it. Some interesting numerical results and our observations on the same are put forward in the last two sections.

2. Annular Walsh Functions

To define annular Walsh function $\Psi_n^\varepsilon(r)$ of index $n \geq 0$ and argument r over an annular region with ε and 1 as inner and outer radii, respectively, it is necessary to express the integer n in the form

$$n = \sum_{m=0}^{\bar{n}-1} k_m 2^m, \tag{1}$$

where k_m are the bits, 0 or 1 of the binary numeral for n, and $(2^{\bar{n}})$ is the power of 2 that just exceeds n. For all r in $(\varepsilon, 1)$, $\Psi_n^\varepsilon(r)$ is defined as

$$\Psi_n^\varepsilon(r) = \prod_{m=0}^{\bar{n}-1} \text{sgn}\left\{\cos\left[k_m 2^m \pi \frac{(r^2 - \varepsilon^2)}{(1 - \varepsilon^2)}\right]\right\}, \tag{2}$$

where

$$\text{sgn}(x) = \begin{cases} +1, & x > 0, \\ 0, & x = 0, \\ -1, & x < 0. \end{cases} \tag{3}$$

The orthogonality condition implies that

$$\int_\varepsilon^1 \Psi_m^\varepsilon(r) \Psi_n^\varepsilon(r) r\, dr = \frac{1-\varepsilon^2}{2} \delta_{mn}, \tag{4}$$

where δ_{mn} is the Kronecker delta defined as

$$\delta_{mn} = \begin{cases} 0, & m \neq n, \\ 1, & m = n. \end{cases} \tag{5}$$

Figure 1 shows the first four annular Walsh functions for central obscuration $\varepsilon = 0.3$ in two dimensions. Figure 2 presents values of the functions $\Psi_n^{0.3}(r)$, $n = 0, \ldots, 3$, along the radius in an azimuthal direction. It should be noted that the order of the functions n is equal to the number of zero crossings, or sign changes of the function in the interval $(0.3, 1)$, and

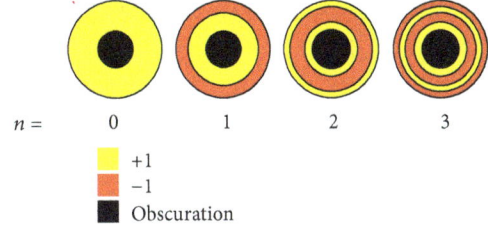

FIGURE 1: Annular Walsh functions $\Psi_n^\varepsilon(r)$ in two dimensions of order $n = 0, 1, 2, 3$ for central obscuration $\varepsilon = 0.3$.

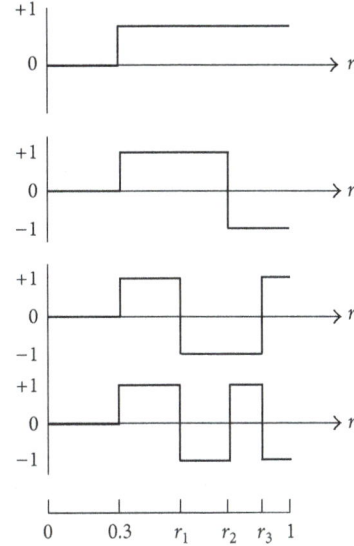

FIGURE 2: Annular Walsh functions $\Psi_n^\varepsilon(r)$ of order $n = 0, 1, 2, 3$ along radius r for central obscuration $\varepsilon = 0.3$.

locations of the points of zero crossings for members of the set of functions $\Psi_n^\varepsilon(r)$, $n = 0, \ldots, 3$ are given by

$$r_i = \sqrt{\frac{[(4-i)\varepsilon^2 + i]}{4}}, \quad i = 1, 2, 3. \tag{6}$$

In general, for the first N, where $N = 2^b$ and b is a positive integer, and annular Walsh functions $\Psi_n^\varepsilon(r)$, $n = 0, 1, \ldots, (N-1)$, the zero crossings are located at

$$r_i = \sqrt{\frac{[(N-i)\varepsilon^2 + i]}{N}}, \quad i = 1, 2, \ldots, (N-1). \tag{7}$$

The inner and outer radii of the annulus is $r_0 = \varepsilon$ and $r_N = 1$. Note that the set of $(N-1)$ zero crossing locations, r_i, $i = 1, 2, \ldots, (N-1)$, consists of all zero crossing locations required for specifying members of this particular set of Walsh functions. An individual member of this set of Walsh functions will have the same number of zero crossings as its order.

For computational purposes it is often convenient to express an annular Walsh function $\Psi_n^\varepsilon(r)$ as

$$\Psi_n^\varepsilon(r) = \sum_{j=1}^N h_{nj} \zeta_j^N(r), \tag{8}$$

TABLE 1: The Hadamard matrix for $N = 2^2$.

+1	+1	+1	+1
+1	+1	−1	−1
+1	−1	−1	+1
+1	−1	+1	−1

where $N = 2^{\tilde{n}}$. For a particular value of n, the integer $\tilde{n}$ is taken such that N just exceeds n. $\zeta_j^N(r)$ are zero-one functions, also known as Walsh block functions, defined as

$$\zeta_j^N (r) = \begin{cases} 1, & r_{j-1} \le r \le r_j, \\ 0, & \text{otherwise.} \end{cases} \tag{9}$$

For all annular Walsh functions of order $n < N$, the same values of r_j, $j = 0, N$ are to be used. They are unique for the particular value of N, and their values are given in (7). h_{nj} are the elements of a $(N \times N)$ Hadamard matrix whose elements are +1 or −1 [13]. Table 1 gives the Hadamard matrix for $N = 2^2$. The Hadamard matrices for other values of N can be obtained from the defining relation (2) of annular Walsh functions.

3. Far-Field Amplitude Distribution

Figure 3 shows the image space of an axially symmetric imaging system. It is well known that the complex amplitude distribution on the image plane corresponding to an axial object point is given by the far-field diffraction pattern of the pupil function over the exit pupil. The complex amplitude $F(p)$ at a point Q' in the far-field due to a point object on the axis is, apart from the multiplicative constant, given by the Hankel transform of order zero of the pupil function [20, 21]:

$$F(p) = \int_0^1 f(r) J_0 (pr) r \, dr, \tag{10}$$

where $f(r)$ is the circularly symmetric pupil function. The variable r is the fractional coordinate for a point P' on the pupil sphere of center O' and radius $O'E'$, where E' is the axial position of the exit pupil. r is equal to the ratio (r'/h'), where r' is the perpendicular distance of the point P' from the axis and h' is the maximum value of r', so that $0 \le r \le 1$. p is the reduced diffraction variable given by

$$p = \frac{2\pi}{\lambda} \left(n' \sin \alpha' \right) \xi', \tag{11}$$

where α' is the semiangular aperture of the imaging system, n' is the refractive index of the image space, $2\pi/\lambda$ is the propagation constant, and $\xi' (= O'Q')$ is the actual geometrical distance of point Q' on the image plane from the center O' of the diffraction pattern.

For a specific central obscuration ratio ε, annular Walsh filters of various orders can be obtained from the corresponding annular Walsh functions by realizing transmission values +1 and −1 by zero and π phase filters, respectively. In presence

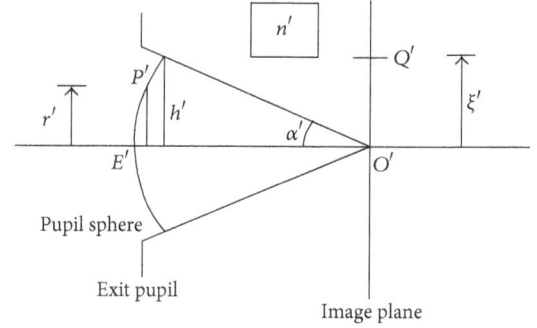

FIGURE 3: Schematic representation of the image space parameters.

of an annular Walsh filter on a uniform pupil, the pupil function $f(r)$ is given by

$$f(r) = \begin{cases} 0, & 0 \le r < \varepsilon, \\ \Psi_n^\varepsilon(r) & \varepsilon \le r < 1. \end{cases} \tag{12}$$

It may be noted that $f(r)$ is binary (value either 0 or +1) only in the case of zero order annular Walsh function $\Psi_0^\varepsilon(r)$; for all other orders, $f(r)$ is ternary with value either 0, +1, or −1.

The far-field amplitude pattern $F_n^\varepsilon(p)$ of an annular Walsh filter $\Psi_n^\varepsilon(r)$ is given by

$$F_n^\varepsilon(p) = \int_\varepsilon^1 \Psi_n^\varepsilon(r) J_0(pr) r \, dr = \sum_{j=1}^N h_{nj} \Gamma_j(p), \tag{13}$$

where

$$\Gamma_j(p) = \int_\varepsilon^1 \zeta_j^N(r) J_0(pr) r \, dr$$
$$= \left[\frac{r_j^2 J_1(pr_j)}{pr_j} - \frac{r_{j-1}^2 J_1(pr_{j-1})}{pr_{j-1}} \right]. \tag{14}$$

4. Numerical Results and Discussion

Figures 4–7 give the far-field amplitude distributions for annular Walsh filters $\Psi_n^\varepsilon(r)$, with central obscuration $\varepsilon = 0.3$, 0.5, 0.7, and 0.9, respectively. In each figure, the amplitude distribution is shown for the annular Walsh filters of orders 0, 1, 2, and 3. Note that the amplitude values are normalized by the central amplitude for an unobstructed uniform aperture. For the sake of underscoring amplitude variations, the scales along the ordinate and abscissa of different curves could not be maintained the same; they were tailored to bring forth the specific features.

It is seen in Figures 4(a), 5(a), 6(a), and 7(a) that for the zero order annular Walsh filters, there is a central maximum of amplitude in the diffraction pattern, and the magnitude of the maximum is obviously dependent on the central obscuration ε. For the sake of comparison, amplitude distribution for an unobscured uniform aperture is shown alongside the same for zero order annular Walsh filters. It is well known that with higher values of obscuration, the central lobe narrows down significantly [20].

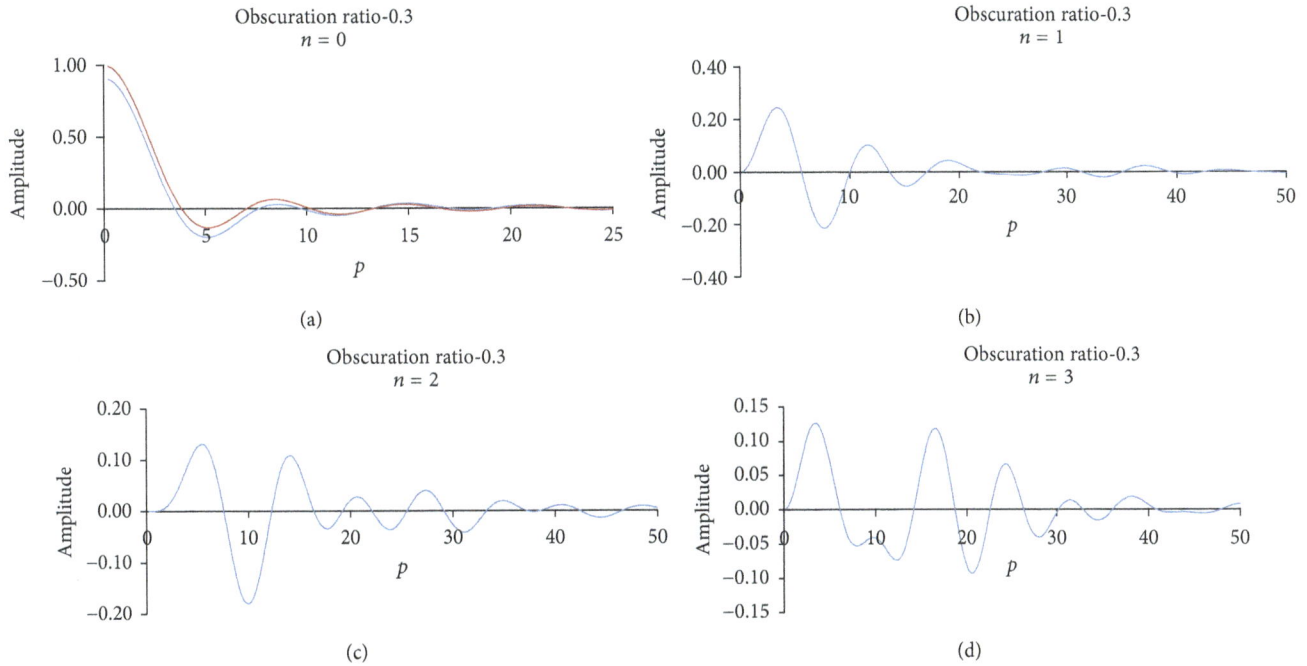

FIGURE 4: Far-field amplitude distributions for annular Walsh filters $\Psi_n^\varepsilon(r)$ on an annular aperture with central obscuration $\varepsilon = 0.3$ for orders (a) $n = 0$, (b) $n = 1$, (c) $n = 2$, and (d) $n = 3$. The red graph in (a) represents amplitude distribution for an unobscured uniform aperture.

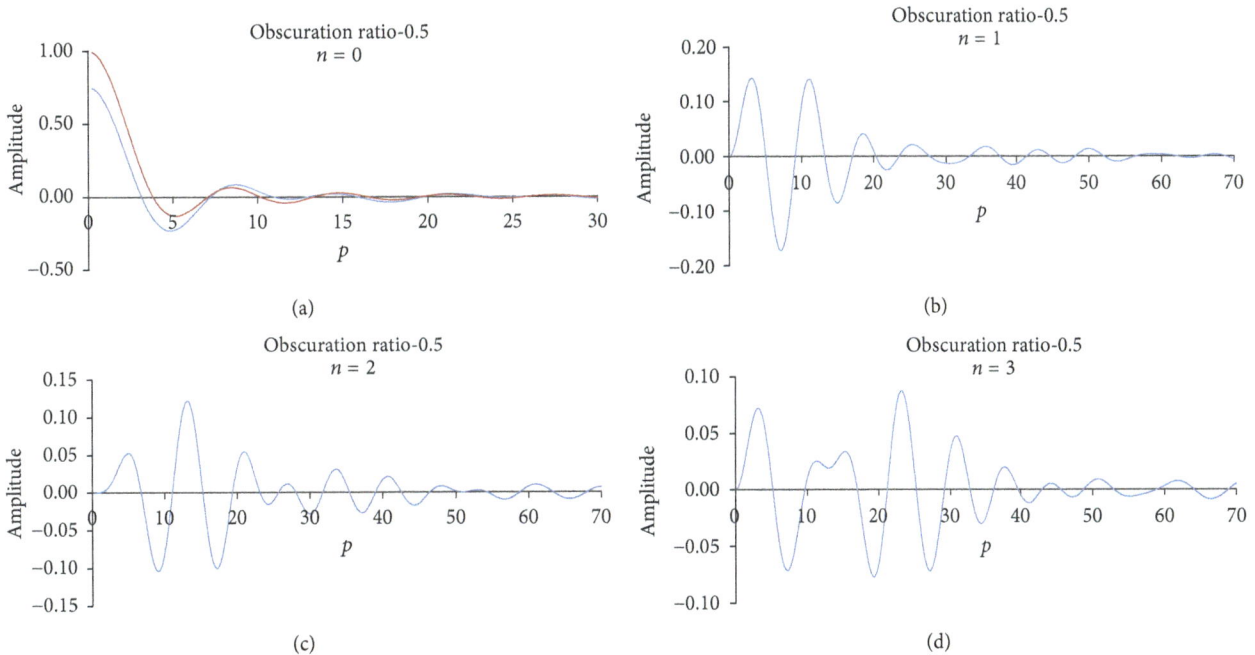

FIGURE 5: Far-field amplitude distributions for annular Walsh filters $\Psi_n^\varepsilon(r)$ on an annular aperture with central obscuration $\varepsilon = 0.5$ for orders (a) $n = 0$, (b) $n = 1$, (c) $n = 2$, and (d) $n = 3$. The red graph in (a) represents amplitude distribution for an unobscured uniform aperture.

Figures 4–7 show that for Walsh filters of all orders other than zero, the central amplitude is zero. This is a consequence of orthogonality property of the Walsh functions. The dark center is surrounded by rings of oscillating amplitudes. For lower orders, this oscillation gradually decays after the first few rings, but for higher orders, the oscillations continue for many more rings. As expected from energy considerations, the amplitude of oscillation is significantly less in case of higher order filters compared to the lower order ones in each case of obscuration. From lower to higher obscuration, the ring with peak oscillatory amplitude gradually shifts away from the centre in case of higher order annular Walsh filters.

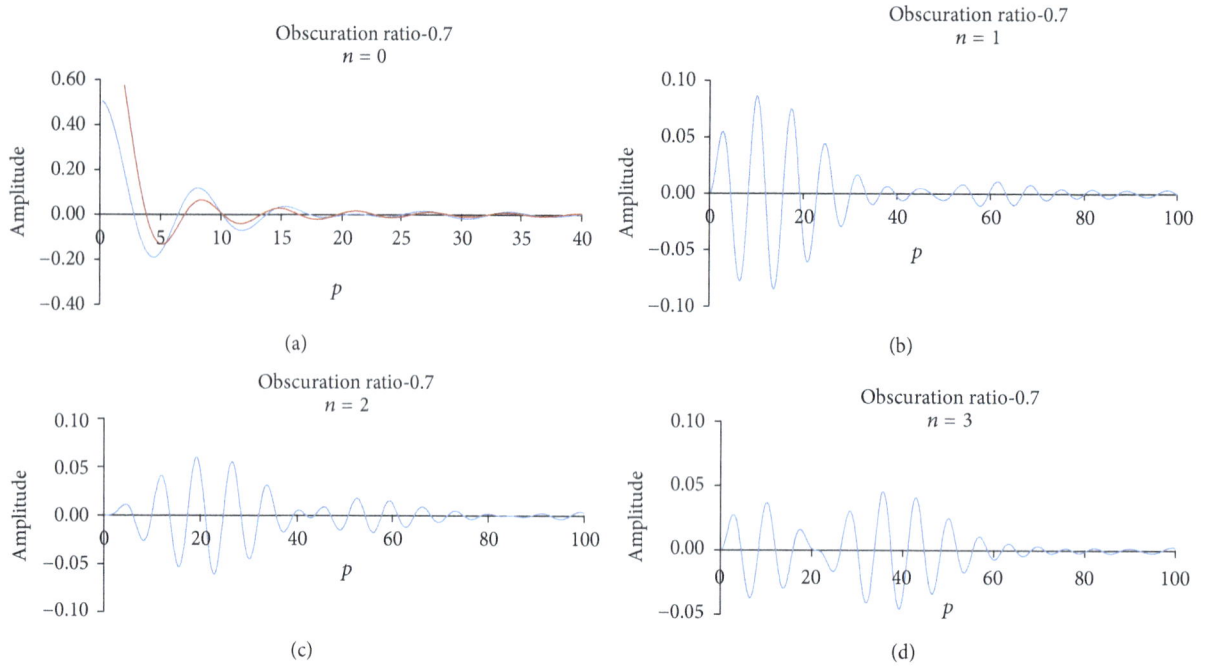

FIGURE 6: Far-field amplitude distributions for annular Walsh filters $\Psi_n^\varepsilon(r)$ on an annular aperture with central obscuration $\varepsilon = 0.7$ for orders (a) $n = 0$, (b) $n = 1$, (c) $n = 2$, and (d) $n = 3$. The red graph in (a) represents amplitude distribution for an unobscured uniform aperture.

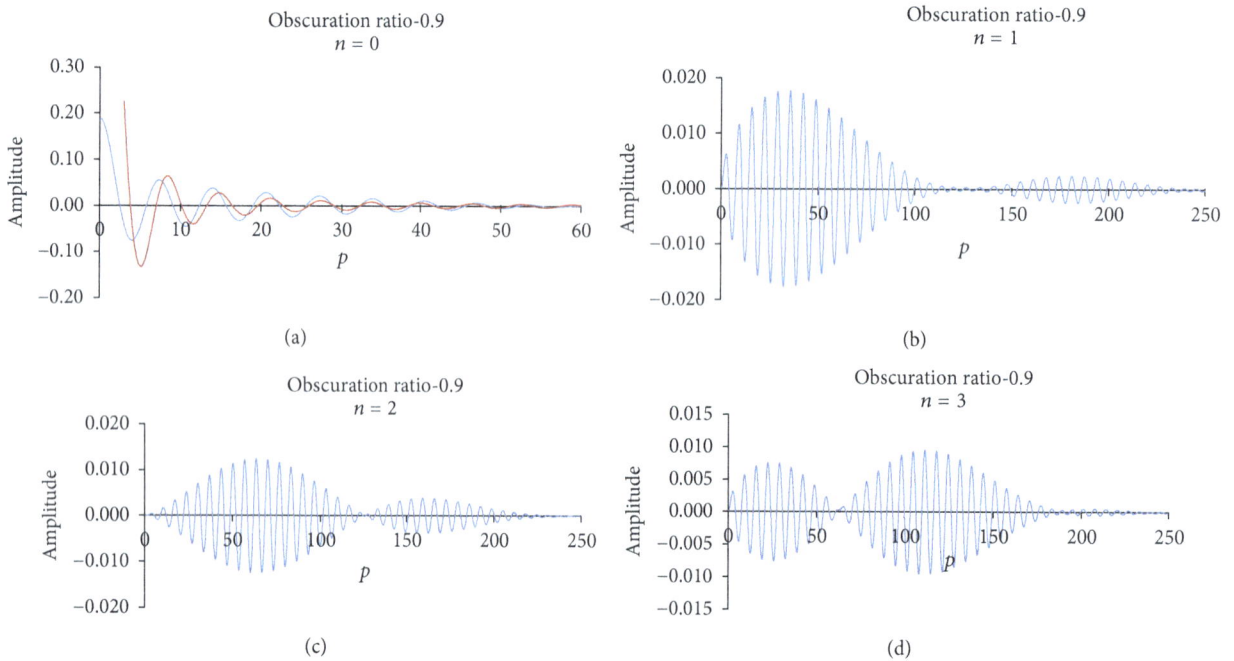

FIGURE 7: Far-field amplitude distributions for annular Walsh filters $\Psi_n^\varepsilon(r)$ on an annular aperture with central obscuration $\varepsilon = 0.9$ for orders (a) $n = 0$, (b) $n = 1$, (c) $n = 2$, and (d) $n = 3$. The red graph in (a) represents amplitude distribution for an unobscured uniform aperture.

Use of ternary values of transmission 0, +1, and −1 in annular Walsh filters has opened up new possibilities for generation of interesting far-field patterns. Amplitude patterns shown in Figures 6(b)–6(d) and 7(b)–7(d) corresponding to annular Walsh filters of orders 1, 2, and 3 on a pupil with large obscuration have distinct characteristics that remain to be explored further for many practical applications, for example, optical encryption. Incidentally, it is also observed that for orders higher than 3, the typical nature of amplitude distribution is more pronounced, with the frequency of oscillations in amplitude becoming increasingly higher with concomitant decrease in peak amplitude.

5. Concluding Remarks

Optimal combinations of annular Walsh filters can help in realizing desired amplitude distributions on the far-field plane for applications using coherent illumination. An important application is to obtain "superresolution" effect by narrowing down of the central lobe and suppression of the side ripples. Solution of the related optimization problem can make effective use of orthogonality of annular Walsh functions. Incidentally, inverse problems on pupil synthesis for prespecified intensity characteristics in the far-field plane were tackled earlier for the case of incoherent illumination by the use of radial Walsh filters for unobscured apertures [18]. It is obvious that the availability of an additional degree of freedom, namely, the obscuration ratio ε, is likely to expand the scope of this inverse problem.

Ready availability of high efficiency spatial light modulators has facilitated the practical realization of phase filters. [22–24]. Nevertheless, it is obvious that, in general, practical implementation of filters with binary or ternary phase values is relatively easier than filters with continuously varying phase. Use of Walsh function-based analysis and synthesis of pupil plane filters has become useful, for it can circumvent the tricky problem of synthesizing continuously varying phase by providing alternative analytical treatments that deal with a finite number of discrete phase levels.

Finally, it remains to investigate the axial resolution characteristics of annular Walsh filters and also their usefulness in the synthesis of three-dimensional light fields [25].

Acknowledgment

Authors are indebted to the anonymous reviewers whose comments have significantly improved the quality of our presentation.

References

[1] G. T. di Francia, "Nuovo pupille superrisolvente," *Atti della Fondazione Giorgio Ronchi*, vol. 7, pp. 366–372, 1952.

[2] E. H. Linfoot and E. Wolf, "Diffraction images in systems with an annular aperture," *Proceedings of the Physical Society B*, vol. 66, no. 2, pp. 145–149, 1953.

[3] A. Boivin, *Théorie et Calcul des Figures de Diffraction de Revolution*, Gauthier-Villars, Paris, France, 1964.

[4] Z. S. Hegedus, "Annular pupil arrays—application to confocal scanning," *Optica Acta*, vol. 32, pp. 815–826, 1985.

[5] C. J. R. Sheppard and A. Choudhury, "Annular pupils, radial polarization, and superresolution," *Applied Optics*, vol. 43, no. 22, pp. 4322–4327, 2004.

[6] T. R. M. Sales and G. M. Morris, "Diffractive superresolution elements," *Journal of the Optical Society of America A*, vol. 14, no. 7, pp. 1637–1646, 1997.

[7] H. Luo and C. Zhou, "Comparison of superresolution effects with annular phase and amplitude filters," *Applied Optics*, vol. 43, no. 34, pp. 6242–6247, 2004.

[8] M. Yun, L. Liu, J. Sun, and D. Liu, "Three-dimensional super-resolution by three-zone complex pupil filters," *Journal of the Optical Society of America A*, vol. 22, no. 2, pp. 272–277, 2005.

[9] M. Yun, M. Wang, and L. Liu, "Super-resolution with annular binary phase filter in the 4Pi-confocal system," *Journal of Optics A*, vol. 7, no. 11, pp. 640–644, 2005.

[10] J. L. Walsh, "A closed set of normal orthogonal functions," *The American Journal of Mathematics*, vol. 55, pp. 5–24, 1923.

[11] K. G. Beauchamp, *Walsh Functions and Their Applications*, Academic, New York, NY, USA, 1985.

[12] C. J. R. Sheppard, "Binary phase filters with a maximally-flat response," *Optics Letters*, vol. 36, no. 8, pp. 1386–1388, 2011.

[13] H. F. Harmuth, *Transmission of Information by Orthogonal Functions*, Springer, Berlin, Germany, 1972.

[14] H. C. Andrews, *Computer Techniques in Image Processing*, Academic, New York, NY, USA, 1970.

[15] M. De and L. N. Hazra, "Walsh functions in problems of optical imagery," *Optica Acta*, vol. 24, pp. 221–234, 1977.

[16] L. N. Hazra and A. Guha, "Farfield diffraction properties of radial Walsh filters," *Journal of the Optical Society of America A*, vol. 3, pp. 843–846, 1986.

[17] L. N. Hazra and A. Banerjee, "Application of Walsh function in generation of optimum apodizers," *Journal of Optics*, vol. 5, pp. 19–26, 1976.

[18] L. N. Hazra, "A new class of optimum amplitude filters," *Optics Communications*, vol. 21, no. 2, pp. 232–236, 1977.

[19] L. N. Hazra, "Walsh filters for tailoring of resolution in microscopic imaging," *Micron*, vol. 38, no. 2, pp. 129–135, 2007.

[20] M. Born and E. Wolf, *Principles of Optics*, Pergamon, Oxford, UK, 1980.

[21] H. H. Hopkins, "Canonical and real space coordinates used in the theory of image formation," in *Applied Optics and Optical Engineering*, R. R. Shannon and J. C. Wyant, Eds., vol. 9, p. 307, Academic, New York, NY, USA, 1983.

[22] E. Fernández, A. Márquez, S. Gallego, R. Fuentes, C. García, and I. Pascua, "Hybrid ternary modulation applied to multiplexing holograms in photopolymers for data page storage," *Journal of Lightwave Technology*, vol. 28, pp. 776–783, 2010.

[23] A. Au, C.-S. Wu, S.-T. Wu, and U. Efron, "Ternary phase and amplitude modulations using a twisted nematic liquid crystal spatial light modulator," *Applied Optics*, vol. 34, no. 2, pp. 281–284, 1995.

[24] S. Mukhopadhyay, S. Sarkar, K. Bhattacharya, and L. N. Hazra, "Polarisation phase shifting interferometric technique for phase calibration of a reflective phase spatial light modulator," *Optical Engineering*, vol. 52, Article ID 035602, 2 pages, 2013.

[25] R. Piestun and J. Shamir, "Synthesis of three-dimensional light fields and applications," *Proceedings of the IEEE*, vol. 90, no. 2, pp. 222–244, 2002.

13

Laser Control of Self-Organization Process in Microscopic Region and Fabrication of Fine Microporous Structure

Yukimasa Matsumura,[1] Wataru Inami,[2,3] and Yoshimasa Kawata[1,3]

[1] Graduate School of Science and Technology, Shizuoka University, Hamamatsu, Shizuoka 432-8561, Japan
[2] Division of Global Research Leaders, Shizuoka University, Hamamatsu, Shizuoka 432-8561, Japan
[3] Japan Science and Technology Agency, CREST, Shiyod-ku, Tokyo 103-0075, Japan

Correspondence should be addressed to Yukimasa Matsumura, f5945008@ipc.shizuoka.ac.jp

Academic Editor: Dieter Schuöcker

We present a controlling technique of microporous structure by laser irradiation during self-organization process. Self-organization process is fabrication method of microstructure. Polymer solution was dropped on the substrate at high humid condition. Water in air appears dropping air temperature below the dew point. The honeycomb structure with regularly aligned pores on the film was fabricated by attaching water droplets onto the solution surface. We demonstrate that it was possible to prevent forming pores at the region of laser irradiation and flat surface was fabricated. We also demonstrated that a combination structure with two pore sizes and flat surface was produced by a single laser-pulse irradiation. Our method is a unique microfabrication processing technique that combines the advantages of bottom-up and top-down techniques. This method is a promising technique that can be applied to produce for photonic crystals, biological cell culturing, surface science and electronics fields, and so forth.

1. Introduction

Microfabrication techniques have been studied by many researchers with the developments of nanotechnologies [1]. Microfabrication techniques are categorized as top-down or bottom-up fabrication techniques. The top-down fabrication technique creates nanoscale and submicron scale devices by using externally controlled equipment to direct their assembly. One of typical top-down methods is photolithography. The advantages of the top-down methods are high precision and high reproducibility in the fabrication process [2]. The top-down method, however, requires expensive equipment, long processing time, and complicated procedures.

On the other hand, the bottom-up fabrication technique builds smaller components to more complex assemblies. The bottom-up techniques are known as self-organization process. The bottom-up methods have the advantages of short process time and low cost. They easily produce a uniform pattern in large area, while the arbitrary structures are difficult to be produced by bottom-up techniques.

It has been reported that microporous films are formed by the cast on the glass substrate under high humid condition by self-organization process of solution casting method [3–17]. The microstructures by self-organization process are expected to be used in many applications as follows. The microporous structure of regular aligned pores is made of polystyrene, and the line defect in this microporous structure is applied for photonic crystal [18, 19]. Micropatterned substrates by self-organization process are used as potential scaffolds for regeneration of cardiac and fibrous tissues [20–25]. Pincushion films are prepared by peeling off the top layer of the self-organization honeycomb film, and metalized pincushion films are fabricated by electroless plating. These metalized pincushion structures are expected to application of field emitter arrays [26].

We propose controlling self-organization process by laser irradiation in order to produce arbitrary structures. It is a

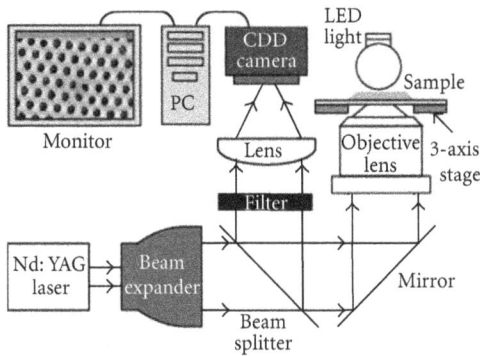

FIGURE 1: Optical setup for controlling of self-organization process.

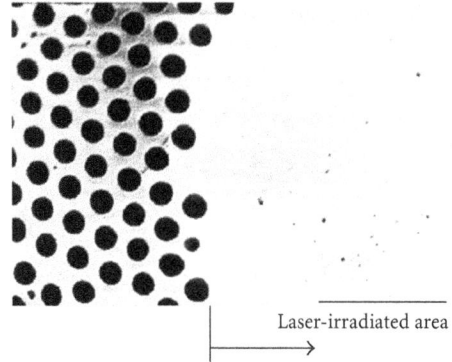

FIGURE 2: SEM image of controlling self-organized microporous polystyrene film by laser irradiation from below. Scale bar: 20 μm.

unique technique that combines top-down and bottom-up techniques.

2. Laser Control of Self-Organized Process

2.1. Fabrication of Self-Organized Microporous Films. It has been reported that honeycomb microporous films are fabricated by self-organization process [3–17]. The mechanism of fabrication process of honeycomb structure on a film is explained as follows.

Polymers with hydrophilic group are dissolved in highly volatile solvent. This solution is dropped on a glass substrate in high humid air. Air near the solution surface is cooled because the solvent draw heat during evaporation. Because the air temperature drops below the dew point, the water droplets are generated near the solution surface and they attach on the surface. Water droplets on the solution surface are packed as honeycomb structure by capillary force. Finally, the solvent and the water droplets are completely evaporated and microporous droplets are produced at the position that the water droplets are attached on the solution surface.

We used a solution of polystyrene which was mono-carboxy-terminated (weight-average molecular weight: 50,000, Scientific Polymer Products, Inc.) in chloroform (Kanto chemical co., Inc.). This polymer has carboxyl group as hydrophilic group. The polystyrene chloroform solution of 25–50 μL was dropped onto the cover glass (24 × 40 mm, Matsunami glass Ind., Ltd.) at high humid condition of relative humidity 75%. Herewith, the polystyrene layer of about 15 μm thickness with microporous structure was fabricated. The fabricated microporous films were observed with scanning electron microscope (SEM; JSM-6390, JEOL Ltd.) after sputtering of gold of 20 nm thickness.

2.2. Laser Control of Self-Organization Process. Figure 1 shows the optical setup to control self-organization process by laser irradiation during fabrication of microporous film. A Q-switched Nd: YAG laser is used as a light source. Wavelength, pulse width, and repetition frequency of the laser are 532 nm, 3–5 ns, and 1 Hz, respectively. Laser light is focused into the solution on the cover glass by an objective lens. The solution thickness decreases because of solvent evaporation, and the solution surface is moved in the optical

axis direction during self-organization process. Therefore we tracked the solution surface with moving the glass substrate in the optical axis direction by the 3-axis stage. The solution droplet surface is observed by a CCD camera with the illumination of an LED light.

3. Experimental Demonstration of Laser Control of Self-Organization Process

3.1. One Pulse Laser Irradiation from Bottom Side of the Solution. The surface of solution was laser-irradiated from bottom side after 53 seconds from the solution being dropped on the cover glass. The solution concentration is 40 g/L and drip amount is 25 μL. Numerical aperture of objective lens is 0.4. Laser pulse energy is 15 μJ. Figure 2 shows SEM images of self-organization microporous polystyrene film with irradiation of a laser pulse. The figure shows that the plane surface without pores was fabricated by the laser irradiation in the microporous structure during self-organization process. If laser light is irradiated after finishing self-organization process, the self-organization microporous films are damaged by the laser pulse.

We can recognize clearly that plane surface without pores area was produced by the laser irradiation. The result shows that we can prevent forming pores at the laser irradiated area. In fact, we may say that self-organization process can be controlled by a pattern of the laser irradiation and planar surface of arbitrary pattern such as line is fabricated in the microporous structure.

We discuss the mechanism that the water droplets on the solution surface are removed by the laser irradiation. Figure 3 shows the illustration of mechanism of laser controlling of self-organization process. We believe that the shock wave is main factor to remove the water droplets. When the laser light is irradiated to a solution and the irradiation intensity exceeds the threshold value, the shock wave is induced with the laser ablation pressure [27]. The shock wave exceeds the solution surface tension, and the water droplets are removed at the laser-irradiated region. Because the solution is irradiated by a laser pulse just before it is completely evaporated, there no enough time water droplets are attached

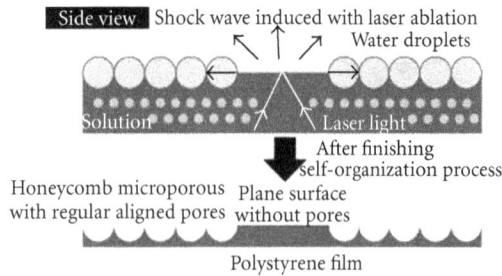

FIGURE 3: Side view pattern diagram of solution surface with laser irradiation.

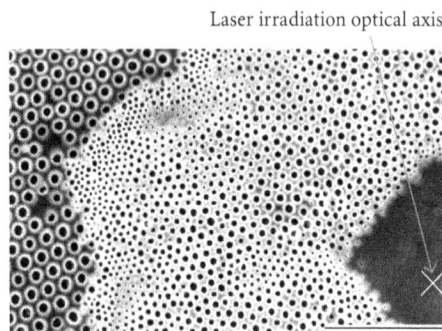

FIGURE 4: SEM image of two pore sizes and plane surface polystyrene film by laser irradiation from below focusing inside solution. Scale bar: 50 μm.

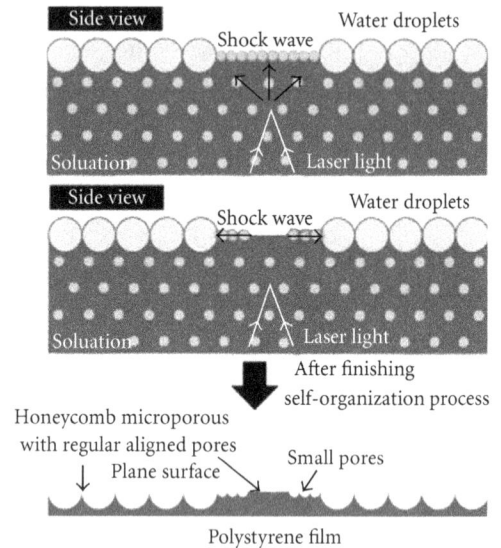

FIGURE 5: Side view pattern diagram of inside solution with laser irradiation.

again in this area and the plane surface without pores is generated.

3.2. Fabrication of Microporous Structure by Focusing Laser Light inside Solution Droplet.

We found that it was possible to produce a combination of large and small pore sizes and plane surface without pores by focusing laser light inside the droplet solution. The solution thickness is about 50 μm when the laser light is irradiated. The laser light is focused at the point of below about 15 μm from the solution surface after 110 seconds from solution being dropped on the cover glass. The solution concentration is 35 g/L and drip amount is 25 μL. Laser pulse energy is 300 μJ.

Figure 4 shows SEM image of the combination structure produced by focusing laser light inside of solution and the optical axis is also shown in this figure. In the area near the optical axis plane surface without pores was produced. The small pores were fabricated at the region surrounding the plane surface area and regularly aligned pores with honeycomb structure also surround the area of small pores.

Figure 5 shows the illustration of mechanism of fabricating small pores and plane surface by the laser irradiation. The reason that the small water droplets were produced is described as follows. On self-organization process, polystyrene chloroform solution is evaporated and temperature at the solution surface is decreased by solvent evaporation. Because of temperature decrease at the solution surface, the convection occurrs in the solution by temperature difference between surface and inside of the solution.

Some water droplets on the solution surface are entered inside the solution by convection flow. The water droplets on the solution surface are grown due to constant moisture supplying, but the water droplets in the solution are not grown because no moisture was supplied to the droplets and small water droplets are kept in small size.

When the laser was focused inside the solution, the small water droplets inside the solution were pushed to the surface by the shock wave of the laser pulse irradiation. All water droplets were removed near the optical axis because the shock wave was strong enough. Because these two phenomena occurred at about the same time, the central area of the laser irradiation was the plane surface without pores and the surrounding area is small pores structure. Unaffected area by the laser irradiation is large pores structure.

3.3. One Pulse Laser Irradiation from the Top Side of the Solution.

The surface of solution is irradiated by the laser from the top side after 90 seconds from the solution being dropped on the cover glass. Solution concentration is 40 g/L and drip amount is 50 μL. Numerical aperture of objective lens is 0.42. Laser pulse energy is 15 μJ. Figure 6 shows SEM images of self-organization microporous polystyrene film with irradiation of a laser pulse. This picture shows that the plane surface without pores was fabricated in the microporous structure. Compared to the result of Figure 2, the plane surface area of Figure 6 is narrower. The plane surface area is reduced to about one-sixth. We may say that the irradiation of laser from the top side reduces the laser-controlled area.

The reduction of controlled area by the laser irradiation from the top side was attributed to spherical aberrations. In the case of the laser irradiation from the bottom side, the laser light transmits the polystyrene chloroform solution, so the laser spot is aberrated during the propagation in the solution [28]. However, in the case of the laser irradiation

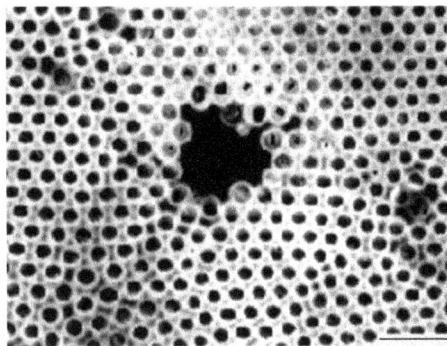

FIGURE 6: SEM image of controlling self-organized microporous polystyrene film by laser irradiation from above. Scale bar: 20 μm.

from the top side, there are no aberrations and the laser spot is well-focused on the solution surface.

4. Conclusion

We demonstrated that the microporous structure was able to be controlled by the laser light irradiation during self-organization process. The plane surface area without pores was produced by removing the attached water droplets from the surface with a shock wave induced with a laser pulse. When the laser light was irradiated the polystyrene chloroform solution from the top side of the solution, the area where the water droplets were removed was reduced. We also demonstrated that in the case of the laser irradiation from the bottom side of the solution, the combination structure of two pore size and plane surface was also fabricated by focusing inside the solution. In this case, the small water droplets inside the solution were pushed to the surface with the shock wave and the small water droplets on the solution surface were removed. Our method has a potential that combines the processes of bottom-up and top-down methods. It is a promising technique that can be applied to produce photonic crystals, biological culturing process, water repellent materials, electronic, and so on.

References

[1] M. Madou, *Fundamental of Microfabrication*, CRC Press LLC, 1997.

[2] T. Kawai, *Graphic Explanation All of Application of Nanotechnology*, Kogyochosakai, 2002.

[3] G. Widawski, M. Rawiso, and B. Francois, "Self-organized honeycomb morphology of star-polymer polystyrene films," *Nature*, vol. 369, no. 6479, pp. 387–389, 1994.

[4] N. Maruyama, T. Koito, J. Nishida et al., "Mesoscopic patterns of molecular aggregates on solid substrates," *Thin Solid Films*, vol. 327–329, no. 1-2, pp. 854–856, 1998.

[5] B. de Boer, U. Stalmach, H. Nijland, and G. Hadziioannou, "Microporous honeycomb-structured films of semiconducting block copolymers and their use as patterned templates," *Advanced Materials*, vol. 12, no. 21, pp. 1581–1583, 2000.

[6] M. H. Stenzel-Rosenbaum, T. P. Davis, and A. G. Fane, "Porous polymer films and honeycomb structures made by the self-organization of well-defined macromolecular structures

created by living radical polymerization techniques," *Angewandte Chemie International Edition*, vol. 40, no. 18, pp. 3428–3432, 2001.

[7] M. Srinivasarao, D. Collings, A. Philips, and S. Patel, "Three-dimensionally ordered array of air bubbles in a polymer film," *Science*, vol. 292, no. 5514, pp. 79–83, 2001.

[8] L. Song, R. K. Bly, J. N. Wilson et al., "Facile microstructuring of organic semiconducting polymers by the breath figure method: hexagonally ordered bubble arrays in rigid-rod polymers," *Advanced Materials*, vol. 16, no. 2, pp. 115–118, 2004.

[9] L. V. Govor, I. A. Bashmakov, F. N. Kaputski, M. Pientka, and J. Parisi, "Self-organized formation of low-dimensional network structures starting from a nitrocellulose solution," *Macromolecular Chemistry and Physics*, vol. 201, no. 18, pp. 2721–2728, 2000.

[10] Y. Tian, S. Liu, H. Ding, L. Wang, B. Liu, and Y. Shi, "Formation of deformed honeycomb-patterned films from fluorinated polyimide," *Polymer*, vol. 48, no. 8, pp. 2338–2344, 2007.

[11] Y. Xu, B. Zhu, and Y. Xu, "A study on formation of regular honeycomb pattern in polysulfone film," *Polymer*, vol. 46, no. 3, pp. 713–717, 2005.

[12] O. Pitois and B. François, "Crystallization of condensation droplets on a liquid surface," *Colloid and Polymer Science*, vol. 277, no. 6, pp. 574–578, 1999.

[13] H. Yabu, M. Kojima, M. Tsubouchi, S. Y. Onoue, M. Sugitani, and M. Shimomura, "Fabrication of photo-cross linked honeycomb-patterned films," *Colloids and Surfaces A*, vol. 284-285, pp. 254–256, 2006.

[14] X. Zhao, Q. Cai, G. Shi, Y. Shi, and G. Chen, "Formation of ordered microporous films with water as templates from poly(D,L-lactic-co-glycolic acid) solution," *Journal of Applied Polymer Science*, vol. 90, no. 7, pp. 1846–1850, 2003.

[15] O. Pitois and B. François, "Formation of ordered microporous membranes," *European Physical Journal B*, vol. 8, no. 2, pp. 225–231, 1999.

[16] J. Peng, Y. Han, J. Fu, Y. Yang, and B. Li, "Formation of regular hole pattern in polymer films," *Macromolecular Chemistry and Physics*, vol. 204, no. 1, pp. 125–130, 2003.

[17] B. François, Y. Ederlé, and C. Mathis, "Honeycomb membranes made from C60(PS)6," *Synthetic Metals*, vol. 103, no. 1–3, pp. 2362–2363, 1999.

[18] P. Jiang, X. Y. Hu, H. Yang, and Q. H. Gong, "Fabrication of two-dimensional organic photonic crystal microcavity," *Chinese Physics Letters*, vol. 23, no. 7, pp. 1813–1815, 2006.

[19] S. Matsushita, R. Fujiwara, and M. Shimomura, "Calculation of photonic energy bands of self-assembled-type TiO$_2$ photonic crystals as dye-sensitized solar battery," *Colloids and Surfaces A*, vol. 313-314, pp. 617–620, 2008.

[20] A. Tsuruma, M. Tanaka, N. Fukushima, and M. Shimomura, "Morphological changes in neurons by self-organized patterned films," *e-Journal of Surface Science and Nanotechnology*, vol. 3, pp. 312–315, 2005.

[21] H. Yabu, M. Takebayashi, M. Tanaka, and M. Shimomura, "Superhydrophobic and lipophobic properties of self-organized honeycomb and pincushion structures," *Langmuir*, vol. 21, no. 8, pp. 3235–3237, 2005.

[22] M. Tanaka, K. Nishikawa, H. Okubo et al., "Control of hepatocyte adhesion and function on self-organized honeycomb-patterned polymer film," *Colloids and Surfaces A*, vol. 284-285, pp. 464–469, 2006.

[23] H. Sunami, E. Ito, M. Tanaka, S. Yamamoto, and M. Shimomura, "Effect of honeycomb film on protein adsorption, cell

adhesion and proliferation," *Colloids and Surfaces A*, vol. 284-285, pp. 548–551, 2006.

[24] Y. Fukuhira, H. Kaneko, M. Yamaga, M. Tanaka, S. Yamamoto, and M. Shimomura, "Effect of honeycomb-patterned structure on chondrocyte behavior in vitro," *Colloids and Surfaces A*, vol. 313-314, pp. 520–525, 2008.

[25] K. Arai, M. Tanaka, S. Yamamoto, and M. Shimomura, "Effect of pore size of honeycomb films on the morphology, adhesion and cytoskeletal organization of cardiac myocytes," *Colloids and Surfaces A*, vol. 313-314, pp. 530–535, 2008.

[26] Y. Hirai, H. Yabu, and M. Shimomura, "Electroless deposition of zinc oxide on pincushion films prepared by self-organization," *Colloids and Surfaces A*, vol. 313-314, pp. 312–315, 2008.

[27] M. Yoshida, "Study of equation of state using laser-induced shock-wave compression 1. Generation and properties of laser-induced shock waves," *Journal of Plasma and Fusion Research*, vol. 80, no. 6, pp. 427–431, 2004.

[28] M. Born and E. Wolf, *Principle of Optics*, Pergamon, 1964.

14

Emissivity Measurement of Semitransparent Textiles

P. Bison, A. Bortolin, G. Cadelano, G. Ferrarini, and E. Grinzato

ITC, CNR, Corso Stati Uniti 4, 35127 Padova, Italy

Correspondence should be addressed to P. Bison, paolo.bison@itc.cnr.it

Academic Editor: Laura Abbozzo Ronchi

In the textiles production industry it is more and more common to advertise new textiles, especially for sportswear, by claiming their ability to emit IR radiation in the long wave band at a higher degree with respect to normal clothes, that is highly beneficial to improve sporting performances. Three textiles are compared, one normal and two "special," with Ag^+ ions and carbon powder added, with different colors. The emissivity of the textiles has been measured to determine if it is increased in the "special" textiles with respect to the normal one. No substantial increase has been noticed. Nonetheless, the test implied some nonstandard procedures due to the semitransparent nature of the textiles, in comparison with the normal procedure that is commonly used on opaque surfaces.

1. Introduction

Two textiles are made of polypropylene (PP) and charged with Ag^+ ions and Carbon powder. They differ from the color that is green for the first and blue for the second. They are compared with a third "normal" PP textile of green color and the same weft of the previous. The purpose is to demonstrate if the emissivity in the Infrared-Long Wave band (IR-LW) [1–3] of the charged textiles is increased or not due to the presence of the charging elements.

The measurement is carried out in the wavelength interval 8–14 μm; it is an integral measurement; that is, it represents the average value of the spectral emissivity in the considered interval [4].

The measurement is carried out by means of a microbolometric camera that exhibits an almost flat spectral response at the various wavelengths. Therefore, it is not taken into account any spectral response of the detector even because of the comparative nature of the measurement [5, 6].

Any evaluation of the taking angle is neglected and so the dependence of the emissivity with the view angle. The measurement is performed with a normal view with respect to the textiles [7].

The measurement technique consists in laying down the charged textiles, to be measured, and the "normal" (the reference) one, side by side and in contact with a thick aluminum plate that is assumed to be as much isothermal as possible. Observing the two textiles (that own the same temperature) by an IR camera allows to evaluate the IR radiation emitted by their surfaces. The possible difference in the radiation collected by the IR camera in correspondence of the "measured" textiles and the "reference" one is due to the emissivity difference. In this case the measurement is more difficult than the one for an opaque surface in so far the textiles are semitransparent and the radiation collected by the camera is due to the contribution of the textiles themselves plus the background, each one emitted with its own emissivity and weighted by the surface fraction that it covers. Therefore, a preliminary assessment of the transmittance coefficient of the textiles is done in such a way to determine the percentage of radiation emitted by the textiles themselves and the percentage that they transmit, coming from the background.

2. Equations Related to the Emissivity Measurement

In case of an opaque object at temperature T_o and emissivity ε_m and with a surrounding environment at temperature T_a the radiance measured by the IR camera is given by [8]

$$I_m = \varepsilon_m I_o + (1 - \varepsilon_m)I_a, \tag{1}$$

where $\varepsilon_m I_o$ is the radiance emitted by the object surface and $(1 - \varepsilon_m)I_a$ is the radiance generated by the environment and reflected by the object surface. In (1) the effect of the absorption and emission of radiation by the atmosphere is neglected due to the small distance between the camera and the object. In case of comparative measurement, in which one material is the reference (subscript r) and the other is the measured one (subscript m) the ratio between the two emissivities is given by

$$R = \frac{\varepsilon_m}{\varepsilon_r} = \frac{I_m - I_a}{I_r - I_a}, \qquad (2)$$

where I_r, I_m, and I_a are the radiances measured by the camera in correspondence of the reference material, the measured material, and the environment, respectively.

In case of semitransparent material, as it is the case of the textiles, (1) is transformed in

$$I_m = (1 - \tau_m)[\varepsilon_m I_o + (1 - \varepsilon_m)I_a] + \tau_m I_s, \qquad (3)$$

where τ_m is the transmittance coefficient of the textiles (that is the one's complement of the surface percentage covered by the textiles) and I_s is the radiance emitted by the surface of the metallic plate on which the textiles are laid down and that is heated at a constant temperature during the measurement. In case of the comparative measurement of semitransparent materials with transmittance coefficients τ_m and τ_r, respectively, for the measured and reference textiles, one obtains

$$R_{st} = \frac{\varepsilon_m}{\varepsilon_r} = \frac{1 - \tau_r}{1 - \tau_m}\frac{I_m - I_a - \tau_m(I_s - I_a)}{I_r - I_a - \tau_r(I_s - I_a)}. \qquad (4)$$

3. Measurement Uncertainty Analysis

The evaluation of the ratio R_{st} by means of (4) requires the measurement of six quantities. Each one is measured many times (some tenths for transmittances to some thousands for radiances). The statistical evaluation of (4) is therefore necessary, together with its uncertainty by means of the uncertainties propagation:

$$\delta R_{st} = \sqrt{\begin{aligned}&\left[\frac{\partial R_{st}}{\partial \tau_r}\delta\tau_r\right]^2 + \left[\frac{\partial R_{st}}{\partial \tau_m}\delta\tau_m\right]^2 + \left[\frac{\partial R_{st}}{\partial I_r}\delta I_r\right]^2 \cdots \\ &\cdots \left[\frac{\partial R_{st}}{\partial I_m}\delta I_m\right]^2 + \left[\frac{\partial R_{st}}{\partial I_a}\delta I_a\right]^2 + \left[\frac{\partial R_{st}}{\partial I_s}\delta I_s\right]^2\end{aligned}}, \qquad (5)$$

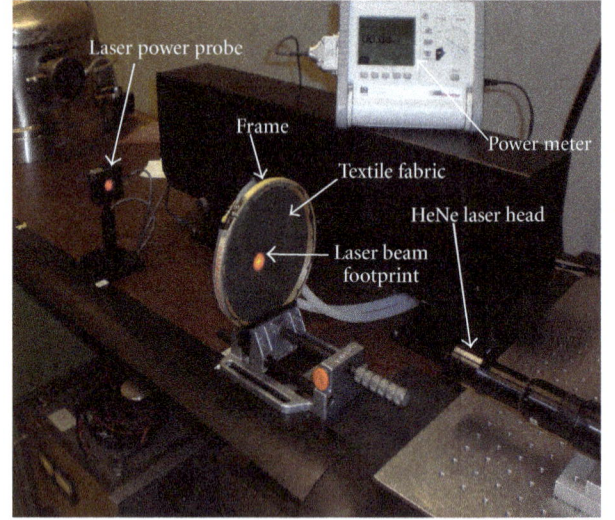

FIGURE 1: Experimental layout for the transmittance measurement in the optical wavelength range.

where the uncertainties $\delta\tau_r, \delta\tau_s, \ldots, \delta I_s$ have been estimated by their standard deviation and the partial derivatives of R_{st} are given by

$$\frac{\partial R_{st}}{\partial \tau_r} = R_{st}\left[\frac{I_s - I_a}{I_r - (1 - \tau_r)I_a - \tau_r I_s} - \frac{1}{1 - \tau_r}\right],$$

$$\frac{\partial R_{st}}{\partial \tau_m} = R_{st}\left[\frac{1}{1 - \tau_m} - \frac{I_s - I_a}{I_m - (1 - \tau_m)I_a - \tau_m I_s}\right],$$

$$\frac{\partial R_{st}}{\partial I_r} = R_{st}\frac{-1}{I_r - (1 - \tau_r)I_a - \tau_r I_s},$$

$$\frac{\partial R_{st}}{\partial I_m} = R_{st}\frac{1}{I_m - (1 - \tau_m)I_a - \tau_m I_s}, \qquad (6)$$

$$\frac{\partial R_{st}}{\partial I_a} = R_{st}\left[\begin{aligned}&\frac{1 - \tau_r}{I_r - (1 - \tau_r)I_a - \tau_r I_s} - \cdots \\ &\cdots \frac{1 - \tau_m}{I_m - (1 - \tau_m)I_a - \tau_m I_s}\end{aligned}\right],$$

$$\frac{\partial R_{st}}{\partial I_s} = R_{st}\left[\begin{aligned}&\frac{\tau_r}{I_r - (1 - \tau_r)I_a - \tau_r I_s} - \cdots \\ &\cdots \frac{\tau_m}{I_m - (1 - \tau_m)I_a - \tau_m I_s}\end{aligned}\right].$$

4. Experimental Apparatus for Transmittance Measurement

The measurement apparatus for the transmittance evaluation consists of a Continuous Wave HeNe laser with emission at the wavelength $\lambda = 632.8$ nm, at a power of 10 mW. The beam is expanded to an FWHM of almost 2 cm. The power of the beam is measured by a suitable detector connected to a power meter, once without any obstacle in the optical path and then with the textiles in between. From the ratio of the measured power with and without the textiles, the transmittance is obtained. See Figure 1.

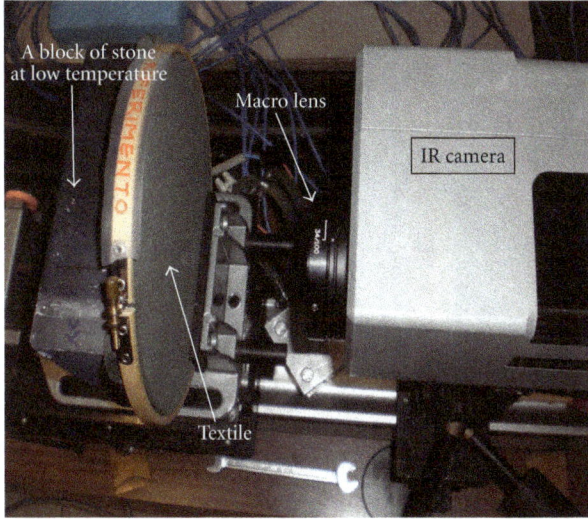

FIGURE 2: Experimental layout for the transmittance measurement in IR wavelength range.

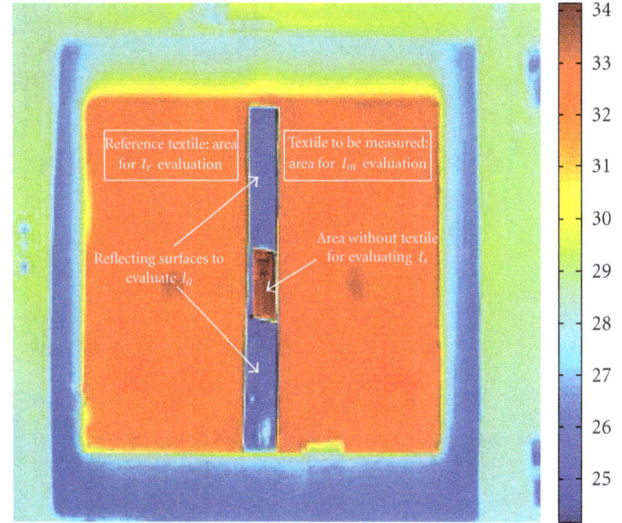

FIGURE 4: IR image relative to Figure 3. See the indication of the areas utilized to evaluate (4).

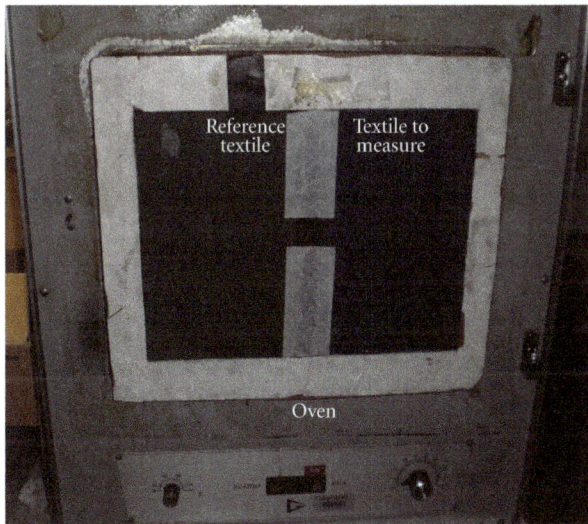

FIGURE 3: Experimental layout for the emissivity measurement.

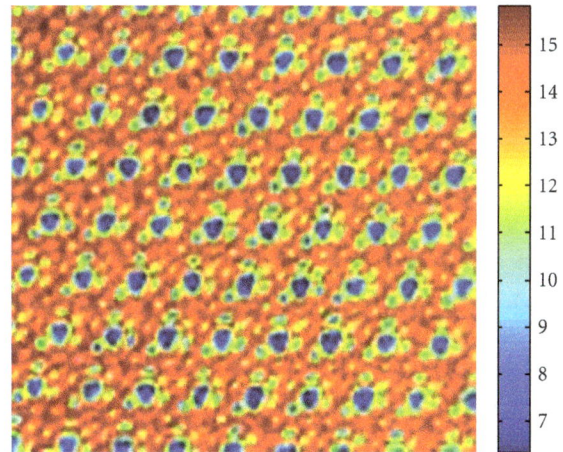

FIGURE 5: Results of the transmittance measurement in the IR. Notice the regular appearance of the cold background areas.

TABLE 1: Results of the transmittance measurement by laser attenuation.

Textiles	Transmittance Mean $\pm$ standard dev.
RV	0.13 ± 0.02
CV	0.14 ± 0.02
CB	0.15 ± 0.02

A possible objection to the assessment of the transmittance as described above is connected to the wavelength of the laser source that is 20 times smaller of the radiation considered in the emissivity evaluation. On purpose a second experimental layout has been prepared. An IR camera with a macro lens ($\sim 100\,\mu$m/pixel) observes a cold background through the textiles at ambient temperature. The background object has no contact with the textiles, avoiding any conduction effect. See Figure 2.

5. Experimental Apparatus for Emissivity Measurement

The measurement apparatus is composed of an aluminum plate 2 cm thick, on which the reference and measured textiles are laid down side by side. The plate is successively inserted on the opening of an oven from which it receives heat. The aluminum plate guarantees the uniformity of the temperature of the textiles. An IR camera observes the textiles that are heated at a temperature of 35-36°C, around the human body temperature. See Figure 3.

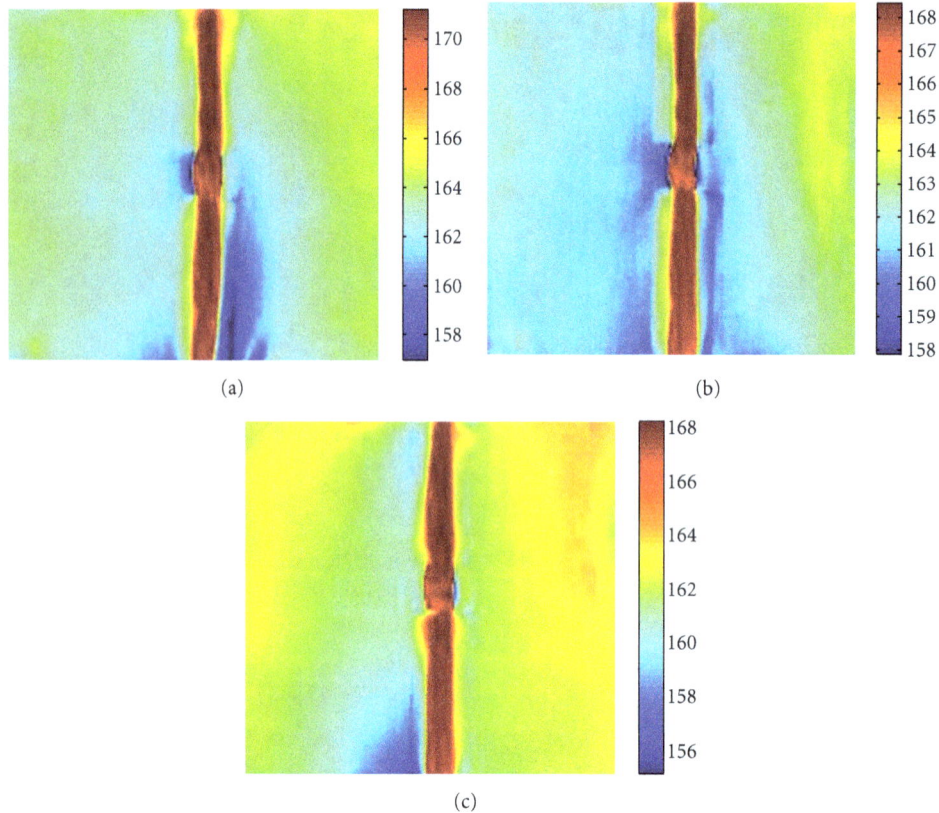

FIGURE 6: (a) Textiles RV on left and CV on right; (b) textiles RV on left and CB on right; (c) textiles CV on left and CB on right.

In the successive Figure 4 the IR image relative to Figure 3 is shown. The areas utilized for the radiance measurements, according to (4) are indicated.

6. Measurements of Transmittance by Laser

The measurements have been carried out on three textiles: the reference one of green color RV, the charged one of green color CV, and the charged one of blue color CB. Each textile has been positioned at 1/4, 1/2, and 3/4 of the laser beam optical path between the laser head and the detector. For each position the measurement has been repeated four times, for a total of 12 measurements. The results with mean values and standard deviations are reported in Table 1.

A second transmittance measurement has been carried out in the IR. Figure 5 shows the IR image of the textiles with some cold areas regularly disposed in the image. The measurement in this case consists in determining the percentage of cold areas in comparison with the total area. The weakness of this technique is related to the arbitrary threshold that should discriminate between cold and hot areas. By varying the threshold between 10.0 and 11.0°C one obtains a transmittance value between 0.116 and 0.170. By choosing a threshold at 10.5°C one obtain a transmittance of 0.142 close to the values obtained with the laser attenuation technique.

TABLE 2: Results of the emissivity ratio according to (4).

Textiles	R_{st} Mean ± standard dev.
RV-CV	1.036 ± 0.065
RV-CB	1.061 ± 0.053
CV-CB	1.059 ± 0.065

7. Emissivity Measurement

Three experiments have been carried out:

(i) in the first, the reference textile RV is on the left and the charged green textile CV is on the right, see Figure 6(a);

(ii) in the second, RV is on the left while the charged textile of blue color is on the right, see Figure 6(b);

(iii) in the third, two charged textiles are directly compared: CV on the left and CB on the right, see Figure 6(c).

The IR images and the computations are done in Object Signal, units furnished directly by the IR camera software, before the transformation in temperature unit, and proportional to radiance.

In Table 2 the results are reported.

8. Conclusion

The CV textile shows an emissivity 3.6% higher than the RV. The CB textile presents an emissivity 5.3% higher than RV. The measurements are affected by an uncertainty of about 5%, that means the differences in emissivity are comparable to the measurement error. The last test, that directly compares the two charged textiles, shows that the blue textile has an emissivity higher than the green one. From these considerations one could guess that the colors used in the textiles affect emissivity more than the charging elements.

Acknowledgment

This work has been financed by the fabrics company IDEE PER IL TESSILE.

References

[1] Wikipedia Contributors, *Emissivity—Wikipedia, the Free Encyclopedia*, 2010.

[2] X. P. V. Maldague, *Theory and Practice of Infrared Technology for Nondestructive Testing*, Wiley-Interscience, 2001.

[3] X. P. V. Maldague, Ed., *Infrared and Thermal Testing*, American Society for Nondestructive Testing, 3rd edition, 2001.

[4] R. Siegel and J. R. Howell, *Thermal Radiation Heat Transfer*, Hemisphere, 1981.

[5] F. Bertrand, J. L. Tissot, and G. Destefanis, "Second generation cooled infrared detectors state of the art and prospects," in *Proceedings of the 4th International Workshop on Advanced Infrared Technology and Applications*, L. Ronchi Abbozzo, Ed., Fondazione Vasco Ronchi, 1997.

[6] W. D. Rogatto, Ed., *Electro-Optical Components*, vol. 3 of *The Infrared and Electro-Optical Systems Handbook*, ERIM and SPIE Optical Engineering Press, 1993.

[7] H. C. Hottel and A. F. Sarofim, *Radiative Transfer*, McGrew-Hill Book Company, 1967.

[8] R. P. Madding, "Emissivity measurement and temperature correction accuracy considerations," in *Thermosense XXI*, vol. 3700 of *Proceedings of SPIE*, pp. 393–401, April 1999.

Meta

15

Experimental Investigation of an Index-Mismatched Multiphase Flow Using Optical Techniques

H. Coronado Diaz and Ronald J. Hugo

Mechanical & Manufacturing Engineering, University of Calgary, 2500 University Drive NW, Calgary, AB, Canada T2N 1N4

Correspondence should be addressed to H. Coronado Diaz; hcoronado@gmail.com

Academic Editor: Mina Hoorfar

An experimental investigation of multiphase flow involving a liquid (water) and a gas (air) is performed. The results for three different scenarios are presented: fixed bubble, ascending bubble, and dispersed-bubble turbulent pipe flow. This study involves a comparison of statistical data collected using two sensing systems, a wavefront sensor and a high-speed video camera. A signal analysis technique based on signal attenuation is developed for data collected using the wavefront sensor. The three experiments performed provide experimental evidence that the Shack-Hartmann wavefront sensor, operating on signal attenuation, is a viable method for the study of multiphase bubble flows.

1. Introduction

Multiphase flow is defined as the simultaneous flow of several phases, with the simplest case being two-phase flow. The flow of two phases is found in many industrial processes including chemical and nuclear reactors, distillation towers, pipeline transport, injection of fluids for secondary recovery of oil and geothermal power plants.

One form of two-phase flow is gas-liquid dispersed-bubble turbulent pipe flow. This flow is of significant importance in chemical and petroleum process industries where the interfacial area between phases needs to be large for improved process efficiency. Bubble size, velocity of bubbles, and their distribution along the pipe are all important parameters for optimizing the efficiency of processes that rely on dispersed-bubble turbulent pipe flow.

Noninvasiveness is a desired attribute in any multiphase flow measurement technique given that the interference between the flow and the measurement device can affect the measured values. Many of the existing multiphase flow measurement techniques rely on the ability of the sensing apparatus to discriminate between certain physical properties that vary between phases, such as electromagnetic radiation attenuation or electrical impedance.

A large number of investigations reported over the last fifteen years describe the application of nonintrusive flow sensing techniques applied to the measurement and the study of multiphase fluid flows [1]. Nonintrusive measurement techniques typically provide either integral or line-integrated information, with the most common consisting of capacitance [2], conductance [3], ultrasound [4], and gamma-radiation absorption [5] measurements. Both integral and line-integrated information collected from multiple sensors is often transformed into spatial information through the use of tomographic reconstruction algorithms.

The long-term goal of a larger research program is to apply optical tomography to the investigation of multiphase pipe flow. The multiple projection line-integrated data collected in the optical tomography system is obtained using Shack-Hartmann wavefront sensors (SH-WFS) [6]. The SH-WFS is able to provide information about both strong index-of-refraction variations (gas-liquid or solid-liquid interfaces), as performed in the current investigation, as well as weaker liquid-liquid interfaces. An example of a flow with weaker liquid-liquid interfaces is a continuous-phase water-glycerine mixture with an immiscible dispersed-phase mineral oil. Through experimental design, the water-glycerine concentration can be altered so that the index mismatch between the water-glycerine mixture and the mineral oil is small. This results in a two-component system with weak index-of-refraction variations. Consequently, the SH-WFS operates in a manner that is close to its original design [6].

A weak index-of-refraction system is not considered in the current investigation. What is performed here, rather, is an examination of the response of a single SH-WFS to a flow for which the SH-WFS has not been designed: a gas-liquid multiphase flow involving strong index-of-refraction variations ($n \approx 1.33$ for water versus $n \approx 1.00$ for air). If the SH-WFS can be demonstrated to provide meaningful quantitative data for a flow with strong index-of-refraction variation, then the use of SH-WFS sensors in optical tomography systems for the study of flows with both strong and weak index-of-refraction variations is supported. In order to investigate this further, it is first necessary to examine the performance of the SH-WFS when applied to a flow with strong index-of-refraction variation. In order to do this, three different but relatively well-understood gas-liquid multiphase bubble flows are considered.

This paper begins by describing the experimental apparatus and optical equipment used during this investigation. The response of the SH-WFS is investigated for the detection of both a single stationary air bubble and multiple ascending bubbles. This investigation is followed by a comparison of the performance of the SH-WFS to a high-speed video camera (HSVC) for detection of multiple ascending bubbles and for dispersed-bubble turbulent pipe flow. To conclude, the overall performance of the SH-WFS for application to index-mismatched multiphase flows is discussed.

2. Experimental Apparatus

Two different experimental systems are used during the investigation to examine the performance of the Shack-Hartmann wavefront sensor. In the first setup, a small vertical half-pipe section is used to examine the aberrating effects of both a single stationary bubble and of multiple ascending bubbles. The second system, a multiphase flow facility, is used to study the horizontal bubble flow under turbulent flow conditions. In both experimental systems, an acrylic optical contour is used to correct for the optical aberrations induced by the curvature of the water-filled pipe section [7].

2.1. Optical Diagnostics Equipment. The optical diagnostics used in the investigation consist of a Shack-Hartmann wavefront sensor (SH-WFS), and a high-speed video camera (HSVC). The HSVC provides the advantage of two-dimensional images. Consequently, it also provides a good method by which the performance of the SH-WFS system can be examined when applied to a gas-liquid multiphase flow involving strong index-of-refraction variations.

2.1.1. Shack-Hartmann Wavefront Sensor (SH-WFS). The general arrangement of the SH-WFS with respect to both the half-pipe and the full-pipe is illustrated in Figure 1. A laser diode light source is expanded and collimated prior to transmission through the pipe section containing the aberrating

multiphase flow. The emerging aberrated light transmits to the SH-WFS, which consists of a one-dimensional array of 64 cylindrical microlenses that focus the light onto a standard 2048 pixel line-scan camera (Dalsa Model CL-C4 2048 M). The analog readout of the line-scan camera is digitized using an Echotek VME-based A/D system with 8-bit resolution. Consequently, the readout data is quantified in terms of the A/D converter counts that range from 0 to 255. Each microlens is 0.448 mm wide with a focal length of 37.45 mm. The center of the focal spot generated by each microlens is quantified by examining the intensity profile of the 32 pixels that fill each subaperture (pixel size = 14 μm). The one-dimensional SH-WFS used in the investigation is capable of acquiring optical wavefronts at a temporal rate of 5,000 frames per second (fps).

Traditional Shack-Hartmann wavefront sensing applications involve the comparison of focal-spot centroidal locations produced by both a reference wavefront and an aberrated wavefront. The reference wavefront is often an unaberrated wavefront, produced by transmission through a medium without index-of-refraction variations. The change in the focal-spot centroidal location between the aberrated and unaberrated wavefronts is then computed for each subaperture, quantifying the distortion of the aberrated wavefront [8]. This method is based on the assumption that the focal-spot deflection in each subaperture will be less than one half of the number of pixels in a subaperture.

Given that the current investigation involves optical transmission through water continuous media ($n \approx 1.33$) with dispersed air bubbles ($n \approx 1.00$), strong index-of-refraction variations are present resulting in large focal-spot deflections. This paper demonstrates that the study of disperse bubble flow of water and air requires the use of an extinction-based approach rather than the standard method of SH-WFS processing.

2.1.2. High-Speed Video Camera (HSVC). The HSVC used during the experiments is a Photron Ultima 1024, capable of acquiring images at frame rates ranging from 60 to 16,000 fps. This camera is able to operate at a full 1024 × 1024 resolution at frame rates of up to 500 fps. The size of the pixel for this camera is 12 μm.

The HSVC enables images of the flow to be collected at high rates and then postprocessed for statistical quantities that can be compared to statistical data derived using the SH-WFS.

2.1.3. Optical Contour (OC). When using optical diagnostics to investigate pipe flow, special care needs to be taken to ensure that the strong optical aberrations induced by the fluid-filled pipe wall do not negatively impact the optical measurement being performed.

Prior to commencing this investigation, a custom-made corrective optic (referred to as an optical contour) was designed and manufactured using a computer numerically controlled (CNC) 5-axis machining station and used to correct for the aberrations induced by the water-filled acrylic pipe. The original design and manufacture of the optical

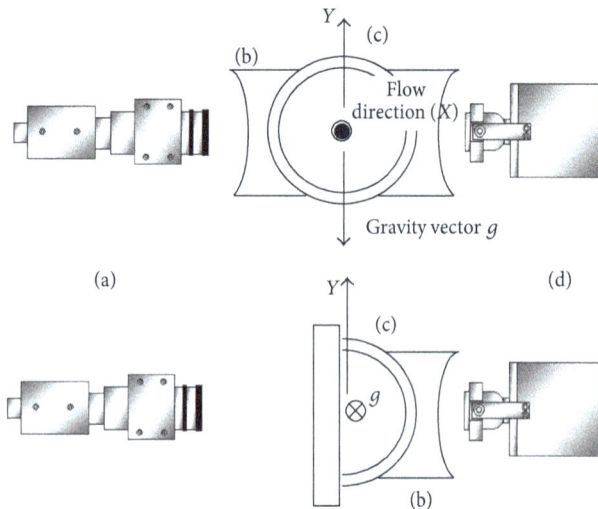

FIGURE 1: Schematic of the Shack-Hartmann wavefront sensor around the pipe and half-pipe: (a) laser-diode collimated light source, (b) optical contour, (c) pipe/half-pipe, and (d) 1D Shack Hartmann wavefront sensor.

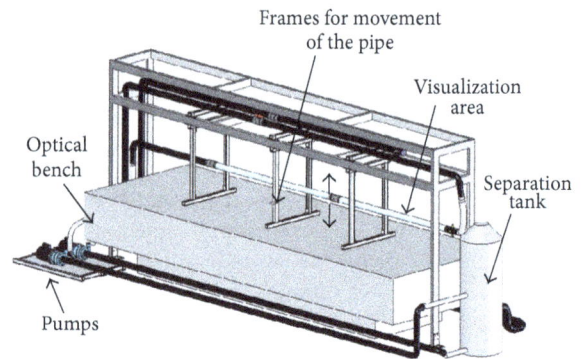

FIGURE 2: Description of the components of the flow facility.

contour is described in de Witt et al. [7]. The optical contour was designed to be clamped to the external pipe surface, making it relatively easy to attach it to different pipe locations. The optical contour was found to correct for the strong aberrations that would have otherwise been present while imaging through the water-filled pipe. The optical contour allows for approximately 80% of the central pipe to be investigated as shown in de Witt et al. [7].

The camera and the beam source are positioned perpendicular to the pipe axis as shown in Figure 1, with the spatial extent of flow investigated approximately 24.5 mm in the y-direction of the 57.15 mm internal diameter cast-acrylic pipe. As noted in Figure 1 (upper), two optical contours are required when applying the SH-WFS to the investigation of pipe flow. A replica of the CNC-manufactured optical contour described by de Witt et al. [7] is produced by first creating a mold of the CNC-manufactured optical contour and then pouring a time-hardening transparent urethane into the mold. The transparent urethane optical contour shows good optical characteristics when compared with the original acrylic lens. Moreover, it can be produced at a fraction of the cost and in less time than the CNC-manufactured optical contour.

2.2. Half-Pipe Apparatus. The half-pipe test section shown in Figure 1 (lower) enables the response of the optical diagnostic equipment to be examined while detecting air bubbles in water and in the absence of a net bulk velocity. The inner diameter of the acrylic half-pipe section is 57.15 mm with a wall thickness of 12.7 mm. The half-pipe section is attached to an acrylic wall 11.4 mm in thickness. The optical contour is clamped to the outside of the acrylic half pipe, and given that one external surface was planar, the half-pipe apparatus requires only one optical contour.

Air bubbles are introduced into the half-pipe apparatus using two different methods. A single stationary bubble is introduced by injecting a known volume of air through a syringe into a small-diameter hose. A small-diameter hose is placed in the water-filled acrylic half pipe in a manner that ensures that the end of the hose is aligned with the sensing area of the SH-WFS.

For the generation of multiple ascending bubbles, a piece of porous material (sand packing of 2.54 cm length and 1.27 cm diameter) is connected to a regenerative blower (EG&G Rotron Blower, with a maximum flow of 0.008684 m^3/s and a maximum pressure of 61.0 cm of water) and placed at the bottom of the half pipe. The air supply hose connecting the blower to the sand packing has an internal diameter of 0.5 mm and is secured internally to the half pipe so as not to interfere with the measurement area (2.54 cm wide) within the pipe.

2.3. Optically Active Multiphase Flow Facility. The multiphase flow facility was designed and built specifically to study dispersed multiphase flow using optical methods. The facility can operate using air, water, and oil; however, only air and water are considered in the current investigation.

A schematic of the multiphase flow facility is shown in Figure 2 with water and air as the operating fluids. The cast acrylic pipe section shown in Figure 2 runs the length of a 4.572 m by 1.829 m optical bench, enabling a wide variety of optical sensing methods to be applied to the study of multiphase flow. The facility can be broken down into five main sections as follows.

(1) Visualization Section. A 3.05 m long section of 5.72 cm inner diameter cast acrylic pipe with corrective optical contour forms the main diagnostic section.

(2) Pumping System. A centrifugal pump capable of delivering 7.25 L/s at open flow is used to circulate the liquid phase. The air is supplied by a 3.7 kW 2-stage air compressor with 300 L vertical tank capable of delivering 8.5 L/s at 689 kPa. The flow from the compressor is measured using a gas turbine flowmeter capable of sensing 1.9–28.3 L/s.

(3) Piping. ABS black plastic pipe with 50.8 mm inner diameter is used for the fluid stream. Flow metering is performed on two straight section runs at the top of the flow facility. Flow rate for both water and oil is measured using a Panametrics UFP-1000 portable ultrasonic liquid flowmeter.

(4) Separation and Storage. A 196 L vertical cylindrical tank with a diameter of 45.72 cm is designed for the separation of water and air.

(5) Air Injection. A pipe of 38.1 mm diameter is connected to a 25 mm compressed air supply, and air injection is performed through a series of four 12.7 mm holes, each separated by 90° around the perimeter of the 5.72 cm pipe.

3. Shack-Hartmann Wavefront Sensor

The performance of the SH-WFS is first examined by conducting experiments with a single bubble and with multiple ascending bubbles in the water-filled half-pipe.

3.1. Single Bubble Experiment. A single stationary bubble is introduced into the half-pipe apparatus by injecting a known volume of air (average volume = 0.0344 ± 0.0035 cm^3) through a syringe into a small-diameter hose. A small-diameter hose is placed into the water-filled acrylic half-pipe so that the end of the hose is aligned with the sensing area of the SH-WFS.

The left part of Figure 3 shows a schematic of the incoming wavefront passing through a bubble where light encountering the bubble is shown to be scattered away from the detector. Results from a ray-tracing model (not shown) independently confirm this assumption. The right part of the same figure shows the corresponding intensity profile collected by the SH-WFS for a subset of the 2048 pixel array. Each 0.448 mm wide lens functions to focus the light into a relatively well-defined and quantifiable focal spot. Overlaid on the intensity profile is the reference measurement for the case of no bubble. By comparing the reference measurement to the single bubble measurement, it becomes apparent that the air bubble serves to scatter light away from the detectors, as depicted in the schematic to the left in the figure. The main consequence of this result is that it was not possible to apply the standard Shack-Hartmann data processing techniques that were described earlier. Rather, an alternate method of evaluating the data from the SH-WFS is required.

Due to low-signal levels, an evaluation that computes intensity attenuation is performed. To compare the intensities, the attenuation for each lens is calculated as the summed absolute difference between the reference intensity profile and the intensity profile collected with the air bubble in place. The attenuation is then passed through a thresholding algorithm that only accepts attenuations larger than 50 counts. A threshold of 50 counts is determined by quantifying noise as the summed absolute difference for multiple reference intensity profiles.

This thresholding algorithm is found to reject variations caused by electronic noise and results in clearer plots. The maximum value of attenuation is well above the threshold level and is greater than 1000 counts. The resulting plot in Figure 4 shows the existence of an aberration in lenses 37 through 46 (pixels 1168 to 1456). Given the 32 pixels per lenslet and a pixel width of 14 μm, the radius of the single bubble aberration is quantified to be 0.4032 cm. Knowing the average volume injected by the syringe and assuming that the bubble remains spherical, the average radius is computed to be 0.4038 ± 0.0136 cm or a difference of 0.15%.

In order to ascertain the benefits of using the SH-WFS for bubble measurements, Figure 5 shows the intensity profile obtained for a single stationary bubble while using a standard line-scan camera without lenslet array. Once again, data from a reference measurement collected for the case of no bubble is overlaid. Although an attenuation region caused by the bubble is evident, it is more difficult to visually perceive this region than for the case with the lenslet array presented in Figure 3. The pixel-by-pixel absolute difference between the Single Bubble Measurement and the reference measurement is calculated, thresholded, and plotted as a contour plot in Figure 6. Although a region of strong attenuation is noted from pixel 814 to pixel 1096, artifacts of the differencing and thresholding routine appear at pixels that are not aligned with the bubble, as noted specifically near pixels 725 and 1400.

The bubble diameter can be calculated from Figure 6 as the length from pixel 814 to pixel 1096, resulting in a diameter of 0.3948 cm. When compared with the 0.4038 ± 0.0136 cm average bubble diameter, an error of 2.23% can be calculated. This error is slightly larger than that obtained while using the lenslet array. The main difference between the two results is attributed to the increased difficulty in the thresholding routine for the case without lenslet array.

A more severe threshold is applied to the case without lenslet array in order to filter the data. The contour plot of the differences between intensity profiles (Figure 6) results in increased noise near the edges of the bubble. The diameter of the single bubble is estimated with increased error from Figure 6 due to the inconsistency in light attenuation along the pixel array. In comparison, the results for the experiments with the lenslet array (cf. Figure 4) lead to a more uniform region for the single bubble experiment.

3.2. Multiple Ascending Bubbles Experiment. Data for multiple ascending bubbles in the half-pipe apparatus is collected at a line rate of 2.5 kHz, yielding 256 lines of data with 0.0004 seconds between samples. When multiple bubbles are studied, it is noted that clustering between two or more bubbles may be observed leading to an uncertainty in the response of the SH-WFS.

For the case of multiple ascending bubbles, the subaperture spot centroidal location does not shift significantly when compared to the unaberrated reference signal. Furthermore, the signal for multiple bubbles is not fully attenuated as it is for a single bubble. It is suspected that scattered light from one bubble may be detected by other subapertures not directly aligned with that bubble. Even though the amount of light extinction is less than that for the case of a single bubble,

Intensity plot for the case of the SH-WFS with lenslet array

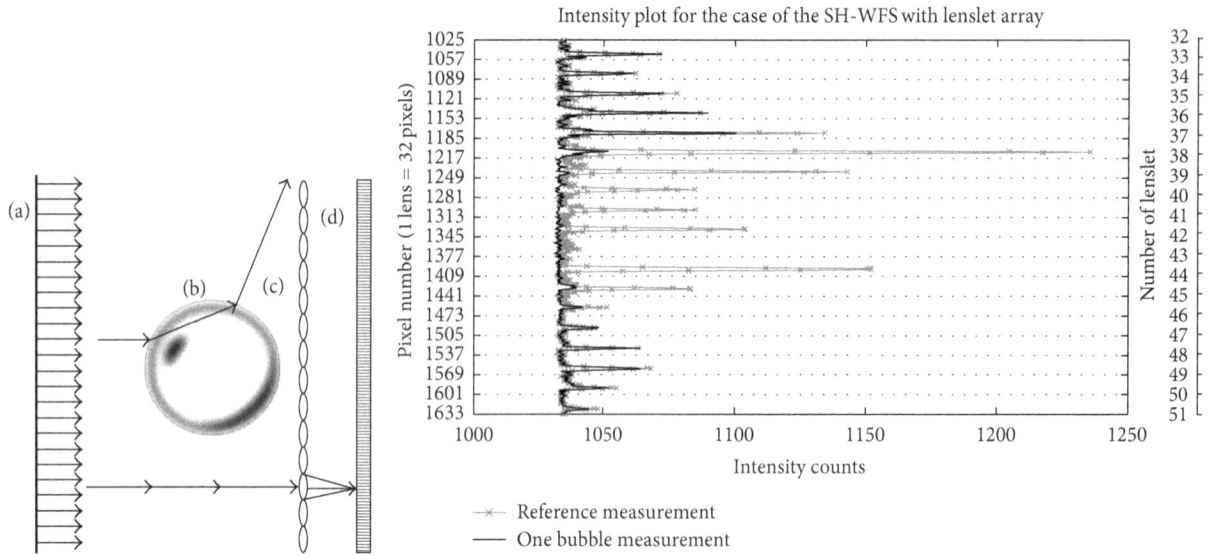

FIGURE 3: Scheme of the wavefront passing through the bubble in the half-pipe and plot with the intensity response from the wavefront sensor, (a) incoming wavefront, (b) single bubble, (c) lenslet array, and (d) pixel array.

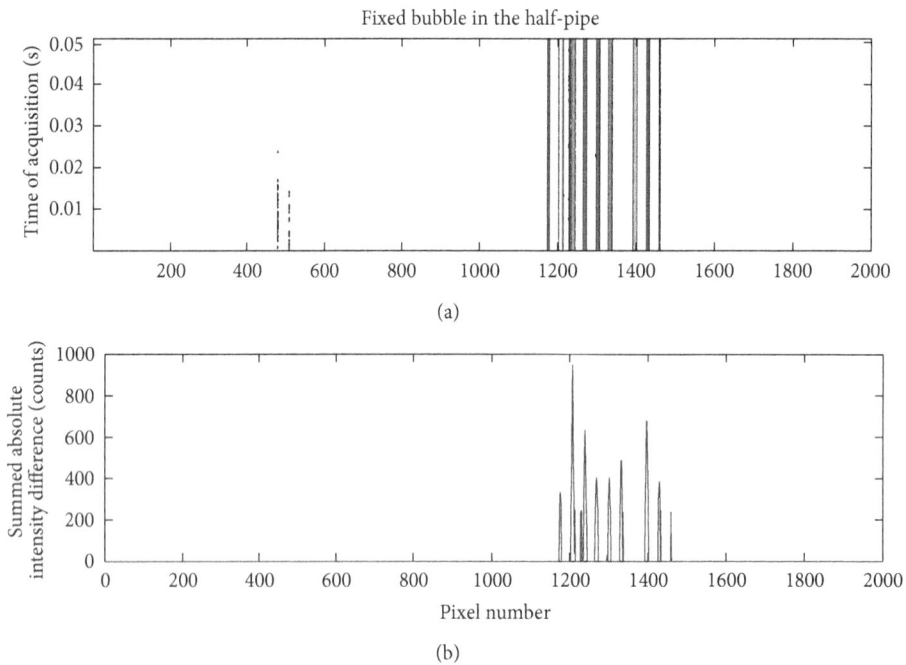

FIGURE 4: Contour of intensity difference with time for one bubble (a); profile of intensity difference at one time sample (b).

the extinction is still more quantifiable than a shift in spot centroidal location. Consequently, the same signal analysis routine as performed for the single bubble experiment is used, with results shown in Figure 7.

When the lenslet array is removed from the SH-WFS, the data collected is more difficult to interpret. Figure 8 shows an example of the datasets obtained for ascending bubbles without the lenslet array. It becomes evident that the lenslet array improves the ability to detect bubbles, especially while analyzing bubbles in motion. The behavior of the SH-WFS is

examined in further detail in the next section when the SW-WFS is compared to a HSVC.

4. Shack-Hartmann Wavefront Sensor versus High-Speed Camera

Two main experiments are performed involving the application of both the SH-WFS and the HSVC to the study of air bubbles in water. The first experiment involves ascending

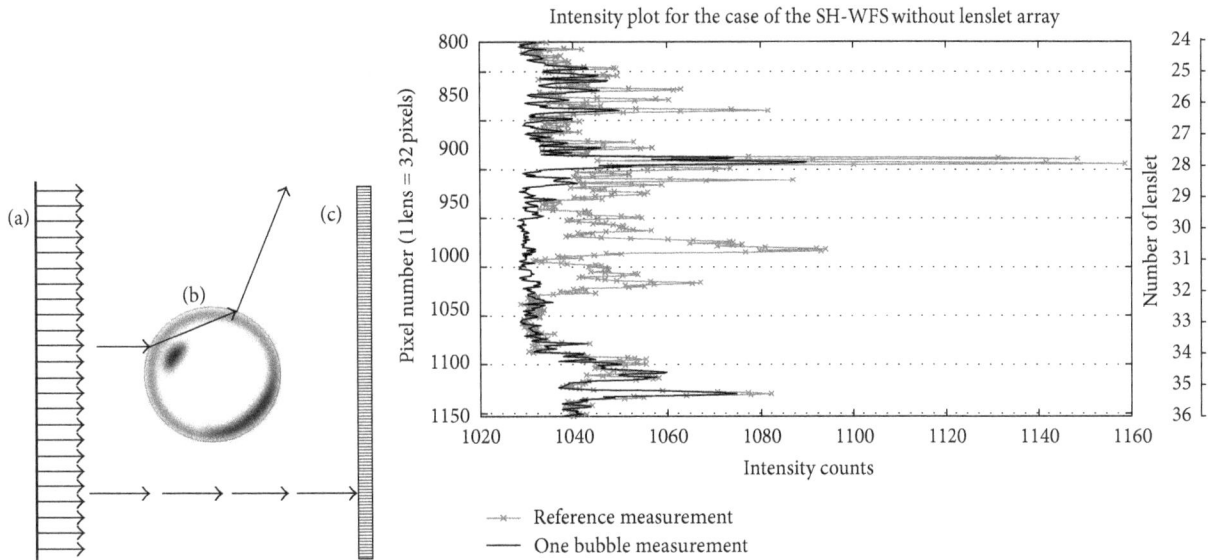

FIGURE 5: Schematic of the wavefront passing through the bubble in the half pipe and plot with the intensity response from the wavefront sensor, (a) incoming wavefront (b) single bubble (c) pixel array.

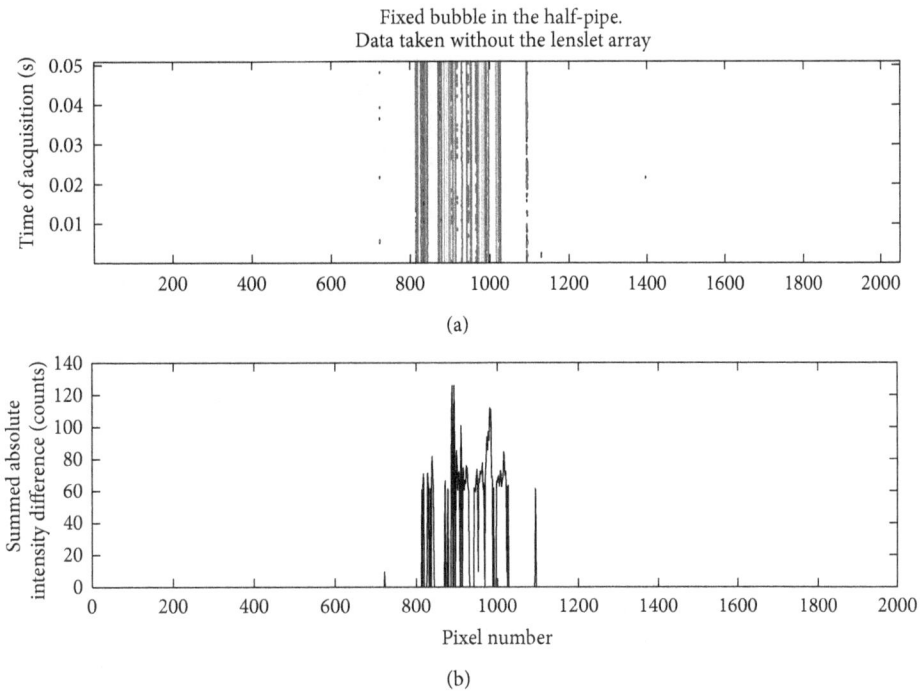

FIGURE 6: Contours of intensity difference with time for one bubble without lenslet array (a) and profile of intensity difference at one time sample without lenslet array (b).

bubbles in the water-filled half-pipe, and the second involves dispersed-bubble turbulent pipe flow.

4.1. Half-Pipe Experiment: Multiple Ascending Bubbles. The flow is analyzed using both the SH-WFS and the HSVC. The high-speed video data is collected at a frame rate of 2 kHz with a resolution of 256 by 512 pixels. Figure 9 shows an example of the data taken with the HSVC for the ascending bubbles experiment. Each dataset collected using the HSVC consists of approximately 70 to 120 bubbles ascending through the viewing area, from which it is possible to extract statistical information for comparison with the SH-WFS. Bubble diameter is extracted from HSVC frames displaying spherical bubbles, while velocity is extracted for bubbles moving vertically upward without collision. By using the HSVC, the average diameter of the bubbles is estimated

Bubble flow in the half-pipe
Difference between the intensity profiles

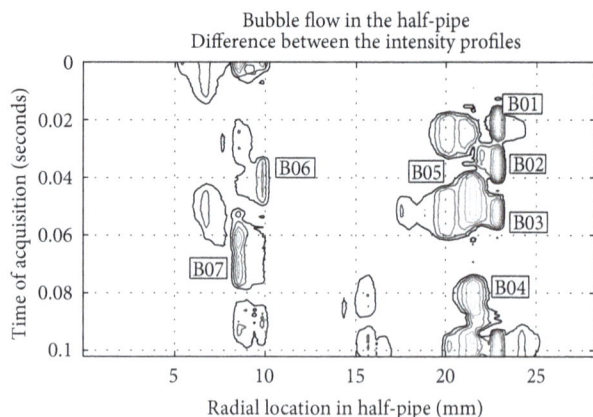

FIGURE 7: Variation in intensity measured for the bubble flow in the half-pipe.

Bubble flow in the half-pipe. Case without the lenslet array.
Difference between the intensity profiles

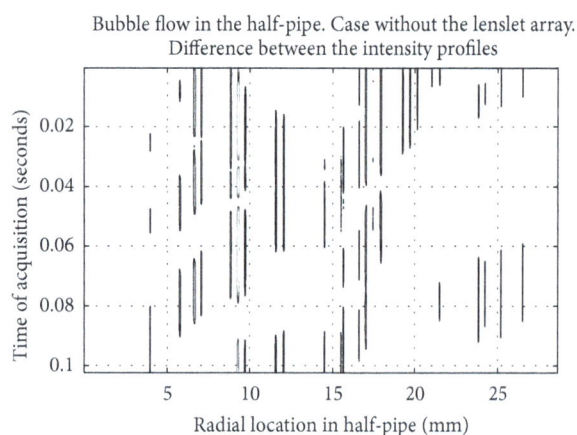

FIGURE 8: Variation in intensity measured for the bubble flow in the half-pipe without the lenslet array.

to be 2.33 ± 0.5 mm and the average velocity of the bubbles is estimated to be 0.250 m/s.

Analysis of the SH-WFS data reveals the average diameter of the bubbles as 2.08 ± 0.224 mm and the average velocity as 0.236 ± 0.041 m/s. Consequently, both the average size and velocity collected using the SH-WFS falls within the range of values estimated using the HSVC. A more detailed comparison involving probability density functions for both sensors is shown in Figure 10 from which it is noted that similar distributions result for both sensing systems.

Processing data collected during the multiple ascending bubble experiment reveals the difficulties associated with either sensing system when multiple bubbles are present. For instance, in the case of results from the SH-WFS it is difficult to say what is happening when two contour peaks are very close together. This behavior may be attributed to two bubbles, with one behind the other, or to interactions between bubbles, such as coalescence. Data collected using the HSVC presents similar challenges, as denoted by the circles in Figure 9. Multiple bubbles challenge automatic bubble detection routines, forcing one to manually inspect images.

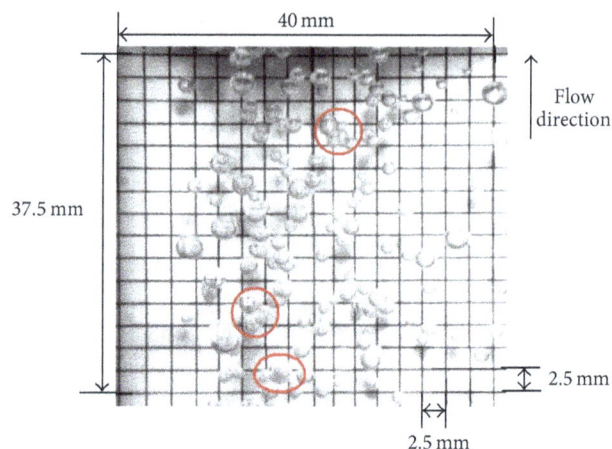

FIGURE 9: Data taken with the HSVC for the case of ascending bubbles in the half-pipe. The circles show clusters of bubbles observed.

4.2. Optically Active Multiphase Flow Facility. As in the case for the water-filled half-pipe, the bubble flow investigated using the multiphase flow facility is performed by collecting data using both the HSVC and the SH-WFS. The multiphase flow facility enables the horizontal turbulent pipe flow to be investigated. This type of flow behaves in a fundamentally different manner from the previously discussed ascending bubble flow. Due to the presence of strong shear forces, bubbles in the horizontal dispersed-bubble turbulent pipe flow are more likely to be ellipsoid in shape rather than spherical. Also, the flow of bubbles is affected by the velocity field of the continuous phase (carrying fluid), with bubbles moving slower along the walls and faster towards the centerline of the pipe. While comparing the two measurement methods, an analysis of bubble size, velocity, and position within the pipe is performed.

The water velocities that resulted in dispersed-bubble flow using the multiphase flow facility ranged from 1.31 m/s to 1.82 m/s. Seven separate water velocities are investigated. Figure 11 shows an example of the bubble distributions in the pipe for each investigated velocity using the HSVC. A total of 1.024 seconds of data is collected with the HSVC for each water velocity at a frame rate of 1000 frames per second (1024 frames). Data collected using the HSVC is taken across the entire pipe diameter.

It is not possible to quantify the amount of air injected during these tests due to the air flow rates being below the sensing range of the gas turbine flow meter. The flow rate of injected air is, however, assumed to be constant for all water velocities investigated given that a fixed flow restriction is imposed to the air inlet hose for all cases.

Figure 12 shows data collected using the SH-WFS for each water velocity. A total of 15 SH-WFS data sets (0.1024 seconds each) are collected for a total of 1.54 seconds of data at each flow velocity.

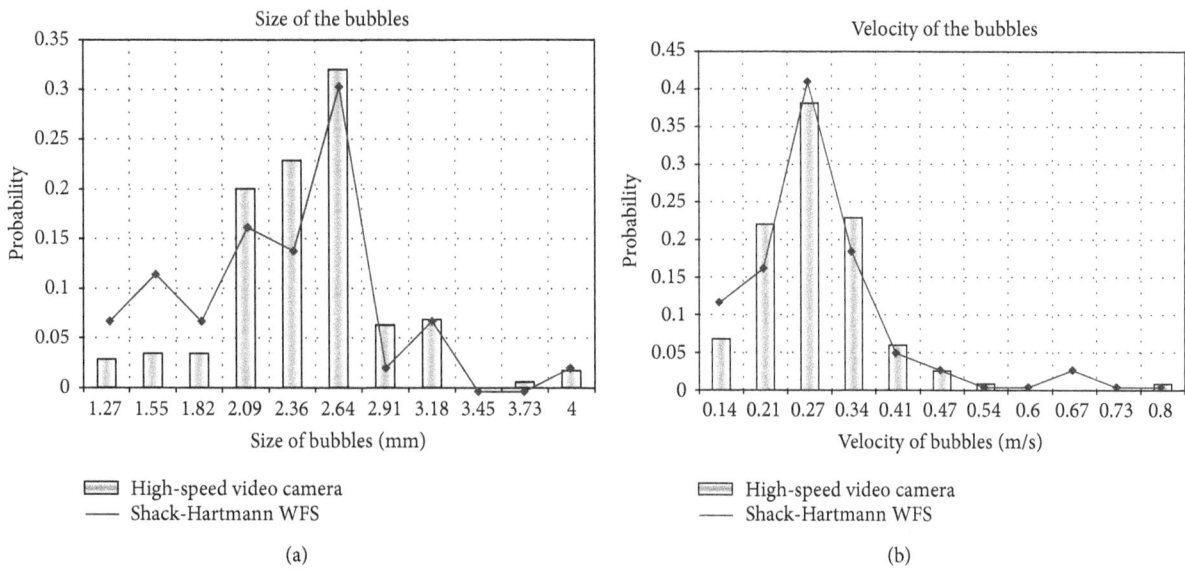

(a)   (b)

FIGURE 10: Statistical analysis for the diameter of spherical particles in the flow and velocity of the ascending bubbles.

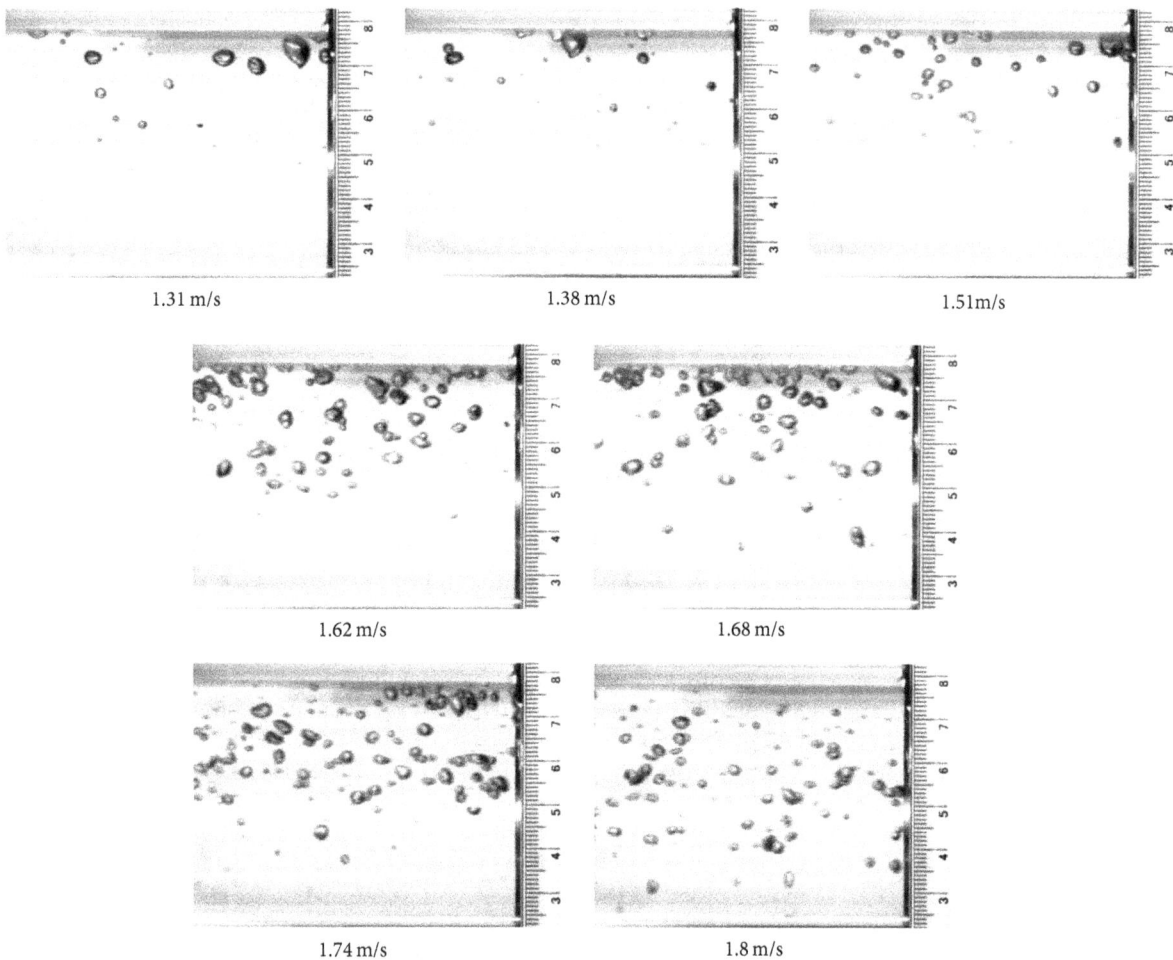

1.31 m/s   1.38 m/s   1.51m/s

1.62 m/s   1.68 m/s

1.74 m/s   1.8 m/s

FIGURE 11: Typical results from the HSVC for each water velocity sample.

The SH-WFS is only able to detect across 24.5 mm of the 57.15 mm inner diameter cast acrylic pipe, and consequently it is necessary for the frame of reference of the SH-WFS detector to be aligned with the physical space of the pipe. This is performed by placing a dark object along the center of the pipe and by noting the subaperture (subaperture 18) where a loss of signal occurred. This subaperture is then referenced as the pipe centerline and labeled by a dark solid line in all plots. Even though the sensing length of the SH-WFS is 28.67 mm, only 24.5 mm of the sensing length results in meaningful intensity measurements. The optical contour causes a reduction of the collimated light at the entry of the pipe and magnification at the exit; this effect reduces the width of the collimated beam within the pipe by 5%. For more details about the behavior of light through the optical contour, the interested reader is referred to de Witt et al. [7].

4.2.1. Data Processing. In order to compare HSVC data to SH-WFS data, HSVC data is extracted only when a bubble crosses a reference line. Each bubble-crossing event yields data with regards to the following characteristics: the vertical position of the bubble in the pipe, the bubble size (both horizontal and vertical major axes), and the number of frames for the bubble to move through the reference line (bubble horizontal velocity). The use of a reference line to analyze data from HSVC enables a direct comparison to velocity data measured using SH-WFS.

Bubble size is measured with the help of a grid that is superimposed on each image. The grid that is used has a minimum size of 1 mm per cell. Approximately 350 bubble-crossing events are detected at the highest water velocities, while only 50 to 80 bubble-crossing events are detected at the lowest velocities.

Image analysis is not required for the SH-WFS data; however, as with the HSVC data, the SH-WFS data is also analyzed manually. The sizes and velocities are obtained from the summed absolute difference of intensity data. A similar grid to the one used with the HSVC is used to detect the size of the objects (bubbles) in the contour plots. The size of the bubbles is calculated by multiplying the size of the pixel in the sensor (14 μm) by the maximum number of pixels in the contour plot. The time for a crossing event is calculated with the maximum length of the body in the contour plot (vertical axis). The convective velocities are computed by dividing each diameter by the corresponding time for a crossing. The position of each bubble inside the pipe is calculated by referencing the pipe centerline to the middle point of each bubble.

Given that the line-scan camera is a one-dimensional sensor, it is only possible to quantify bubble size in the cross-flow direction. Consequently, the data collected with the SH-WFS includes the vertical position of the bubble in the pipe, the size of the bubble (vertical axis dimension), and the number of frames (time) that it would take for each bubble to move through the sensing plane.

The image distortion effects due to the presence of the optical contour are also taken into account, and the data presented here is the corrected physical dimension. The magnification used for the optical contour is as reported in de Witt et al. [7].

4.2.2. Bubble Shape. Of the more than 1000 bubbles analyzed using the HSVC data, only 26% of the bubbles reveal spherical shape. The remaining bubbles are ellipsoidal in shape. An analysis of the shape of the bubbles is made using the data acquired with the HSVC and a ratio of bubble size in the flow direction (X) to bubble size in the cross-flow direction (Y) computed. A probability density function of bubble aspect ratio is shown in Figure 13. Three distributions are plotted, one with data for all water velocities and the other two with data from both the lowest and highest velocities. The bubbles become more ellipsoidal at the higher flow velocities, with the long axis oriented in the flow direction. The mean bubble size in the X direction is 2.44 mm versus 2.14 mm in the Y direction.

Investigations by Crowe et al. [9] and Hesketh et al. [10, 11] have reported a similar segregated distribution for bubble size in horizontal dispersed-bubble turbulent pipe flow.

This variation in bubble aspect ratio will affect velocity calculations made using the SH-WFS measurements given the assumption that bubbles are spherical. Bubbles are more likely to be elongated in the flow direction, resulting in a reduction in measured bubble velocity using the SH-WFS. The data presented in this paper is taken very close to the injection point which results in increased bubble shape variations. The assumption of sphericity has been made by other researchers investigating the horizontal bubble flow [12, 13] with the assumption most valid for fully developed flow with bubble diameter less than 2.0 mm.

4.2.3. Bubble Size and Position. Figure 14 shows the data collected for the lowest and highest water velocities of bubble size versus pipe position using both the HSVC and the SH-WFS. When comparing the behavior of bubble size versus pipe position at the lowest water velocities, the largest bubbles are observed towards the top of the pipe. As water velocity increases, the bubbles are seen to distribute more evenly across the entire pipe diameter.

A pattern is observed in the size information computed using the SH-WFS data where bubble size falls into discrete bands. These discrete bands result as the size calculation is performed based on the width of each lenslet, resulting in a resolution of 0.448 mm with an uncertainty of ±0.224 mm. When two bubbles are very close to each other or when the difference in intensity shows a non-regular shape towards the edge (due to noise from scattered light from another subaperture), a minimum size of half of a lenslet is assigned.

With the exception of the largest water velocity, the maximum bubble size for all data collected with the SH-WFS is smaller than the maximum size calculated using the HSVC. The SH-WFS is also able to detect smaller bubbles than the HSVC. While small bubbles can be seen in the

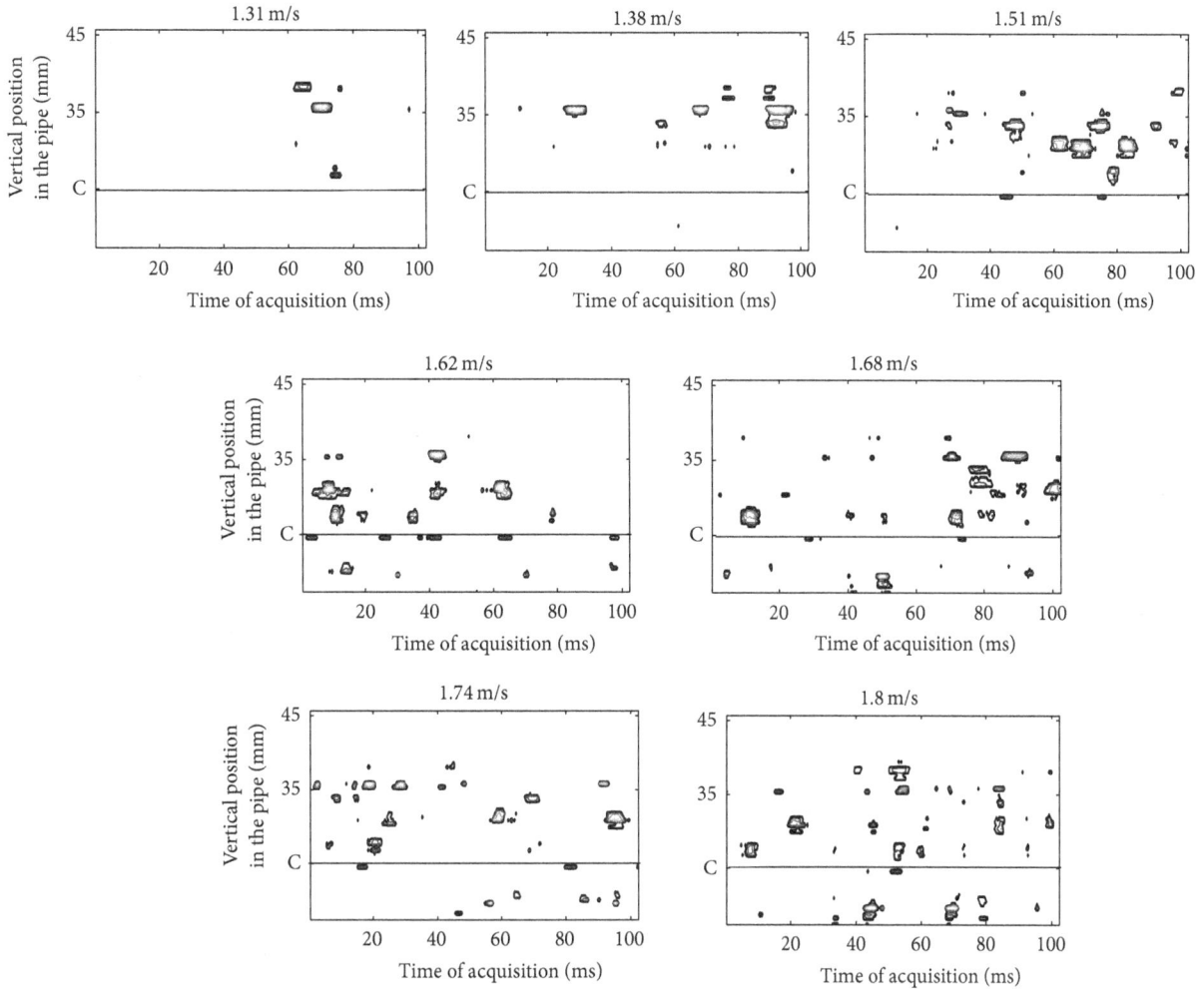

FIGURE 12: Typical results from the SH-WFS for each water velocity sample. Letter "C" in each plot represents the center of the pipe.

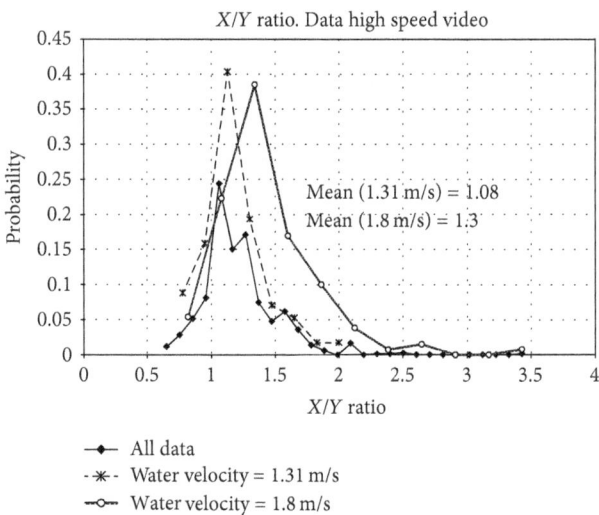

FIGURE 13: Analysis of the sphericity of bubbles with the high-speed video.

high-speed video, it is very difficult to obtain an accurate size calculation due to a limit in sensor resolution. The minimum

size acquired for the SH-WFS was 0.224 mm while for the HSVC the minimum size is 0.54 mm, more than double that of the SH-WFS. The cause for this difference is believed to be attributed to the depth of field of the HSVC. Bubbles outside the depth of field of the HSVC would be blurred, making it more difficult to resolve them. The SH-WFS, on the other hand, involves collimated laser light. Consequently, it does not have a depth of field but rather is able to detect bubbles independent of where they are located within the pipe. If the HSVC had a smaller field of view like the SH-WFS, the resolving capability of the HSVC would have improved. The number of pixels in the SH-WFS is also larger than that for the HSVC, which also results in the noted difference. Had a higher resolution HSVC been used, the results would have been more comparable.

4.2.4. Bubble Velocity and Position. A comparison of bubble velocity data with pipe position is shown in Figure 15. The distributions of velocities for both sensors show good agreement, with similar maximum velocities for both sensors. The analysis of longer data sets would help to reduce the scatter observed in the distribution for both size and velocity.

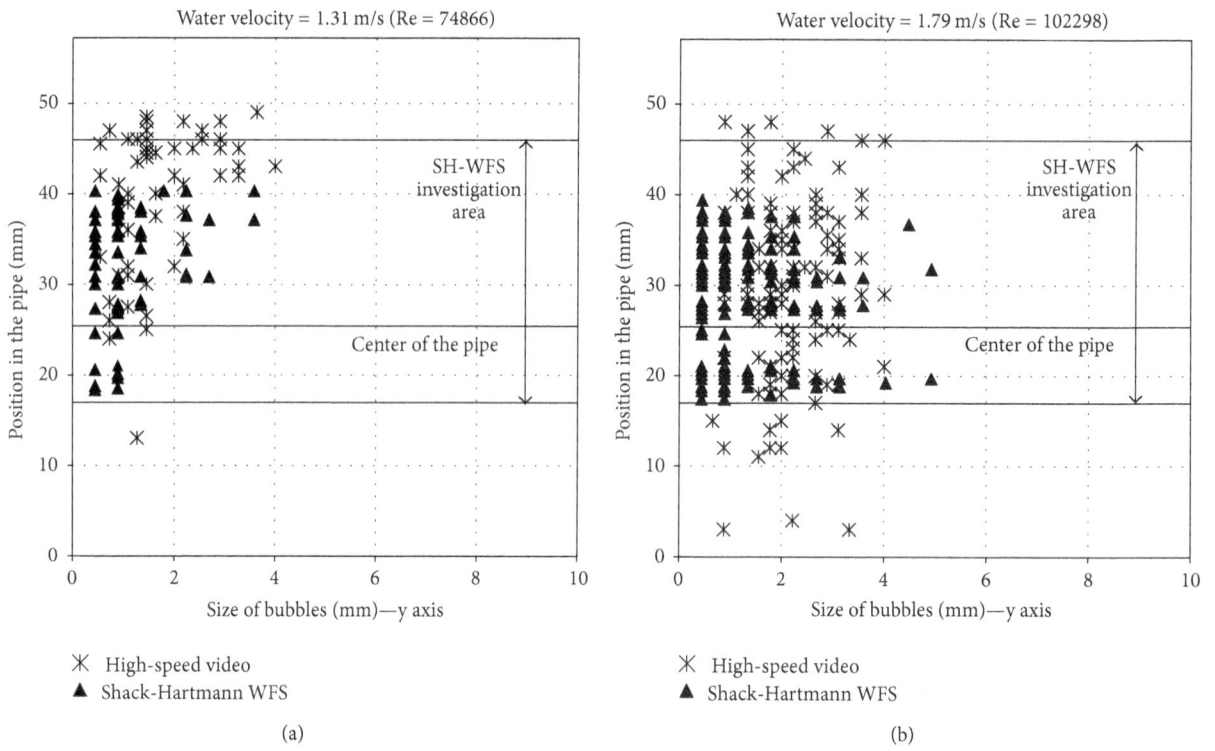

FIGURE 14: Comparison of position of bubbles in the pipe versus size of the bubbles.

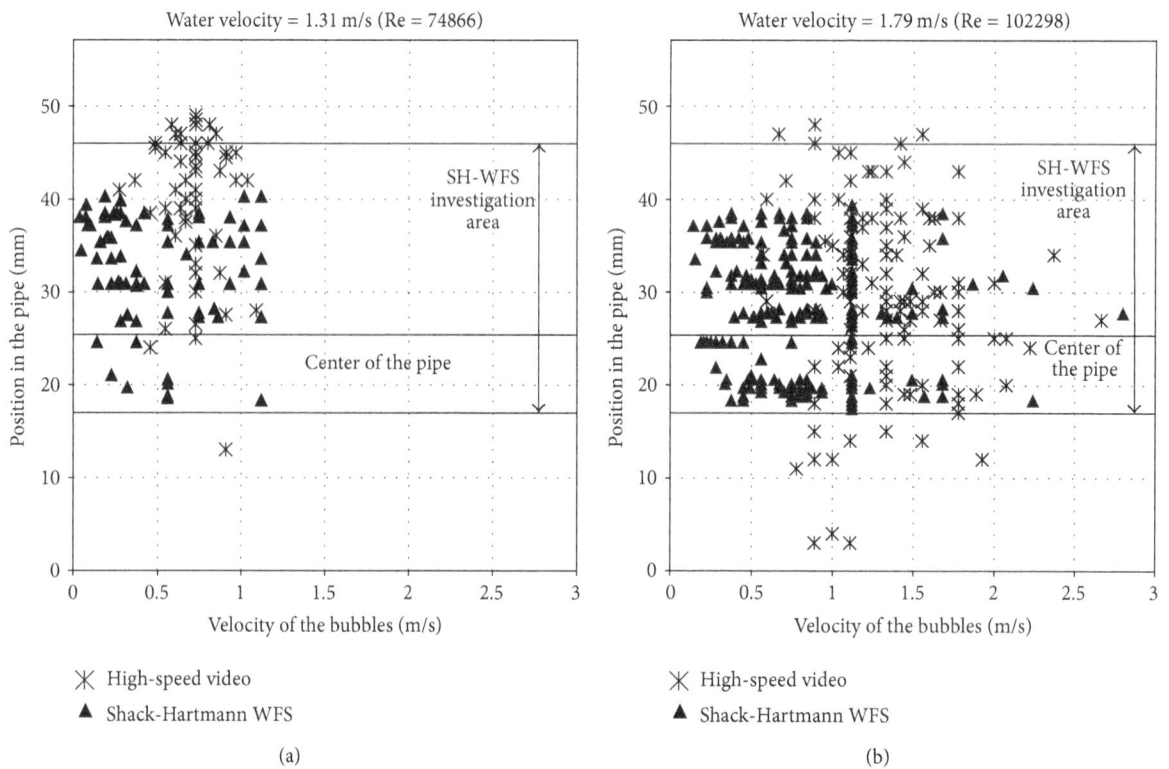

FIGURE 15: Comparison of position of the bubbles in the pipe versus velocity of the bubbles.

5. Conclusions

This paper has investigated the performance of a SH-WFS for the analysis of dispersed-bubble turbulent flow. The presence of dispersed bubbles in the continuous water media is found to cause extinction of the collimated light source for the SH-WFS.

A statistical comparison between the SH-WFS and HSVC reveals good agreement for two different experiments: ascending bubble flow and horizontal pipe flow. The depth of field of the HSVC makes it more difficult to detect small bubbles (less than 0.55 mm) when compared to the SH-WFS, and consequently the SH-WFS has a better resolution than the HSVC for the optical configuration investigated. Had the HSVC had a smaller field of view or larger number of pixels, its resolution would have improved.

The three individual experiments that were performed-single fixed bubble, multiple ascending bubbles, and horizontal dispersed-bubble turbulent pipe flow, all provide strong experimental evidence that the SH-WFS is a reliable method for the study of multiphase bubble flows. The SH-WFS has certain advantages/disadvantages when compared to other measurement systems.

In comparison to the HSVC, the SH-WFS has the advantage of being able to detect bubbles at any plane inside the pipe whereas bubble measurement using the HSVC was more dependent on the depth of field of the camera lens system. The bubble detection procedures for the SH-WFS are simpler than those for the HSVC, providing a computational advantage during postprocessing operations. Given that the SH-WFS is a one-dimensional sensor, it has the advantage of being able to store more frames and to collect them at higher frame rates. The one-dimensional nature of the SH-WFS does pose a disadvantage when compared to the HSVC, given that it is difficult to measure convection velocity without first having to make assumptions about bubble dimension in the flow direction (i.e., the spherical bubble assumption).

The SH-WFS has been shown to perform as an extinction-based sensing technique when applied to a flow containing components with strong index-of-refraction variations, such as the liquid-gas flow investigated here. Additional sensing capability exists when the system is applied to flows with weak index-of-refraction variations, such as two immiscible fluids with minor index of refraction differences. This would enable the SH-WFS to detect both dispersed bubbles and oil droplets in a water-glycerin continuous flow, for instance. Exploration of this research topic has been left as a future investigation.

Acknowledgments

The authors would like to acknowledge the support of the Natural Sciences and Engineering Research Council of Canada (NSERC) through its Discovery Grant Program and to acknowledge the National Council of Science and Technology of Mexico (CONACyT) for its financial support.

References

[1] T. Dyakowski, "Process tomography applied to multi-phase flow measurement," *Measurement Science and Technology*, vol. 7, no. 3, pp. 343–353, 1996.

[2] S. M. Huang, A. B. Plaskowski, C. G. Xie, and M. S. Beck, "Tomographic imaging of two-component flow using capacitance sensors," *Journal of Physics E*, vol. 22, no. 3, pp. 173–177, 1989.

[3] F. J. Dickin, R. A. Williams, and M. S. Beck, "Determination of composition and motion of multicomponent mixtures in process vessels using electrical impedance tomography-I. Principles and process engineering applications," *Chemical Engineering Science*, vol. 48, no. 10, pp. 1883–1897, 1993.

[4] J. Wolf, "Investigation of bubbly flow by ultrasonic tomography," *Particle and Particle Systems Characterization*, vol. 5, no. 4, pp. 170–173, 1988.

[5] S. B. Kumar, D. Moslemian, and M. P. Duduković, "A γ-ray tomographic scanner for imaging voidage distribution in two-phase flow systems," *Flow Measurement and Instrumentation*, vol. 6, no. 1, pp. 61–73, 1995.

[6] R. V. Shack and B. C. Platt, "Production and use of a lenticular Hartmann screen," *Journal of the Optical Society of America*, vol. 61, pp. 656–660, 1971.

[7] B. J. de Witt, H. Coronado-Diaz, and R. J. Hugo, "Optical contouring of an acrylic surface for non-intrusive diagnostics in pipe-flow investigations," *Experiments in Fluids*, vol. 45, no. 1, pp. 95–109, 2008.

[8] R. K. Tyson, *Principles of Adaptive Optics*, Academic Press, New York, NY, USA, 2nd edition, 1991.

[9] C. Crowe, M. Sommerfeld, and Y. Tsuji, *Multiphase Flows with Droplets and Particles*, CRC Press, New York, NY, USA, 1998.

[10] R. P. Hesketh, F. Russell, and A. W. Etchells, "Bubble size in horizontal pipelines," *AIChE Journal*, vol. 33, no. 4, pp. 663–667, 1987.

[11] R. P. Hesketh, A. W. Etchells, and T. W. F. Russell, "Experimental observations of bubble breakage in turbulent flow," *Industrial and Engineering Chemistry Research*, vol. 30, no. 5, pp. 835–841, 1991.

[12] P. Andreussi, A. Paglianti, and F. S. Silva, "Dispersed bubble flow in horizontal pipes," *Chemical Engineering Science*, vol. 54, no. 8, pp. 1101–1107, 1999.

[13] M. M. Razzaque, A. Afacan, S. Liu, K. Nandakumar, J. H. Masliyah, and R. S. Sanders, "Bubble size in coalescence dominant regime of turbulent air-water flow through horizontal pipes," *International Journal of Multiphase Flow*, vol. 29, no. 9, pp. 1451–1471, 2003.

16

Remote Colorimetric and Structural Diagnosis by RGB-ITR Color Laser Scanner Prototype

Massimiliano Guarneri, Mario Ferri de Collibus, Giorgio Fornetti,
Massimo Francucci, Marcello Nuvoli, and Roberto Ricci

ENEA, Via E. Fermi 45, Frascati 00044 (Roma), Italy

Correspondence should be addressed to Massimiliano Guarneri, massimiliano.guarneri@enea.it

Academic Editor: Joseph Rosen

Since several years ENEA's Artificial Vision laboratory is involved in electrooptics systems development. In the last period the efforts are concentrated on cultural heritage remote diagnosis, trying to develop instruments suitable for multiple purposes concerning restoration, cataloguing, and education. Since last five years a new 3D (three-dimensional) laser scanner prototype (RGB-ITR) based on three amplitude-modulated monochromatic laser sources mixed together by dichroic filters is under development. Five pieces of information per each sampled point (pixel) are collected by three avalanche photodiodes and dedicated electronics: two distances and three target reflectivity signals for each channel, red, green, and blue. The combination of these pieces of information opens new scenarios for remote colorimetry allowing diagnoses without the use of scaffolds. Results concerning the use of RGB-ITR as colorimeter are presented.

1. Introduction

One of the biggest problems concerning cultural heritage environment is the study and then the reproduction of pigment color. This difficulty is mainly caused by the tendency to consider the color almost exclusively as subject of visual perception and individual sensibility. It is important to underline that the color is also a physical propriety of the object itself and not only an observer experience that can be measured [1] yet in a nontrivial way.

The first work of colorimetry applied to cultural heritage field was in 1953 by Istituto Centrale per il Restauro di Roma for a preliminary study on the restoration of "Maestà" by Duccio di Boninsegna [2]. After that, 40 years elapsed before colorimetry found new applications in this field, mainly in Italy [3]. This is also the consequence of finding in commerce tested and optimized electronic instruments which are mainly developed for industrial inspection. The migration of this type of instruments, suitable for the measure of uniform colored surfaces, in cultural heritage environment is limited by the complexity and continuous

variation of polychromatic surfaces. For this reason it is important to have a colorimeter, which is able to measure very small areas better if punctual.

Traditional colorimeters [4] work in a very closed range, 0–30 cm, which can be acceptable for canvas or small frescos placed at few meters. For investigating large surfaces in a range of several meters, that is, painted chapels, commercial colorimeters are time-consuming systems and the use of scaffolds is necessary.

For all these reasons a remote and punctual system is desirable, but a new class of problems concerning color measurement is introduced: one among all is the independence of color measurement by distance.

Since several years a new optoelectronic device is developed in the Research Center ENEA of Frascati (Rome): this instrument can be considered a combination of a 3D scanner and a tristimulus colorimeter. The feature of collecting five pieces of information for each pixel, two distances, and three back-reflected signals from target, introduces the opportunity to carry out remote diagnosis of both the structure and the pigment of investigated surfaces [5].

2. Objectives

The main and ambitious aim of methodology and technology described in the paper is to fill the empty let by the use of commercial instruments in cultural heritage environment.

The color 3D laser scanner prototype, called RGB-ITR (Red Green Blue Imaging Topological Radar), developed in ENEA laboratory and used in several national and international campaigns, tries to respond at the demand of punctual and remote analysis of polychromatic surfaces. The submillimetric spatial resolution and beam size ensure a continuous cover of the investigating surface.

The color calibration methods, presented in the next paragraph, demonstrate the possibility to operate remotely for the detection and prevention of pigment modifications and have to be considered a starting point for the definition of this scanner prototype as colorimeter. In fact the feature of operating remotely in a range of 3–30 meters constrains to put in front of each detector an interferential filter whose transmittance spectra do not mimic the CIE color matching function [1]—so to guarantee the acquisition of measures free by external light sources interference. The calibration methods ensure repeatability of pigment measure during the time, that is, years, and permit to have an objective measure of polychromatic surfaces.

3. Methodology

The technique adopted for the simultaneous collection of distance and color information is based on amplitude modulation [6–8] of three monochromatic laser sources (680 nm, 532 nm, and 460 nm), which operate as carrier waves. The distance and color information are extracted from modulating waves by phase-shift and reflectivity responses of the target at the three wavelengths.

3.1. Distance Information. The main formula for distance detection is expressed by [6]

$$D = \frac{v\Delta\varphi}{4\pi f_m},\tag{1}$$

where f_m is the modulation frequency, $\Delta\varphi$ the phase difference between the reference modulating wave and the signal back-reflected by the target, and v the light speed in the medium. For laser optical powers such that the shot noise dominates over all other noise sources in the detection process (typically, a few nW at the output of the detection optical fiber), the accuracy σ_R of distance (i.e., range) measurements can be showed to increase with the modulation frequency f_m according to the following formula [6]:

$$\sigma_R = \frac{v}{2\pi\sqrt{2}m f_m \text{SNR}_i},\tag{2}$$

where m is the modulation depth and

$$\text{SNR}_i = \sqrt{\frac{P\eta\tau}{hf\Gamma}}\tag{3}$$

the current signal-to-noise ratio, which depends on the laser optical frequency f, the integration time τ, the detector's quantum efficiency η, the merit factor Γ of the receiver, and the collected optical power P. Here h denotes the Planck constant.

3.2. Color Information. The color information is elaborated by the detection of the amplitude of back-reflected modulating signal from the target. The raw data are expressed in Volt and are the result of elaboration by a lock-in amplifier Stanford SR-844, directly connected to an avalanche photodiode by a low-noise amplifier. The ranges for each channel are about 0.1–100 mV, which are digitally converted by three 16-bit ADC. Raw data are normalized in a range of 0-1 by minimum/maximum measurement of the scene, if no calibration is applied.

3.3. Color Calibration by Distance Information. The color calibration method as a function of distance is based on the measurement of a white target, Spectralon STR-99-020, placed at different distances and illuminated by the three lasers of RGB-ITR system. Figure 1 shows the three curves acquired during a measurement campaign for the 3D digitization of the S. Peter Martyr Oratory, a 15th century chapel located at Rieti (Italy): each point was collected by moving the white target of 50 cm, in a range of 2.5–21 meters. The curves are fitted by Hermite polynomials, which ensure the continuity in the working range.

These curves can be considered as a sort of RGB-ITR's fingerprint: once optical parameters are fixed, like the spot and optical receiving focalization, the color information can be univocally represented. The color correction of data acquired on investigated surface is given by the ratio between the three colors acquired by the instrument and the amplitude value on the three calibration curves selected by the corresponding distance.

3.4. Color Calibration by Spectrophotometer. For a standardization of color information detected by RGB-ITR system, respecting CIE standards, a further calibration was introduced.

This calibration method was executed in laboratory by using fourteen color Lambertian targets: these targets were acquired by both RGB-ITR system and a MINOLTA spectrophotometer-CM-2600d.

The first column on the left of Figure 2 shows the RGB-ITR data collected by acquiring the fourteen targets, whereas the last column shows the same targets data collected by spectrophotometer. In the central column, instead, the RGB-ITR data after calibration method obtained by MINOLTA spectrophotometer have been displayed. The differences between two instruments are evident but, while spectrophotometer data are acquired by contact, the RGB-ITR data are collected at a distance of 10 meters.

The matching of two color spaces was executed by minimizing the distance between the couple of measures collected

(a)

(b)

(c)

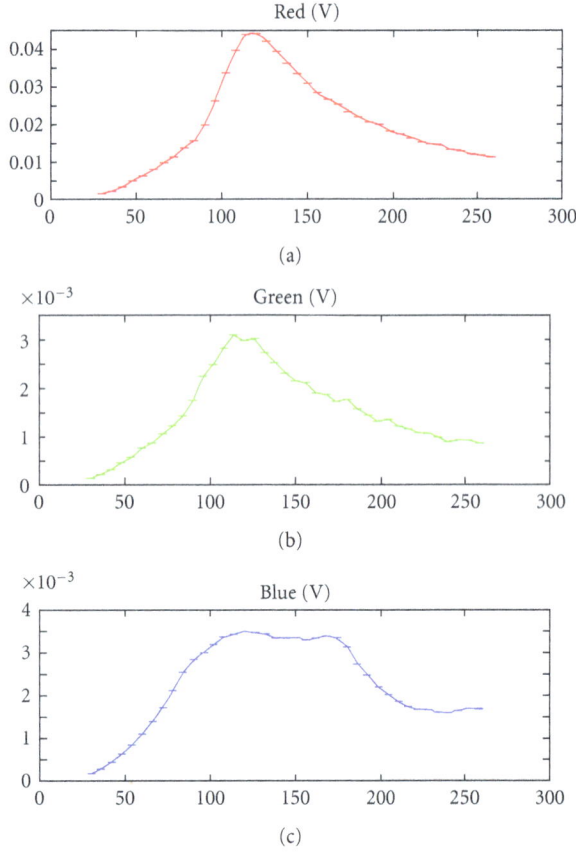

FIGURE 1: Single channel calibration curves, obtained striking a white target with the three laser beams of RGB-ITR system. Each curve represents the amplitude (V) of the back-reflected signal as a function of the distance, expressed as phase shift (degrees) between reference and back-reflected signals. (a) red, (b) green, and (c) blue curves.

FIGURE 2: On the left, RGB-ITR raw data; in the center, RGB-ITR data after calibration method obtained by MINOLTA spectrophotometer; on the right, MINOLTA spectrophotometer data.

by the two instruments. The resulting transformation matrix is obtained solving the equations:

$$
\begin{bmatrix} sR_M \\ sG_M \\ sB_M \\ s \end{bmatrix} = \begin{bmatrix} t_{11} & t_{12} & t_{13} & t_{14} \\ t_{21} & t_{22} & t_{23} & t_{24} \\ t_{31} & t_{32} & t_{33} & t_{34} \\ t_{41} & t_{42} & t_{43} & t_{44} \end{bmatrix} \cdot \begin{bmatrix} R_{\text{ITR}} \\ G_{\text{ITR}} \\ B_{\text{ITR}} \\ 1 \end{bmatrix}, \quad (4)
$$

FIGURE 3: RGB-ITR modules assembled all together on a mobile platform. From bottom, the first two boxes are composed by lock-in amplifiers, for laser modulation and signals detection, and a motion controller of the scanning motors; inside the next box, laser sources and detectors; on the top, the optical head with the scanning mirror.

where (R_M, G_M, B_M) are the colors triplet in spectrophotometer space, while $(R_{\text{ITR}}, G_{\text{ITR}}, B_{\text{ITR}})$ are the colors triplet in RGB-ITR space.

Arranging the terms and groupings, a homogeneous system of three equations with 16 unknowns from t_{11} to t_{44} is obtained, as shown in

$$
\begin{aligned}
& t_{11} \cdot R_{\text{ITR}} + t_{12} \cdot G_{\text{ITR}} + t_{13} \cdot B_{\text{ITR}} + t_{14} - t_{41} \cdot R_{\text{ITR}} \cdot R_M \\
& \quad - t_{42} \cdot G_{\text{ITR}} \cdot R_M - t_{43} \cdot B_{\text{ITR}} \cdot R_M - t_{44} R_M = 0 \\
& t_{21} \cdot R_{\text{ITR}} + t_{22} \cdot G_{\text{ITR}} + t_{23} \cdot B_{\text{ITR}} + t_{24} - t_{41} \cdot R_{\text{ITR}} \cdot G_M \\
& \quad - t_{42} \cdot G_{\text{ITR}} \cdot G_M - t_{43} \cdot B_{\text{ITR}} \cdot G_M - t_{44} G_M = 0 \\
& t_{31} \cdot R_{\text{ITR}} + t_{32} \cdot G_{\text{ITR}} + t_{33} \cdot B_{\text{ITR}} + t_{34} - t_{41} \cdot R_{\text{ITR}} \cdot B_M \\
& \quad - t_{42} \cdot G_{\text{ITR}} \cdot B_M - t_{43} \cdot B_{\text{ITR}} \cdot B_M - t_{44} B_M = 0.
\end{aligned}
\quad (5)
$$

It can be seen that for each point three equations and 16 unknown parameters are interested. Because the measures collected by the two instruments are affected by noise, more than necessary measures for solving the unknowns are considered. The parameter estimation problem is solved by computing the vector $\theta = [t_{11}, t_{12}, t_{13}, t_{14}, \ldots, t_{44}]^y$ that minimizes some cost function of the matrix equation $\Lambda \cdot \theta = 0$. A good estimation using Total Least Squares techniques can be found computing the eigenvector corresponding to the smallest eigenvalue of the matrix $\Lambda^y \cdot \Lambda$. The results obtained are shown in the central column of Figure 2.

Because of the measure by spectrophotometer can be executed only by contact with the surface, the transformation matrix found by solving (5) has to be applied for each point collected by the method illustrated in Section 3.3. The combination of two calibrations permits to standardize, at each distance, the RGB-ITR data with the MINOLTA spectrophotometer color measures.

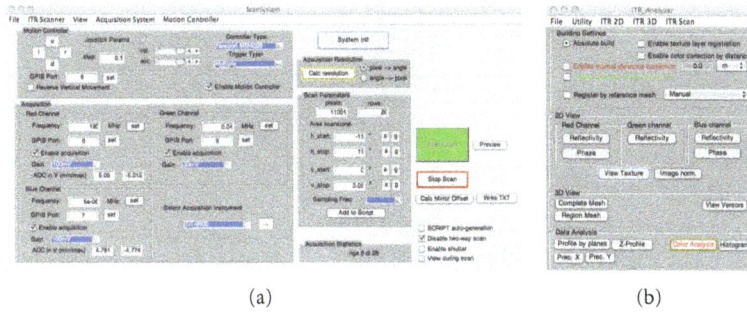

(a)   (b)

FIGURE 4: (a) User interface of ScanSystem software module; (b) user interface of ITRAnalyzer software module.

(a)   (b)

FIGURE 5: (a) RGB-ITR raw data; (b) RGB-ITR data after color calibration.

4. Technology Description

The RGB-ITR is the last offspring of a series of amplitude-modulated (AM), monochromatic 3D laser scanners realized at the ENEA Artificial Vision laboratory (Frascati, Rome), and collectively identified by the acronym ITR (Imaging Topological Radar) [8]. The main new feature improving the last model is the use of three AM laser sources for remote colorimetry.

4.1. ITR's Characteristics and Operating Principle.

The RGB-ITR scanner [5, 9] is essentially composed by two main modules: a so-called passive module, which coincides with the system's optical head, basically including the launching and receiving optics; a so-called active module, which is composed of laser sources, modulators, detectors, and all necessary electronics for collecting and processing data. The two modules are physically separated and optically connected by means of optimized optical fibers. This enables the use of the system in hardly accessible or even hostile environments [10]. Figure 3 shows the entire system assembled as a tower so to guarantee the minimum size and movement facility; the modular configuration permits to adapt the system in the operating environment.

One of the most critical parts in this type of instruments is the movement of the scanning system [11], which a lot of times coincides with the movement of the motorized mirror responsible of the laser beam sweep on the target. The

way that the motors are assembled on the scanning system defines also the uniformity of the resolution and the possible presence of poles in the resulting images on the entire 3D space.

At the moment RGB-ITR scanning system is composed by a mirror mounted on a motor, parallel to the base of the optical head and anchored at a metallic box: the rotational axis is perpendicular at the base of the optical head (horizontal movement). The metallic box can rotate around an axis parallel at the base of the optical head (vertical movement).

Due to construction limits, the actual angle of view of RGB-ITR is $80° \times 310°$, with a point-to-point precision of $0.002°$. This resolution permits to obtain a structured and dense point cloud in the working range of the scanner.

4.2. ITR Software Modules.

All software is assembled and customized in ENEA and it is composed by two main interfaces: the first one, called ScanSystem, controls all the aspects of scanning and data acquisition; the second one, called ITRAnalyzer, is able to build, analyze, and export all ITR data.

ScanSystem permits to change a lot of ITR scanner's parameters, so to guarantee the best solution for the acquisition of the scene under investigation. In Figure 4(a) it is possible to observe the dialog interface: the "Motion Controller" section allows to move manually the scanning mirror so to determine the acquisition range, settable in

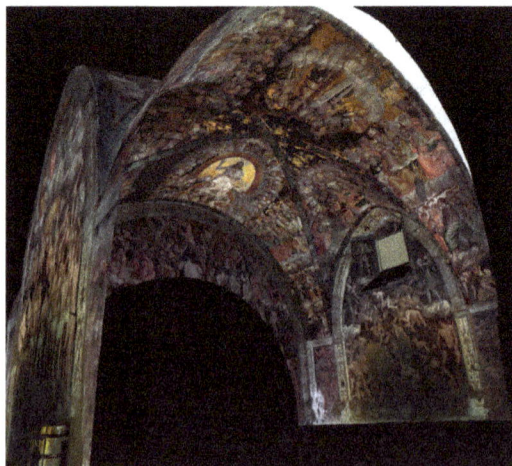

Working range	3–8 m
Spot size	0.5–0.9 mm
Point-by-point distance	0.6 mm
Single-point acquisition time	0.33 ms
Pixel resolution	12000 × 23000
Time for acquisition	1 day
Real site dimension	10 × 6 × 8 m

FIGURE 6: On the left, the entire vault acquired in the S. Peter Martyr church; on the right, a conclusive table of main parameters of the acquisition.

FIGURE 7: A detail of a scanning in progress of "Loggia di Amore e Psiche" in Villa Farnesina (Rome).

"Scan Parameters" module; the "Acquisition" section is useful for setting the modulation frequency, gain filters, and sampling frequency of the three channels; the acquisition starts pressing the button "Start Scan." Another interesting feature of ScanSystem resides in SCRIPT programming that permits to divide the acquisition process in several subscans, so achieving a scanning-parameters customization based on scene complexity. In the last period the color calibration section was added at the software.

ITRAnalyzer manages all data collected by RGB-ITR system, display color, and structure profiles of the investigated target and permits the registration of meshes obtained by different point-of-view scans. This software can be considered the link between RGB-ITR raw data and commercial software for 3D modeling and photo manipulation. In Figure 4(b) it is possible to observe the user interface: the "Building Settings" section allows to set the typology of algorithms to activate before loading the acquired data, like the color calibration

versus distance method; the "2D View" panel allows some previews of each channel information and global colored texture; the "3D View" section allows some previews of entire or partial 3D mesh and versors; the last section, "Data Analysis," is useful for a preliminary on site data analysis of both RGB-ITR data, color, and distance information.

5. Results

In this section some of the results obtained by RGB-ITR system during a few recent field campaigns are presented. Figure 5 shows a scan executed in the church of S. Peter Martyr (Rieti): on the left, raw data obtained directly by the scanner are shown; on the right, color data after the calibration with the methods described above are displayed.

During this campaign the entire vault was acquired. In Figure 6 a screenshot of the entire 3D model and a summary of acquisition main parameters are shown.

Figure 7 shows a detail of a scanning in progress of the ceiling of "Loggia di Amore e Psiche" in Villa Farnesina (Rome). The data shown in this figure are not calibrated by the normalization curves.

6. Business Benefits

The real economic benefit is the sum of several factors, which have an impact in different sectors of cultural heritage environment: the characteristics such as remote diagnosis of both color and structure information, without the use of scaffolds, noninvasiveness, and versatility offer a competitive and cheap solution for a support to restoration and cataloguing. Moreover, the possibility of collecting simultaneous color and distance information decreases drastically time and costs of elaborated data after production and mesh creation, which can also be used for educational purposes.

7. Conclusions

The use of RGB-ITR system for structural and colorimetric remote diagnosis and 3D digitalization was shown. The calibration methods for color acquisition have to be considered a starting point for colorimetric measurements compatible with CIE standards.

In summary the use of three AM laser stimuli and narrow interferential filters permits to concurrently acquire five pieces of information (two distances and three-reflectivity signals back-reflected by target, one for each incoming chromatic laser component) per each sampled point by means of three avalanche photodiodes and dedicated electronics and to perform a discrete color analysis that can be considered a three Delta-function analysis.

The use of supercontinuum lasers [12] is underinvestigation, so to provide a color analysis in the entire visible spectrum.

A new version of ITR system is going to be developed: the main new features are the upgrade of the electronic system, more compact and efficient, and the insertion of new laser sources for infrared and fluorescence analysis.

References

[1] C. Oleari, *Misurare il Colore*, Ulrico Hoepli, 2002.

[2] M. Cordaro, E. Borrelli, and U. Santamaria, "Il problema della misura del colore delle superfici in ICR: dalla colorimetria tristimolo alla spettrofotometria di riflettanza," *Colorimetria e Beni Culturali, Collana Quaderni di Ottica e Fotonica 6*, SIOF, Centro Editoriale Toscano, Firenze, Italy, 2000.

[3] C. Oleari et al., "Colorimetria e Beni Culturali," in *Proceedings of the Firenze 1999 and Venezia 2000*, 1999-2000.

[4] C. L. Hsien, *Introduction to Imaging Color Science*, Foxlink Peripherals, 2005.

[5] M. Ferri De Collibus, L. Bartolini, G. Fornetti et al., "Color (RGB) imaging laser radar," in *International Symposium on Photoelectronic Detection and Imaging 2007: Laser, Ultraviolet, and Terahertz Technology*, vol. 6622, Beijing, China, September 2007.

[6] D. Nitzan, A. E. Brain, and R. O. Duda, "The measurement and use of registered reflectance and range data in scene analysis," *Proceedings of the IEEE*, vol. 65, no. 2, pp. 206–220, 1977.

[7] L. Mullen, A. Laux, B. Concannon, E. P. Zege, I. L. Katsev, and A. S. Prikhach, "Amplitude-modulated laser imager," *Applied Optics*, vol. 43, no. 19, pp. 3874–3892, 2004.

[8] S. Poujouly and B. Journet, "A twofold modulation frequency laser range finder," *Journal of Optics A*, vol. 4, no. 6, pp. S356–S363, 2002.

[9] R. Ricci, L. De Dominicis, M. F. De Collibus et al., "RGB-ITR: an amplitude-modulated 3D colour laser scanner for cultural heritage applications," in *Proceedings of the International Conference LACONA VIII—Lasers in the Conservation of Artworks*, 2009.

[10] L. Bartolini, A. Bordone, A. Coletti et al., "Laser In vessel viewing system for nuclear fusion reactors," in *High-Resolution Wavefront Control: methods, Devices, and Applications II*, vol. 4124, pp. 201–211, San Diego, Calif, USA, August 2000.

[11] G. F. Marshall, *Handbook of Optical and Laser Scanning*, Marcel Dekker, 2004.

[12] S. G. Leon-Saval, T. A. Birks, W. J. Wadsworth, P. S. J. Russell, and M. W. Mason, "Supercontinuum generation in submicron fibre waveguides," *Optics Express*, vol. 12, no. 13, pp. 2864–2869, 2004.

17

Cross-Shaped Terahertz Metal Mesh Filters: Historical Review and Results

Arline M. Melo,[1] Angelo L. Gobbi,[2] Maria H. O. Piazzetta,[2] and Alexandre M. P. A. da Silva[3]

[1] BR Labs Ltd., Rua Lauro Vannucci 1020, 13087-548 Campinas, SP, Brazil
[2] Brazilian Synchrotron Light Laboratory, Rua Giuseppe Maximo Scolfaro, 10000, 13083-970 Campinas, SP, Brazil
[3] Department of Microwave and Optics, University of Campinas, (UNICAMP), Avenue Albert Einstein, 400,
    Cidade Universitária Zeferino Vaz, 13083-970 Campinas, SP, Brazil

Correspondence should be addressed to Arline M. Melo, armame@gmail.com

Academic Editor: Michael Fiddy

Terahertz frequencies experiments has motivated the development of new sources, detectors and optical components. Here we will present a review of THz bandpass filters ranging from 0.4 to 10 THz. We also demonstrate our fabrication process, simulations and experimental results.

1. Introduction and Review

1.1. Basic Concepts. The terahertz, or far infrared range within electromagnetic spectrum placed between microwaves and midinfrared wavelengths, correspond to wavelengths from 3 millimeters up to 3 micrometers. Moreover, the transition between these two spectral bands represents the transition of two different technologies: electronics and photonics (Figure 1). These issues bring us technological challenges to develop new sources, detectors, filters, and other important components [1, 2].

Recently, terahertz range has attracted the interest around the world because its innovative applications including different areas as nondestructive tests, military and civilian security, chemistry, medicine, biology, and others [3–6]. Especially after terrorist attacks on September 11th, terahertz research has increased abruptly because all possible defense applications.

THz radiation can be transmitted through different materials, which became possible to "see" through clothes, shoes, bags, plastics, and paper envelops, allowing the identification of chemical and biological agents like illicit drugs and explosives [3, 5].

1.1.1. Terahertz Metal Mesh Filters. A metal mesh filter, which is a type of frequency selective surfaces (FSS), can be defined by a thin metal film (few to tenths of microns of thickness) which is perforated, using different geometries, in a two-dimensional array. These filters are compact and present an easy and available fabrication process. They also could act as high pass, low pass, band-pass, or reject band filters and this is the most important characteristics of them in other words, their optical behavior can be changed selecting the proper geometry and their parameters dimensions [7].

There are some examples of FSS geometries in the literature presenting different spectral response. In Figure 2, it shows few examples of band pass, high pass, and reject band filters.

The most known band pass geometry is the cross-shaped (see Figure 3) where the frequency profile is determined by the dimensions of the crosses width (K), length (L), and periodicity (G) (Figure 3).

The main fabrication process used to produce metal mesh filters is the photolithography, where a substrate is prepared to receive a photoresist (polymer sensitive to light—UV is the most common radiation used). After the deposition of this polymer, the sample has to be baked and then exposed to a light source using a mask with the geometric patterns. The light will pass through the transparent areas on the mask substrate and interact with the photoresist. In this step, the sample will receive other bake phase followed by the development process of photoresist. Finally the sample

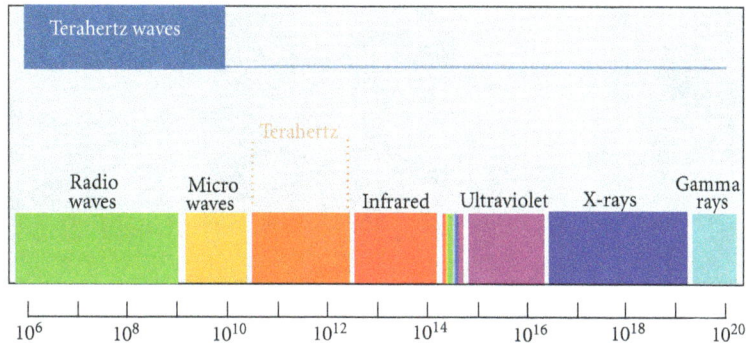

FIGURE 1: Electromagnetic spectrum showing the terahertz range between microwaves and infrared ranges.

is ready to the metal deposition, where the thin film will be evaporated or growed in a metallic bake.

Although conventional photolithography is the most common fabrication process, we can cite few other techniques as direct write using electron beam (Direct Electron Beam Litography, DEBL), nanoimprint and laser ablation [8–10]. Figure 4 shows the results for cross-shaped filters obtained using different fabrication process.

1.1.2. Metal Mesh Filters Applications. The metal mesh filters can be applied in different areas and here we will cite some applications examples cited in the literature:

(i) Astronomy. Infrared astronomy can be possible using satellites and airborne telescopes operating within the whole infrared range and at high-altitude sites, where the observations are affected primary by water vapor atmospheric absorption. There are some atmospheric windows (25 and 38 microns at sites higher than 5000 m) where it is still possible to observe extraterrestrial sources with lower losses. Long midinfrared ranged from 25 up to 40 microns, allows observing reddened sources and different molecular/atomic lines and dusting features [11].

(ii) Free Electron Lasers. Different quasioptical components, including metal mesh filters, are applied as passive selective components (filters) on experiments using terahertz-free electron lasers as the NovoFel at Novosibrisk [12]. These devices should be able to operate over a long period of time under high-power conditions without important degradation of their properties [12].

(iii) Imaging. Other potential applications are related to infrared imaging and spectroscopy, which are related also to the devolpement of high power terahertz sources and high sensitivity detectors within these wavelength range [13].

(iv) Energy-Saving Glasses. In other dimension scale (millimeters), cross-dipole FSS can be used to generate energy-saving glasses, which can be applied on buildings to keep them cooler during summer and warmer during winter. It also allows the reception of useful microwave/RF signals required for mobile phone, GPS, and personal communica-

tion systems. This application is possible using mm-wave metal band-pass filters, which block all infrared wavelengths and transmit the desired mm-waves frequencies [14].

(v) Sensors. Bolometer sensors and cameras are very useful to detect infrared radiation and their performance can be improved by the incorporation of FSS elements with no requirement for external filters to define the absorbing band [15].

1.2. Historical Review and Results. Ulrich [16] is one of the most cited articles about metal mesh filters. In this work, a metal mesh interference filter is used to replace the output coupler of a far infrared molecular laser. This interference filter consists of two parallel reflector grids with a large array of small coupling holes where the reflectance is varied by adjusting the spacing between the grids. Ulrich also was one of the responsibles for showing that the transmission properties of these metallic meshes could be considered as circuit elements on a free space transmission line. Using this theory, it is possible to determine the transmission profile of inductive (square openings) and capacitive (free-standing squares) meshes, high pass and low pass, respectively.

After these first experiments and theoretical approaches, one can find in the literature many other authors who fabricated and tested band pass metal mesh filters at different sub-THz and THz frequencies. It is shown in Table 1 different experimental results found in the literature, which are just a few of the total works about band pass metal mesh filters published until now. In this table, it is also presented the grid parameters of several filters designed to operate within the range from 100 GHz up to 14 THz, exhibiting high-transmission peaks and band pass widths between 13 up to 50% of the central frequency.

The freestanding filters present a high mechanical fragility. Then, some authors have used photolithography process over some support material, p. e. Mylar (polyester film), polyamide, teflon (polytetrafluoroethylene), and others, which are semitransparent within those frequencies range.

Recently, other interesting works have been published using different materials as substrate, such as Polymethylpentene (TPX), TydexBlack, and high-density polyethylene (HDPE). The use of these materials could become a very

FIGURE 2: Examples of filters presented in literature: (a) reject band peak at 6.9 and 20.3 THz; (b) reject band peak at 3.8 and 7.0 THz; (c) reject band peak at 2.7 and 5.9 THz [20]; (d) band pass filter at 1.7 THz; (e) high-pass filter at 1.1 THz [21]; (f) blocking of multiple frequency bands [22]; (g) band pass at 7 microns [23]; (h) band stop sample [24]; (i) band pass sample [12].

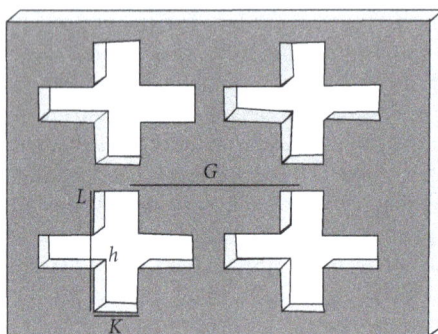

FIGURE 3: Dipole cross-shaped band pass filter parameters.

important way to fabricate useful filters because they can also act as a high-pass filter for wavelengths above near infrared (NIR: 1–20 microns). In this case, it would be very helpful to applications where the background radiation in these wavelength range (visible and NIR) is prohibitive.

The transmission profiles for TydexBlack, TPX, and HDPE are shown in Figure 5.

In Kaufmann et al. [17], different filters at 2.4 THz have been fabricated using THz semitransparent materials as substrate. In this work, it is compared the transmission profile obtained by a suspended metal mesh, a filter on a TydexBlack base, and by a metal mesh sandwiched between TPX layers. The results have shown a high transmittance for the suspended filters, as expected. The measurements resulted in a 40% and 48% transmission for the filter with TydexBlack and the TPX sandwiches, respectively. They also concluded, in the same work, that the transmission is independent from the polarization position angle and the dependence on the incidence angle becomes more important for angles larger than 20 degrees.

Ma et al. [13] also developed similar filters using low-loss substrate material, high-density polyethylene (HDPE). Filters for three different frequencies were fabricated, 1.5, 1.75 and 2.91 THz, using photolithography process above a 1 mm thick HDPE plate. The measured transmission resulted

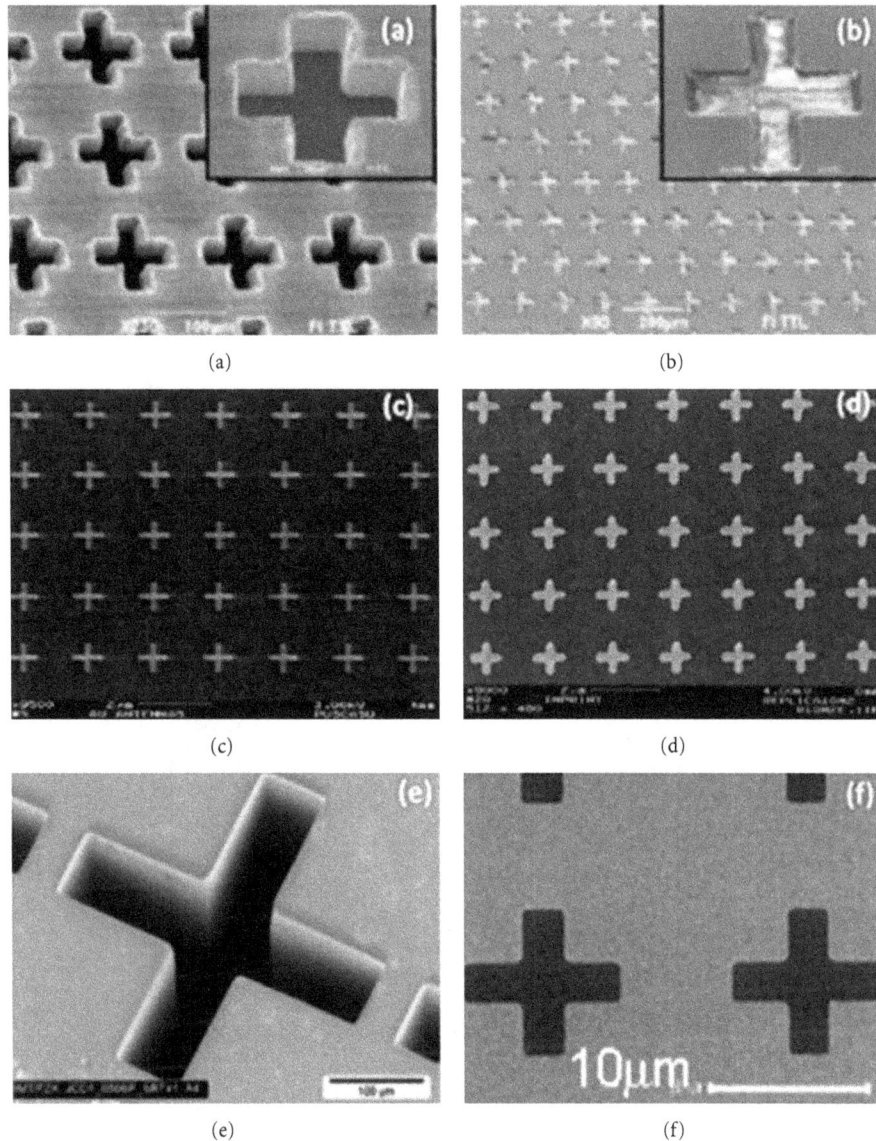

FIGURE 4: Electron microscopy images illustrating the structural results for different fabrication process: (a) laser ablation-free standing [8]; (b) laser ablation-on a support material standing [8]; (c) DEBL [9]; (d) Nanoimprint [9]; (e) X-ray photolithography [10]; (f) UV photolithography [11].

in 51%, 54%, and 52%, for the 1.5, 1.75, and 2.91 THz, respectively.

2. Our Contribution

The second part of this chapter intends to present our contribution to this topic. We will review all important job achievements published in 2008 [18] and continue with more recent developments of our group.

2.1. Project of Different Band Pass Filters between 0.5 and 10 THz. The initial motivation to design band pass filters to these specific frequency range came from an astrophysical application, Solar Physics, and especially solar flares. There are two different ways to observe the Sun, from the ground and using balloons or satellites outside our atmosphere.

Observations through our atmosphere are limited by the water vapor column presented in the observation site. There are, within the terahertz range, three atmospheric windows (which are known as frequency regions within the electromagnetic spectrum presenting absorption coefficient minimums) centered at 405, 670, and 850 GHz. The other higher frequencies filters were designed considering outside-atmosphere observations [19].

All filter parameters designed here are based on Porterfield work. Considering linear variation of the mesh parameters (K, L, and G) we calculate a conversion factor for each parameter which will be used to design other frequencies filters as below, in Table 2.

Using the conversion factors shown in Table 2, we are able to calculate new parameters mesh in order to have

(a)

(b)

(c)

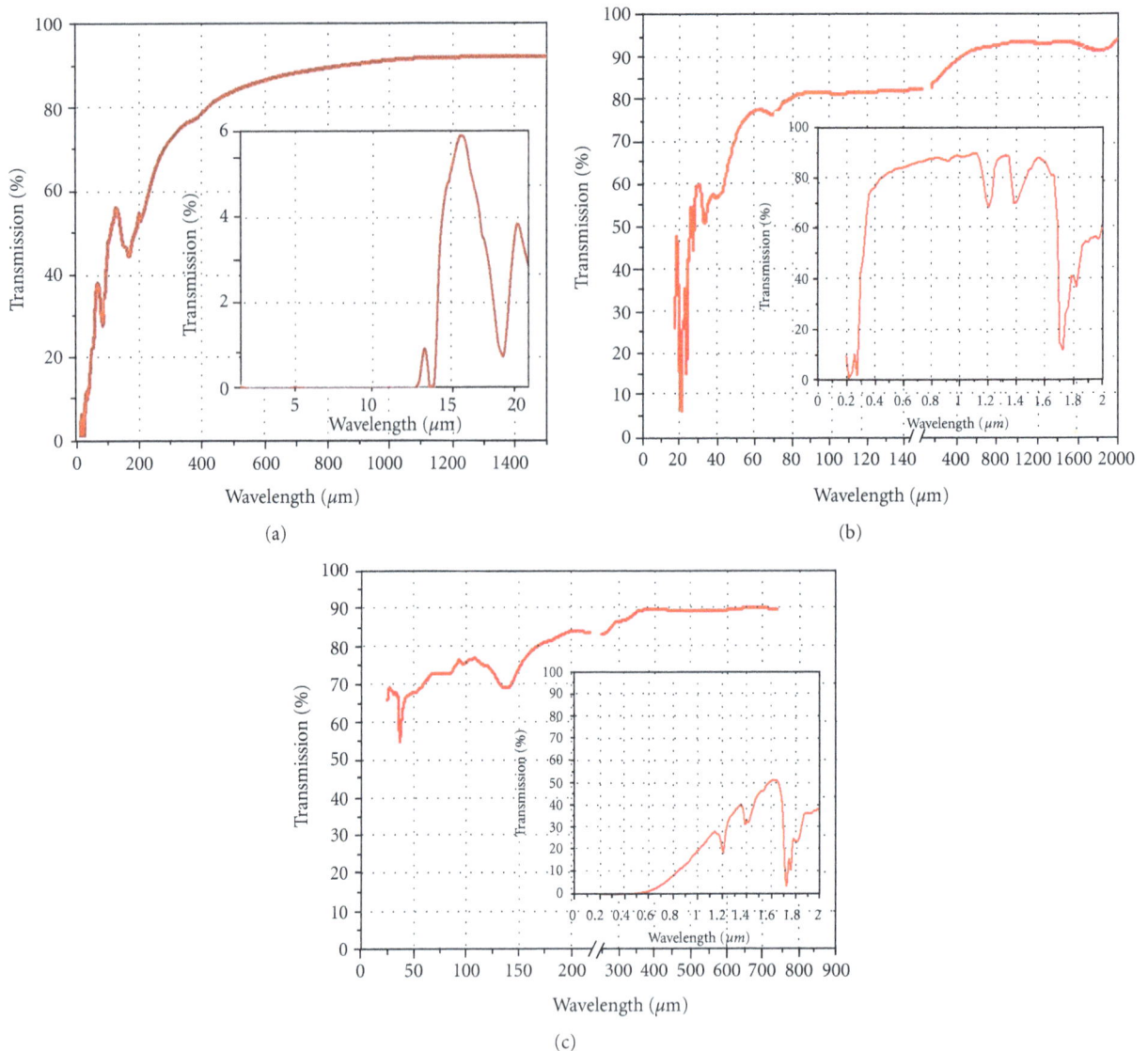

FIGURE 5: Transmission profile for: (a) TydexBlack, (b) TPX, and (c) HDPE by Tydex company measurements.

a new band pass filter just multiplying the factor by the desired central wavelength in microns. In Table 3, it is shown our calculated parameters dimensions for filters operating from 0.4 up to 10 THz. As Porterfield filters presented a band pass width about 15% of the central frequency, all filters design using this method will generate filters with bandwidths around the same value. As you can see in Table 3, we also designed filters to present double bandwidths, 30% of central frequency, and in this case we used simulation methods.

Using a commercial electromagnetic simulator code, CST microwave studio, we could calculate in a good approximation the mesh parameters for filters with larger bandwidths. In this case, we have calculated the initial mesh parameters values for narrow bandwidth (15%) considering the conversion factors presented here. In the optimization process, it was determined, as the software goals, to achieve at least the desired new bandwidth and that the resonant

frequency remained the desired central frequency. At the end of optimization process, CST has returned the new parameters dimensions. We have fabricated and measured these larger bandwidth filters, proving that this can be an easy and effective technique to design filters presenting new bandwidth values (Figure 6).

2.2. Fabrication Process: Photolithography and Metallic Deposition. Filters such as the presented ones can be fabricated using different techniques. One micromachining technique uses polymer film as substrate with a metallic thin layer deposited on one side (e.g., Mylar). However, plastic films add absorption and have thermal and mechanical limitations for certain applications.

Another micromachining procedure, known as LIGA, explores photolithography followed by electroplating techniques, producing a metal grown film with open space cross formats, without absorption, which improves the final filter

TABLE 1: Band pass metal mesh filters results from different authors.

Author	Frequency (THz)	G (microns)	K (microns)	J (microns)	Transmission (%)	Bandpass width (%)	Fabrication process
Page [25]	0.138	972	892	28	91	35	Mylar photolithography
Page [25]	0.174	842	762	35.5	85	41	Mylar photolithography
Page [25]	0.282	504	444	22.5	92	46	Mylar photolithography
Page [25]	0.384	438	288	7.5	72	17	Mylar photolithography
Page [25]	0.480	376	226	11.5	66	16	Mylar photolithography
Page [25]	0.510	286	226	15	78	33	Mylar photolithography
Porterfield [7]	0.587	402	251	66	97	17.3	Free UV photolithography
Porterfield [7]	1.195	201	126	33	100	16.9	Free UV photolithography
Ma [13]	1.5	106.3	81.4	22.6	51	40	HDPE photolithography
Porterfield [7]	1.523	154	98	28	97	18.6	Free UV photolithography
Ma [13]	1.77	76.5	66.4	12.6	54	51	HDPE photolithography
Porterfield [7]	2.084	113	71	19	100	15	Free UV photolithography
Ma [13]	2.91	60	38	10	52	13.7	HDPE photolithography
Kuznetsov [12]	3.75	41.4	29.4	6	59	30	Polyimide LIGA*
Kuznetsov [12]	6.5	28.5	18.5	5	64	30	Polyimide LIGA*
Smith [26]	7.9	22	12.8	3.5	52	15	Polyimide Optical photolithography
Möller [27]	8.3	26.4	23.2	2.4	50	14	Free UV photolithography
Möller [27]	13	16.7	14.7	1.5	60	17	Free UV photolithography
Möller [27]	14	16.4	13.9	2.5	80	23	Free UV photolithography

*LIGA is a german acronym for Lithography, Electroplating and Molding.

TABLE 2: Mesh parameter conversion factors.

Porterfield et al. [7] $K = 261$, $L = 76$, and $K = 402$ micrometers; $f = 0.585$ THz ($\lambda = 512.8$ micrometers)		
K parameter conversion	L parameter conversion	G parameter conversion
$Ck = 261/512.8 = \mathbf{0.50897}$	$Cl = 76/512.8 = \mathbf{0.1482}$	$Cg = 402/512.8 = \mathbf{0.78393}$

transmission. We used the second technique to fabricate filters with frequencies centered between 0.4 and 10 THz.

We used silicon substrates previously prepared with deposited films of SiO_2, Ti, and Au. The lithography has used SU-8, deposited at 1000 rpm during 30 seconds, for filters frequencies below 2 THz, and at 1500 rpm during 30 seconds, for the higher frequency filters.

The obtained samples were submitted to a baking process, then exposed to UV radiation during 30–40 seconds and finally developed. This process sequence prepares the samples for the deposition of the metallic material, nickel in our case, which is finally the material of the filter, Figure 7.

The electroplating process used a WATTS bath of nickel sulfate, nickel chlorine, boric acid, and water, with current density of 3 A/dcm². The last step is the etching, to remove the photoresist and the layers of SiO_2, Ti, and Au using potassium cyanide and HF buffer.

2.3. Microscopic Characterization. After all fabrication process, the next step is an initial characterization of the filters. We usually submit all samples to an optical inspection using a 50x microscope, where we could evaluate defects which affect a region of the filter. As the second step, we are interested to measure the grid parameters to compare with the designed dimensions and evaluate the crosses profiles imperfections. Figure 8 shows good pattern results obtained for 0.67, 1.5, 3, and 4 THz filters.

During the fabrication process, different imperfections could appear in your sample. In Figure 9, we are illustrating a few of them.

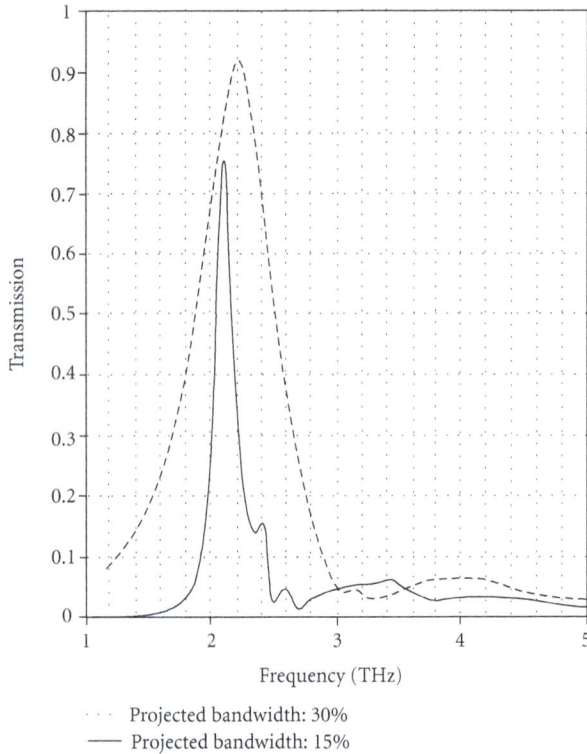

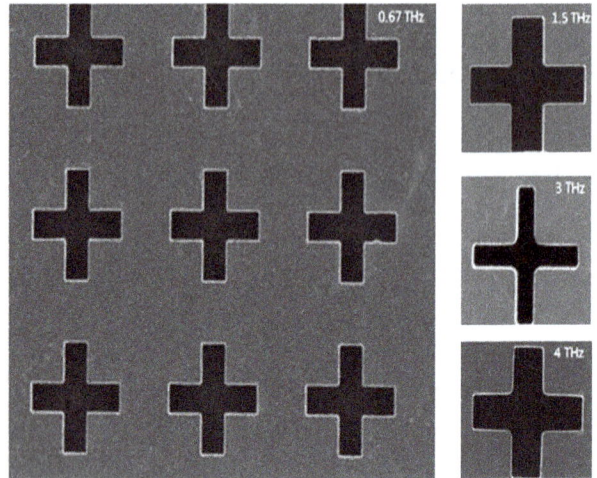

FIGURE 8: SEM images for different band pass filters frequencies.

FIGURE 6: Comparison between measured transmission profiles for the designed band pass filter centered at 2 THz with 30% and 15% bandwidths.

····· Projected bandwidth: 30%
—— Projected bandwidth: 15%

TABLE 3: Mesh parameter designed to band pass filters operating between 0.5 and 10 THz.

Frequency	k (μm)	j (μm)	g (μm)	Expected bandwidth
405 GHz	370	110	590	15%
670 GHz	225	60	350	15%
850 GHz	175	50	302,5	15%
1 THz	153	44.5	235	15%
1.5 THz	102	30	157	15%
2 THz	76	22	120	15%
2 THz	81.4	22.6	106.3	30%
3 THz	66.4	12.6	76.5	30%
4 THz	38	10	60	15%
6 THz	25	7.4	39	15%
8 THz	19	5.6	29.4	15%
10 THz	3.2	15.4	22.5	15%

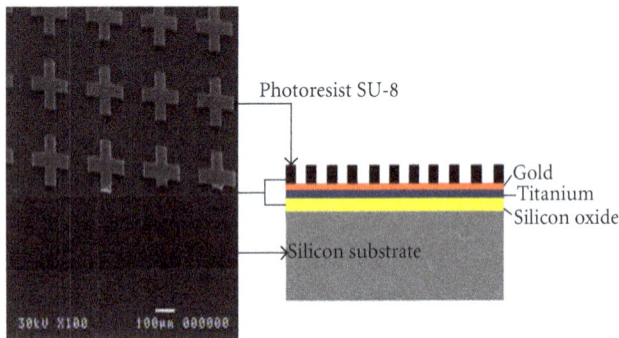

FIGURE 7: SEM micrograph of a sample after the photolithography process. In this image the thick SU8 crosses prepare the sample to the metallic deposition step.

As the filter frequency increase, the metal thickness has to decrease to avoid a waveguide process. For example, considering the 10 THz filter (corresponding to 30 micrometer wavelength), the filter thickness needs to be less than 10 microns. We adopt free standing filters to avoid the high absorption of any support material and in this case the fabrication process of a very thin metallic film is very sensitive to the mechanical stresses. Figure 9(a) depicts the crack of a 7-micron thick film.

Figures 9(b) and 9(d) shows rounded cross shaped profile examples. In these cases, imperfections are related with the photolithography process, especially with the UV exposition time of the SU8 resist and its development process. The resist time exposition has to be very precise and be tested along the process because, if we use less radiation doses the resist corner details could not be sensitized and during the development process the SU8 material will be removed.

Figure 9(c) illustrats a growth metallic film presenting a difference in rounded corner. It could happen if the metallic film is growth until the top of the SU8 crosses. It is explained because the facts that during the development phase the first resist layer started to be removed and the resist profile became rounded at the top. There is a simple empirical rule where the metal film is growing using just 80% of the thickness of the resist crosses.

2.4. THz Transmission Measurements. Figure 10 presents the transmission profiles for seven filter samples measured using a Fourier transform infrared spectroscopy (FTIR). It is shown in Table 4 the measured central frequencies, bandwidths, and transmission percentages for the filters in

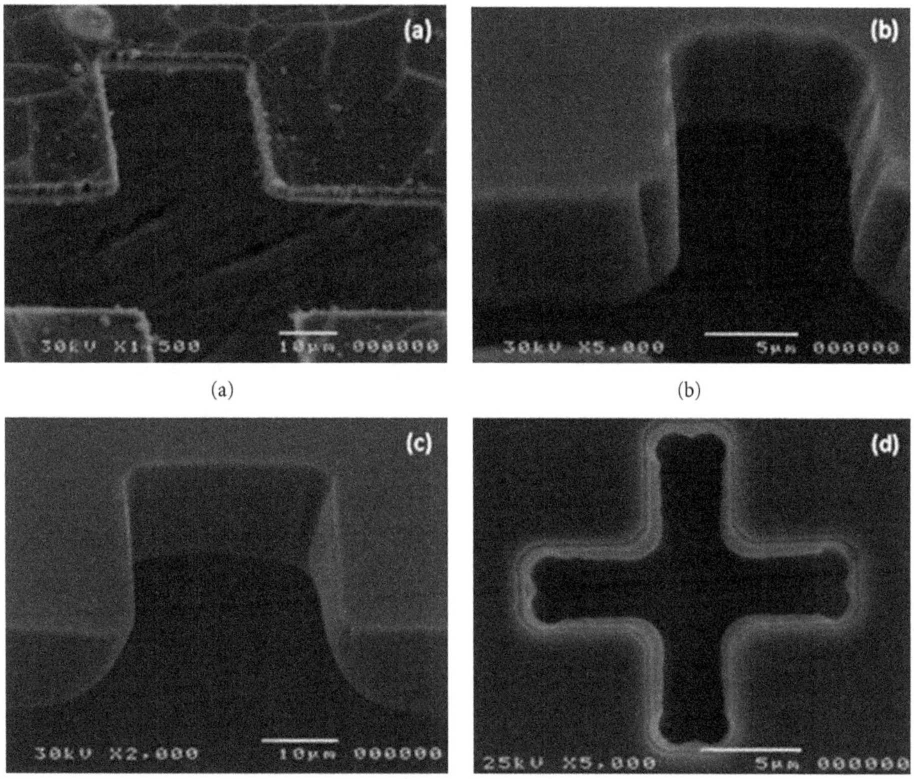

(a)

(b)

(c)

(d)

FIGURE 9: SEM images illustrating different defects by the fabrication process.

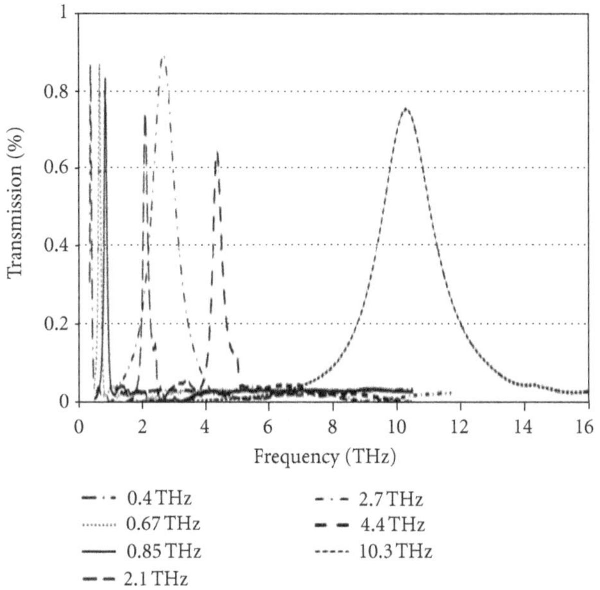

FIGURE 10: Transmission profile for filters designed between 0.4 and 10 THz measured using the Fourier Transform Infrared Spectroscopy (FTIR) technique.

Figure 10. We can see that the results presented are in accordance with the designed central frequency and bandwidth as shown in Table 3.

TABLE 4: Measured central frequency, bandwidth, and transmission for filters presented in Figure 10.

Frequency	Measured frequency	Bandwidth (% Fc)	Transmission (%)
405 GHz	395	17.6	88
670 GHz	674	14.2	87
850 GHz	858	14	84
2 THz	2.1	10.7	75
3 THz	2.7	28.5	90
4 THz	4.4	8.8	67
10 THz	10.3	13.9	76

The Figure 11 presents the transmission profile comparison between FTIR measurements and our simulated data obtained using the CST software. The data show a good agreement between simulated and experimental results, validating the simulation technique to support new filters design.

2.5. *Simulations.* Here we are presenting, using only simulations results, the effect of the individual variation of each mesh parameter on the expected resonant frequency and bandwidth. It will give us an insight about what can happen due to small errors caused by fabrication process. In order to compare the designed filters presented in Table 3 on the same graph, each filter has its frequency range normalized

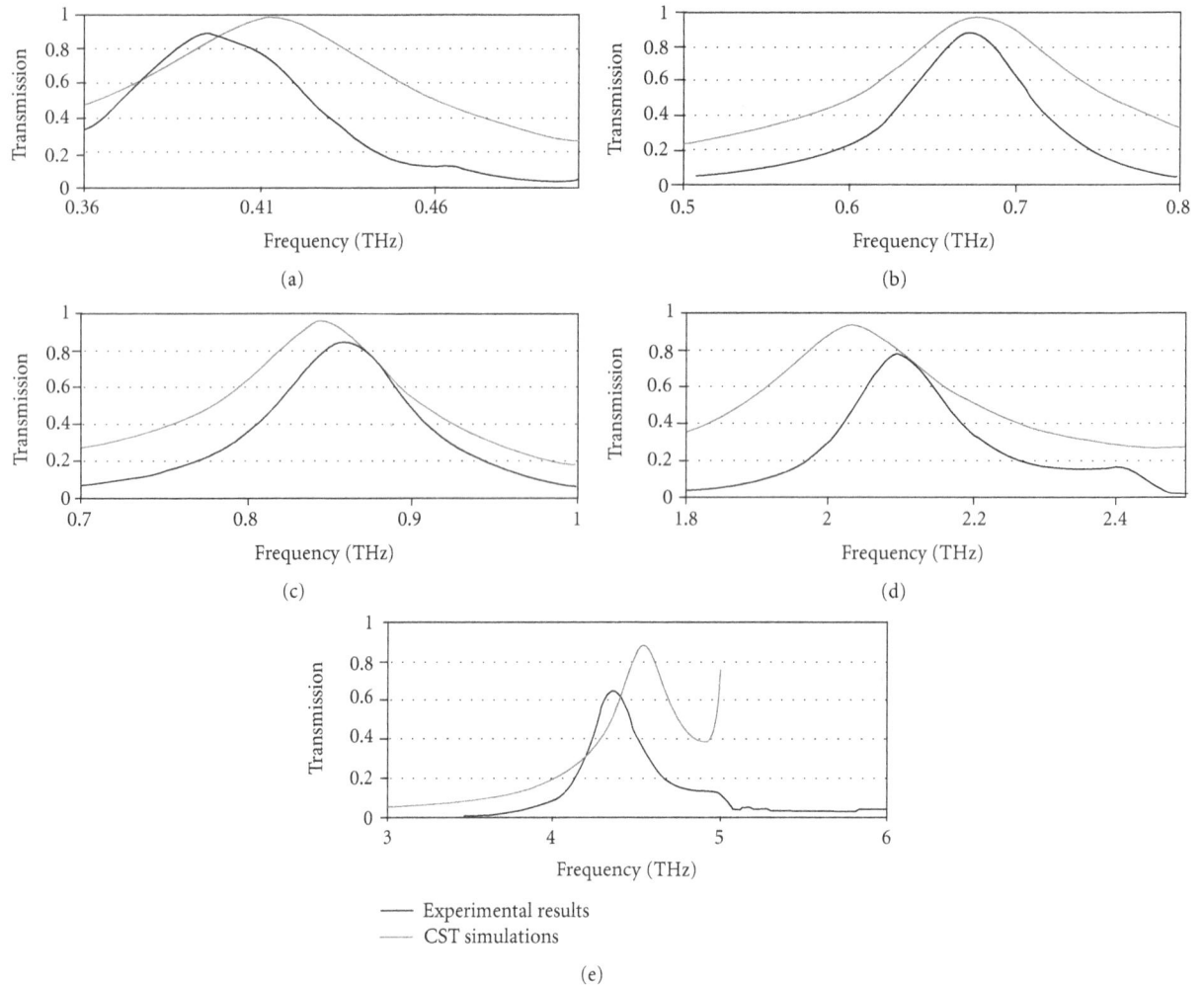

FIGURE 11: Comparison between the measured transmission profile and the simulated data from CST software for filters at 0.4, 0.67, 0.85, 2, and 4 THz.

by its own peak frequency. Following the same method, the range of variation for all parameters is also normalized by the designed values.

In Figure 12(a), it is shown that the variation on the periodicity, parameter G, can linearly shift the peak frequency up or down. In a general way, if parameter G increases, the peak frequency decreases and vice-versa. However, for a variation of up to 20%, the peak frequency suffers a shift no bigger than 10%. In Figure 12(b), one can note that bandwidth is affected in a nonlinear way, increasing by the decrease of parameter G or decreasing by the increase of parameter G.

A different behavior is observed on the variation of parameter K. In Figure 13(a), it is shown that increasing K, the peak frequency is also increased almost linearly. However, for a variation of up to 20%, the peak frequency shifts no more than 2.5%. The same almost linear effect is seen on bandwidth when parameter K is increased.

Parameter L presents the strongest influence on the peak frequency position, as can be seen in Figure 14 (a). For a variation of up to 20%, the central frequency is shifted more

than 10%. Similarly to parameter G, if L is increased, the peak frequency decreases and vice-versa. However, parameters G and L have opposite behaviors regarding bandwidth, when parameter L is increased, bandwidth is also increased but almost linearly.

Another common pattern profile issue caused by fabrication process is rounded corners. In this case, the crosses arms corners are not sharp as designed. Here, this error will be taken into account by the variation in the radius R at the filter's corners as depicted in Figure 15.

In order to compare filters, it is defined as the corner error, the relation of R and parameter L. The maximum corner error is 50%, where $R = L/2$. It is shown in Figure 16 that increasing R, the peak frequency is shifted to higher frequencies up to 10% of the central frequency. The result regarding bandwidth is not conclusive, presenting almost no measurable variation for higher frequencies.

3. Conclusion

3.1. THz Filters Companies. Currently, we find some companies which can fabricate and sell different terahertz metal

TABLE 5: Terahertz metal mesh filters companies.

Company	Filter type	website
BR-labs	Bandpass filter between 0.4 to 10 THz or custom made.	http://www.br-labs.com/
QMC instruments	Low pass filter with edges between 0.3 to 20 THz.	http://www.terahertz.co.uk/
Virginia diodes	Bandpass filter between 0.35 to 5 THz or custom made.	http://www.vadiodes.com/
Mutsumi corporation	Bandpass filter between 0.2 to 2 THz.	http://www.science-mall.co.jp/en

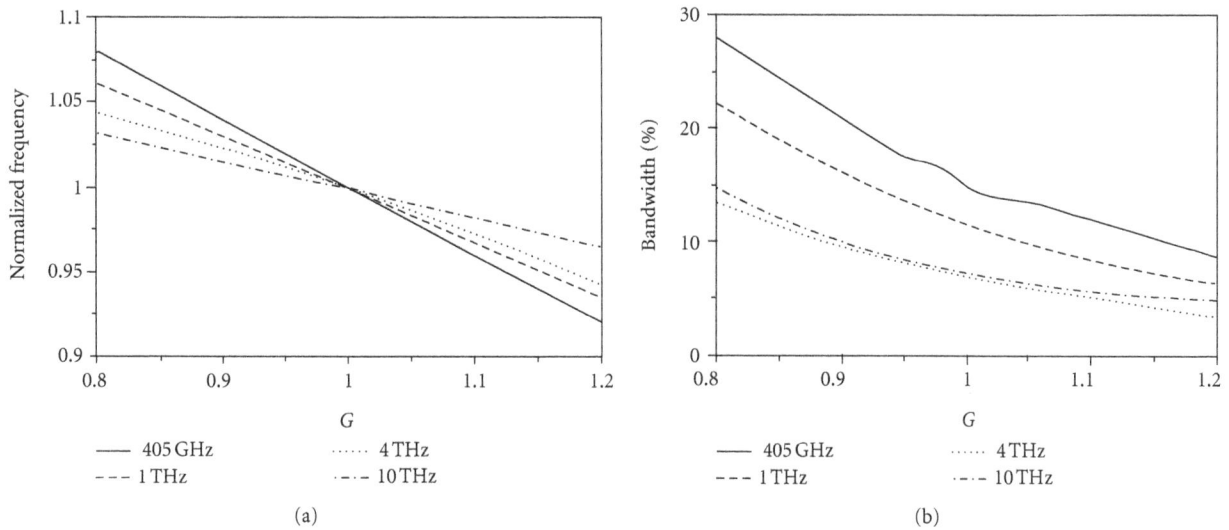

FIGURE 12: Effect of parameter G variation on (a) resonant frequency, (b) bandwidth.

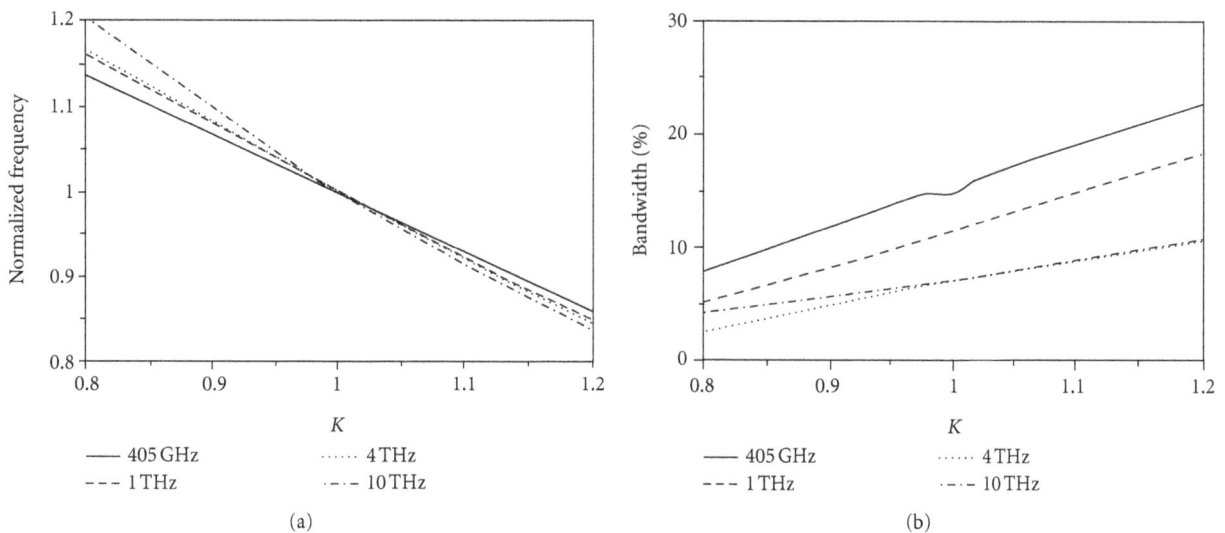

FIGURE 13: Effect of parameter K variation on (a) resonant frequency, (b) bandwidth.

mesh filters (high-pass, low-pass, and band pass transmission). The available frequencies for band pass filters are between 200 GHz up to 10 THz, and the prices are around USD 2.000 up to USD 10.000, varying if the item is in stock or will be custom made. In Table 5 it is presented four examples of companies.

3.2. Conclusions. In this chapter we presented a brief introduction to important concepts and definitions for terahertz and metal mesh filters areas, including the most common applications to both topics. A compact historical review of bandpass filters works is also presented. In the second part of this work we approach our own contributions on

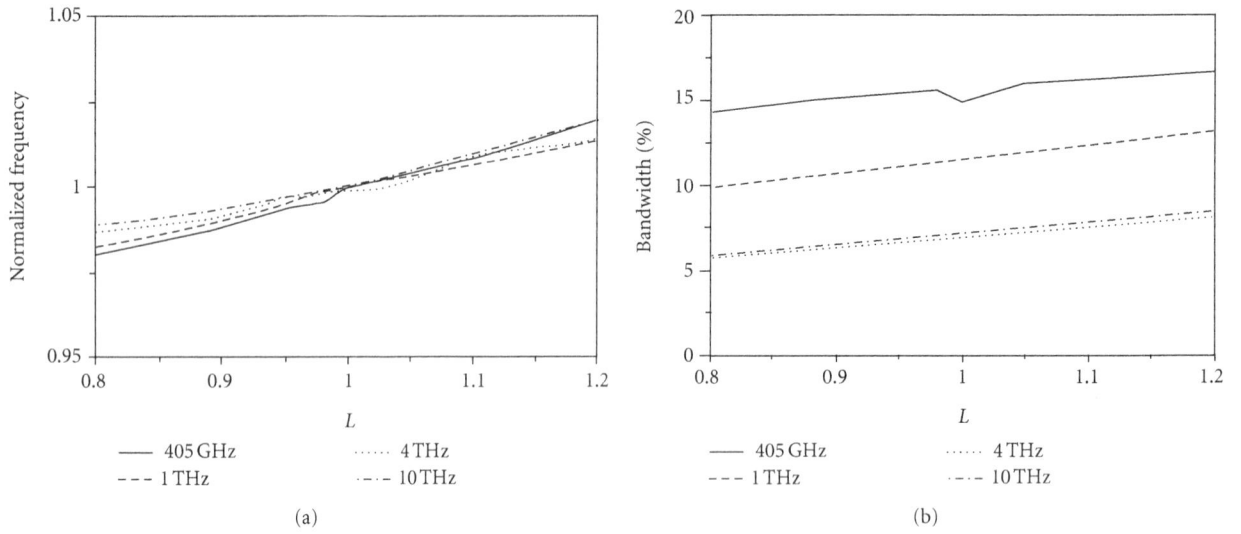

FIGURE 14: Effect of parameter L variation on (a) resonant frequency, (b) bandwidth.

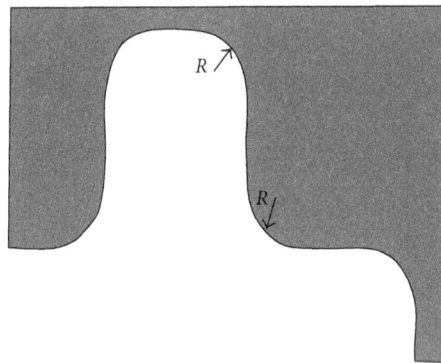

FIGURE 15: Partial view of a filter showing the rounded defect defined by a radius R.

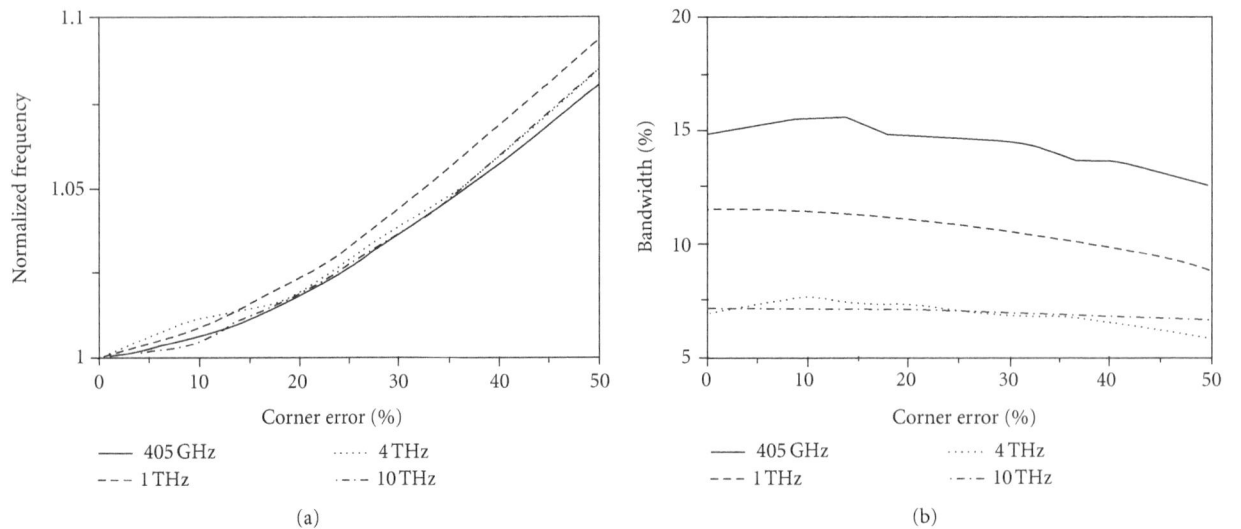

FIGURE 16: Effect of corner radius variation on (a) resonant frequency, (b) bandwidth.

terahertz filters, their design and fabrication process, besides of experimental results. A simple but powerful mathematical technique for designing cross filters is presented. Some useful insights can be taken from the presented parameter variation study in order to understand the mesh parameters and possible process errors influence on the filters frequency response. We conclude featuring some new companies which could provide standard and custom made filters.

Conflict of Interests

All trademarks mentioned in the paper were only cited for technical purposes and do not imply recommendation of the authors. We also indicated no potential conflict of interests.

References

[1] A. Lisauskas, T. Löffer, and G. Hartmut, "Photonics terahertz technology," *Semiconductor Science and Technology*, vol. 20, no. 7, 2005.

[2] G. P. Williams, "Filling the THz gap—high power sources and applications," *Reports on Progress in Physics*, vol. 69, no. 2, pp. 301–326, 2006.

[3] R. Appleby and H. B. Wallace, "Standoff detection of weapons and contraband in the 100 GHz to 1 THz region," *IEEE Transactions on Antennas and Propagation*, vol. 55, no. 11 I, pp. 2944–2956, 2007.

[4] H. B. Liu, H. Zhong, N. Karpowicz, Y. Chen, and X. C. Zhang, "Terahertz spectroscopy and imaging for defense and security applications," *Proceedings of the IEEE*, vol. 95, no. 8, pp. 1514–1527, 2007.

[5] H. Zhong, A. Redo, Y. Chen, and X. C. Zhang, "THz wave standoff detection of explosive materials," in *Terahertz for Military and Security Applications IV*, vol. 6212 of *Proceedings of SPIE*, Orlando, Fla, USA, April 2006.

[6] M. S. Sherwin, C. A. Schmuttenmaer, and P. H. Bucksbaum, *DOE-NSF-NIH Workshop on Opportunities in THz Science*, Arlington, Va, USA, 2004.

[7] D. W. Porterfield, J. L. Hesler, R. Densing, E. R. Mueller, T. W. Crowe, and R. M. Weikle II, "Resonant metal-mesh bandpass filters for the far infrared," *Applied Optics*, vol. 33, no. 25, pp. 6046–6052, 1994.

[8] B. Voisiat, A. Biciunas, I. Kasalynas, and G. Raciukaitis, "Bandpass filters for THz spectral range fabricated by laser ablation," *Applied Physics A*, vol. 104, no. 3, pp. 953–958, 2011.

[9] I. Puscasu, G. Boreman, R. C. Tiberio, D. Spencer, and R. R. Krchnavek, "Comparison of infrared frequency selective surfaces fabricated by direct-write electron-beam and bilayer nanoimprint lithographies," *Journal of Vacuum Science and Technology B*, vol. 18, no. 6, pp. 3578–3581, 2000.

[10] V. Nazmov, E. Reznikova, Y. L. Mathis et al., "Bandpass filters made by LIGA for the THZ region: manufacturing and testing," *Nuclear Instruments and Methods in Physics Research A*, vol. 603, no. 1-2, pp. 150–152, 2009.

[11] S. Sako, T. Miyata, T. Nakamura, T. Onaka, Y. Ikeda, and H. Kataza, "Developing metal mesh filters for mid-infrared astronomy of 25 to 40 micron," in *Advanced Optical and Mechanical Technologies in Telescopes and Instrumentation*, vol. 7018 of *Proceedings of SPIE*, Marseille, France, June 2008.

[12] S. A. Kuznetsov, V. V. Kubarev, P. V. Kalinin et al., "Development of metal mesh based quasi-optical selective components

[13] Y. Ma, A. Khalid, T. D. Drysdale, and D. R. S. Cumming, "Direct fabrication of terahertz optical devices on low-absorption polymer substrates," *Optics Letters*, vol. 34, no. 10, pp. 1555–1557, 2009.

[14] G. I. Kiani, K. P. Esselle, L. G. Olsson, A. Karlsson, and M. Nilsson, "Cross-dipole bandpass frequency selective surface for energy-saving glass used in buildings," *IEEE Transactions on Antennas and Propagation*, vol. 59, no. 2, pp. 520–525, 2011.

[15] D. W. Logan, "A frequency selective bolometer camera for measuring millimeter spectral energy distributions," Access Open Dissertations, paper 70., 2009.

[16] R. Ulrich, "Far-infrared properties of metallic mesh and its complementary structure," *Infrared Physics*, vol. 7, no. 1, pp. 37–55, 1967.

[17] P. Kaufmann, R. Marcon, A. Marun et al., "Selective spectral detection of continuum terahertz radiation," in *Millimeter, Submillimeter, and Far-Infrared Detectors and Instrumentation for Astronomy V*, vol. 7741 of *Proceedings of SPIE*, San Diego, Calif, USA, July 2010.

[18] A. M. Melo, M. A. Kornberg, P. Kaufmann et al., "Metal mesh resonant filters for terahertz frequencies," *Applied Optics*, vol. 47, no. 32, pp. 6064–6069, 2008.

[19] P. Kaufmann, "Emissões da atividade solar do submilimétrico ao infravermelho (SIRA)," FAPESP Project, 2009–2013.

[20] J. A. Bossard, L. Li, J. A. Smith et al., "Terahertz applications of frequency selective surfaces: analysis, design, fabrication and testing," in *Proceedings of the IEEE Antennas and Propagation Society International Symposium with USNC/URSI National Radio Science and AMEREM Meetings*, 2006.

[21] C. Winnewiser, F. Lewer, J. Weinzier, and H. Helm, "Transmission features of frequency selective surface compoments in the far infrared determined by terahertz time domain spectroscopy," *Applied Optics*, vol. 38, no. 18, 1999.

[22] D. H. Kim and J. I. Choi, "Design of a multiband frequency selective surface," *ETRI Journal*, vol. 28, no. 4, pp. 506–508, 2006.

[23] K. E. Paul, C. Zhu, J. C. Love, and G. M. Whitesides, "Fabrication of mid-infrared frequency-selective surfaces by soft lithography," *Applied Optics*, vol. 40, no. 25, pp. 4557–4561, 2001.

[24] G. D. Boreman, *Infrared Antennas & Frequency Selective Surfaces*, CREOL, The College of Optics & Photonics, 2011.

[25] L. A. Page, E. S. Cheng, B. Golubovic, J. Gundersen, and S. S. Meyer, "Millimeter-submillimeter wavelength filter system," *Applied Optics*, vol. 33, no. 1, pp. 11–23, 1994.

[26] H. A. Smith, M. Rebbert, and O. Sternberg, "Designer infrared filters using stacked metal lattices," *Applied Physics Letters*, vol. 82, no. 21, pp. 3605–3607, 2003.

[27] K. D. Möller, J. B. Warren, J. B. Heaney, and C. Kotecki, "Cross-shaped bandpass filters for the near- and mid-infrared wavelength regions," *Applied Optics*, vol. 35, no. 31, pp. 6210–6215, 1996.

and their application in high-power experiments at Novosibirsk terahertz FEL," in *Proceedings of the FEL*, 2007.

18

Detection of Nitroaromatic and Peroxide Explosives in Air Using Infrared Spectroscopy: QCL and FTIR

Leonardo C. Pacheco-Londoño, John R. Castro-Suarez, and Samuel P. Hernández-Rivera

Department of Chemistry, ALERT-DHS Center of Excellence, Center for Chemical Sensors Development, University of Puerto Rico at Mayagüez, P.O. Box 9000, Mayagüez, PR 00681-9000, USA

Correspondence should be addressed to Samuel P. Hernández-Rivera; samuel.hernandez3@upr.edu

Academic Editor: Augusto Belendez

A methodology for processing spectroscopic information using a chemometrics-based analysis was designed and implemented in the detection of highly energetic materials (HEMs) in the gas phase at trace levels. The presence of the nitroaromatic HEM 2,4-dinitrotoluene (2,4-DNT) and the cyclic organic peroxide triacetone triperoxide (TATP) in air was detected by chemometrics-enhanced vibrational spectroscopy. Several infrared experimental setups were tested using traditional heated sources (globar), modulated and nonmodulated FT-IR, and quantum cascade laser- (QCL-) based dispersive IR spectroscopy. The data obtained from the gas phase absorption experiments in the midinfrared (MIR) region were used for building the chemometrics models. Partial least-squares discriminant analysis (PLS-DA) was used to generate pattern recognition schemes for trace amounts of explosives in air. The QCL-based methodology exhibited a better capacity of discrimination for the detected presence of HEM in air compared to other methodologies.

1. Introduction

The detection of highly energetic materials (HEMs) at trace levels in air remains a subject of great importance to national defense and security. In the past few years, most of the published reports have focused on the detection of these important chemical compounds. However, the majority of them require some type of sampling [1, 2]. Obtaining samples in the field is the principal disadvantage of most explosive detection devices because the person doing the sampling is at risk.

Most of the analytical techniques employed for the development of methodologies for HEM detection are based on spectroscopic and chromatographic techniques [1, 2]. Trace amounts of 2,4-dinitrotoluene (2,4-DNT) in air have been detected and discriminated by surface-enhanced Raman spectroscopy (SERS) using a gold surface sensor [3]. These sensors generate a response in the presence or absence of 2,4-DNT and other volatile nitroaromatic HEMs in air. In this case, the sample vapor was introduced to the sensor

with a fan. High-speed fluorescence spectroscopy is another method for the detection of nitroaromatic HEM in the air. This method employs silica microspheres coated with a highly sensitive fluorescent polymer that responds by quenching the fluorescence when HEM molecules attach to the polymer [1, 2, 4–7]. 2,4-DNT can also be detected and quantified by measuring the IR acoustic wave in polymer-coated surfaces [8]. In this method, the presence of 2,4-DNT generates a change in the frequency of the acoustic wave on the surface, and this change is used for detection and quantification.

Air sampling with a solid-phase extraction cartridge to collect a toluene/methyl-*tert*-butyl ether analyte using a modified supercritical fluid extraction apparatus followed by separation and measurement of the extracted analyte by GC has also been used for the analysis of nitroaromatic HEM [9]. The detection of nitroaromatic HEM in air can also be performed by extraction with C-18 solid-phase membranes. In this case, the analyte is desorbed directly in a chromatographic mobile phase [10]. The detection of triacetone triperoxide (TATP) in air has been reported using a gas-washing sampling technique [11]. HPLC with postcolumn UV irradiation and photometric detection following photochemical degradation of TATP has also been used for detection [11, 12].

The use of chemometrics (multivariate analysis) with spectroscopic data in HEM detection has been very valuable because it has allowed for the exploration of very complex ambient matrices [13, 14]. Many chemometrics tools have been developed and tested. However, the most commonly used tools employed to identify, quantify, and classify datasets are those that make use of principal components analysis (PCA), partial least-squares (PLS), and discriminant analysis (DA), as well as their combined usage in PLS-DA and hierarchical cluster analysis (HCA). PCA transforms a set of variables into fewer variables (called dimensions, principal components, or components) that contain most of the information (variance) from the initial dataset. The PCA algorithm seeks to save the information from a large number of variables in a small number of uncorrelated components with minimal loss of information. One of the main reasons for performing a PCA is to reduce the number of variables to a few uncorrelated dimensions that contain as much information as possible (used to avoid multicollinearity in multiple regressions, among other things) [14]. PLS is a linear analysis routine that is used to design and build robust calibration models for quantitative analysis. PLS regression analysis is a quantitative spectral decomposition technique that is closely related to PCA regression [15]. PLS uses the concentration information during the decomposition process, which causes the spectra containing higher constituent concentrations to be weighted more heavily than those with lower concentrations. The main idea of PLS is to obtain as much information about the concentration as possible into the first few latent variables (number of components) [16]. Linear discriminant analysis (LDA or DA) is a multivariate technique that allows for the differentiation of separate objects from distinct populations and allocates new objects into populations previously defined [17]. The usefulness of this methodology is to determine a relationship of belonging or not belonging to a previously defined group.

The application of pattern recognition to infrared spectroscopy can be found in the current literature. PCA was used to analyze the FTIR spectra of mixtures of two monomers [18]. Discrimination between mayonnaise samples that contained different vegetable oils was achieved by PLS-DA of near infrared spectral data [19]. PLS-LDA has also been used in other areas of science and engineering, including biomedical studies such as the classification of tumors [20], early detection of diabetes related to changes in the skin [21], and fault diagnosis in chemical processes [22].

Detection and discrimination of HEM are important in the applications for national defense and security. Being able to detect and prevent a chemical/biological threat long before any damage to civilians, military personnel, and private or public property is a goal of agencies in charge of public security and national defense. To a large extent, remote detection modalities will benefit from chemometrics-enhanced spectroscopic detection.

In this paper, infrared vibrational detection of highly energetic materials, such as 2,4-DNT and TATP, present in the gas phase in air was performed. The detection experiments were performed in the active mode using two source types including a traditional globar IR source for detection using FT-IR spectrometers (for both modulated and unmodulated lights) and a quantum cascade laser (QCL) IR source for detection with a dispersive spectrometer. PLS-DA was used to generate pattern recognition schemes for trace amounts of explosives in air from the obtained IR spectra. Classificatory capacities from different models based on PLS-DA were used to establish the best experimental setup for the detection and classification of these explosives in the gas phase.

2. Experimental Section

2.1. Materials. The reagents used in this investigation included acetone (CH_3COCH_3, 98% w/w, Sigma-Aldrich Chemical Co., Milwaukee, WI, USA), hydrogen peroxide (H_2O_2, 50% in water, Sigma-Aldrich), hydrochloric acid (HCl, 12 M, Merck, VWR, Inc., West Chester, PA, USA), sulfuric acid (H_2SO_4, 18 M, Merck), and dimethyl ether (CH_3OCH_3, Sigma-Aldrich). Standard solutions of 2,4-DNT 1000 parts per million (ppm) GC/MS primary standards were obtained from Restek Corp. (Bellefonte, PA, USA) and from Chem Service, Inc. (West Chester, PA, USA). Crystalline samples of 2,4-DNT were purchased from Chem Service, Inc. (West Chester, PA, USA).

2.2. Synthesis of TATP. TATP cannot be purchased from chemical reagents suppliers. Therefore, it was prepared in small quantities as needed due to the high thermal instability of this powerful explosive. A white crystalline precipitate was obtained using a conventional synthesis method. The crystals were filtered and washed using distilled water followed by recrystallization from methyl ether.

2.3. Instrumentation. The IR equipment used for the experiments was a Bruker Optics (Billerica, MA, USA) model IFS 66v/S Fourier transform IR (FT-IR) interferometer. This system has an evacuable bench equipped with several sources, detectors, beam splitters, and other accessories to perform experiments on solid, liquid, and gas samples from the far IR region (to $50\,cm^{-1}$) to the near IR region (as high as $7500\,cm^{-1}$). For the experiments described, the system was equipped with a deuterated triglycine sulfate (DTGS) detector and a potassium bromide (KBr) beam splitter. An EM-27 open path (OP) FT-IR interferometer (Bruker Optics) was used to obtain the IR spectral information from the TATP samples with a thermoelectrically (TE) cooled mercury-cadmium telluride (MCT) detector.

A LaserTune IR dispersive spectrometer equipped with a TE-cooled MCT detector (Block Engineering, Marlborough, MA, USA) was employed to obtain spectral information of the TATP samples. An Agilent 6890 gas chromatograph (GC) coupled to an Agilent 5893 mass selective detector (MSD) was used for qualitative analyses. An Agilent Technologies model

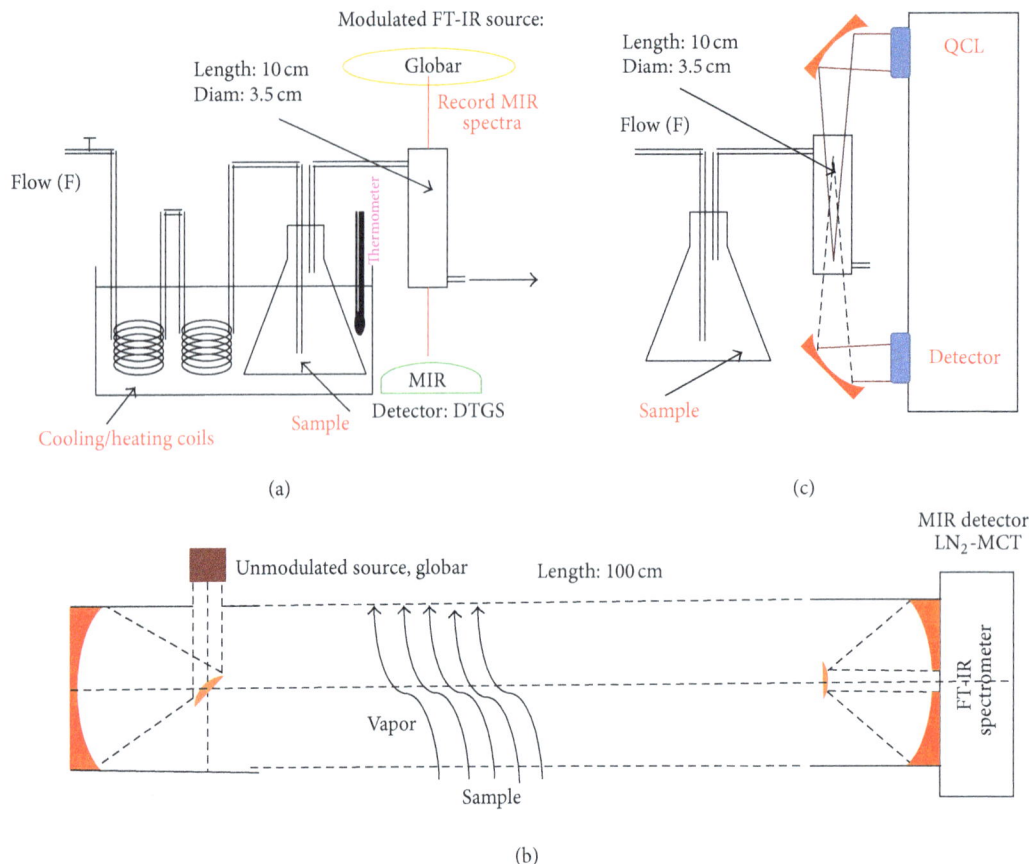

FIGURE 1: Schematic diagram of the experimental setup. (a) FT-IR interferometer with modulated light source. (b) Open Path FT-IR with unmodulated (external) thermal source. (c) Quantum cascade laser (QCL) scanning system.

6890N, Network GC system with microcell ^{63}Ni Electron Capture Detector (μ-ECD), was also used for the quantitative analyses. A capillary column HP-5 MS 5% phenyl methyl siloxane, 30 m, 250 μm in diameter, and 0.25 μm of film thickness, was used.

3. Measurements and Analysis

3.1. Experiment. Figure 1 shows the schematic diagram of the experimental setup used in the investigation. Three types of experiments were performed. The setup for the first experiment is shown in Figure 1(a). The HEM samples were placed on the bottom of 500 mL Erlenmeyer flasks. A flow of dry air (1–16 mL/s) at several temperatures (0–38°C) was used. The temperature was regulated by either scanning in the range of temperature or by using point-by-point fixed temperature measurements. Trace amounts of explosives in the gas phase were dragged from the surface by the air flow and transported to an IR gas cell for detection. Spectra were recorded using the instrument at 4 cm^{-1} of resolution and 25 scans. The spectral range was 400–4000 cm^{-1}.

The number of spectra obtained was 799 for 2,4-DNT/air; 120 for TATP/air; and 1881 spectra of ambient air. Figure 1(b) shows a schematic representation of the EM-27 interferometer setup employed to collect absorption spectra excited by

a globar source. All of the active mode experiments were performed at ambient temperature (25°C) at 30 scans and 4 cm^{-1} resolution. Sets of 100 air spectra, air with TATP, and air with DNT with a spectral range of 650–4000 cm^{-1} were collected.

A third experiment was performed using a quantum cascade laser (QCL) as the source with the Block Engineering LaserTune spectrometer. All of the active mode experiments were performed at the laboratory temperature of 20°C with 1 scan at 4 cm^{-1} resolution. The spectral range was from 830 to 1430 cm^{-1}. Forty-four spectra of air with 2,4-DNT, 25 spectra of air with TATP, and 37 spectra of air were obtained. The presence of TATP in the air was determined by GC-MS, and the concentrations of 2,4-DNT in the air for different fluid conditions were calculated using a calibration curve from GC-μECD, which was only performed for the first experiment.

3.2. Partial Least-Squares (PLS) Discriminant Analysis (DA). PLS-DA runs were performed using the OPUS 6.0 software (Bruker Optics) combined with the Statgraphics Plus v. 15.2 (StatPoint Technologies, Inc., Warrenton, VA, USA) statistical analysis software. The models were evaluated using internal validation, statistical significance (P), and the percentage of cases correctly classified (PCCC). For internal

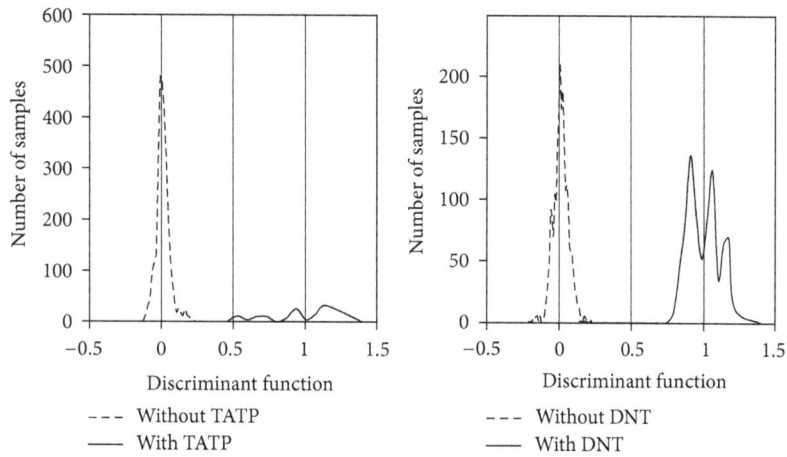

(a)

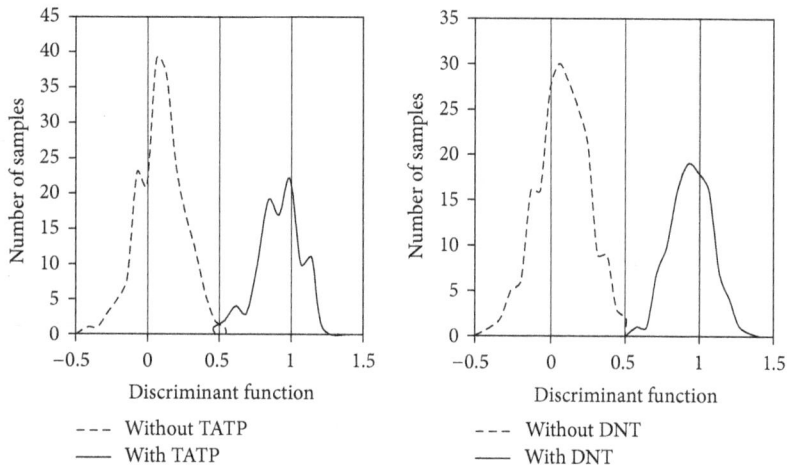

(b)

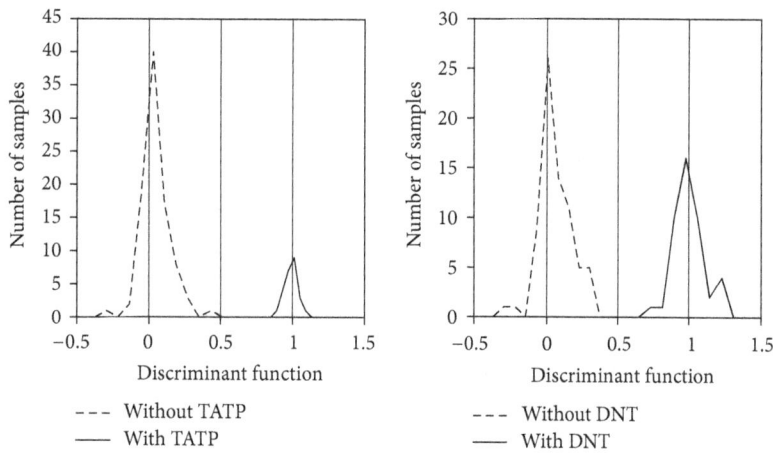

(c)

FIGURE 2: (a) Histogram for discrimination using modulated source FT-IR. (b) Histogram for discrimination using unmodulated thermal source FT-IR. (c) Histogram for discrimination using QCL.

TABLE 1: Validation parameters for the various models constructed.

Parameter	Modulated FTIR (model 1)		Unmodulated FTIR (model 2)		QCL scan (model 3)	
	TATP	DNT	TATP	DNT	TATP	DNT
Wilks' lambda	0.12	0.03	0.09	0.11	0.05	0.06
Canonical correlation	0.94	0.98	0.96	0.94	0.98	0.97
PCCC	100.00%	100.00%	100.00%	100.00%	100.00%	100.00%
PCCCC	100.00%	99.94%	100.00%	99.97%	100.00%	100.00%
Sensibility	100.00%	100.00%	100.00%	99.50%	100.00%	100.00%
Specificity	100.00%	99.75%	100.00%	100.00%	100.00%	100.00%
False alarm	0.00%	0.00%	0.00%	0.50%	0.00%	0.00%
Missed detection	0.00%	0.25%	0.00%	0.00%	0.00%	0.00%
Loadings	7	7	10	10	5	5
Samples (N)	3079	3079	300	300	106	106
P value	<0.0001	<0.0001	<0.0001	<0.0001	<0.0001	<0.0001

validation, each spectrum was successively removed from the dataset and discriminated from a new model built from the remaining spectra. This procedure was performed for each of the spectra in the dataset, and the predicted discrimination was compared with the experimental observations. The generated percentage of cases correctly classified is called the cross-percentage of cases correctly classified (PCCCC) or the leave-one-out cross-validation (LOOCV or CV) procedure. The other statistical indicators that were used included the Wilks' lambda and canonical correlation.

4. Results and Discussion

Figure 2 shows the frequency distribution for the CV of air with TATP and DNT. The solid line represents the data for air with the analyte of interest, and the dotted line represents the data for clean air and air with other analytes. Good discrimination was obtained in all of the statistical experiments. The evaluation is shown in Table 1. The PCCC for all of the models was 100.0%, and complete classification is observed. However, the cross-validation PCCCC was not 100.0% for DNT samples.

In the FT-IR model with a modulated globar source, the PCCCC for TATP was 100% but it was lower for 2,4-DNT. This result can be attributed to the fact that pure air was only analyzed at 25°C, whereas 2,4-DNT was analyzed from 0°C to 38°C. In this model for 2,4-DNT, 0.25% of the sample was not correctly classified. These data missed the detection of or indicated a false negative for air with 2,4-DNT at low temperatures where the sublimation of DNT is very small. For this model of 2,4-DNT, the sensitivity was 100.00%, the specificity was 99.75%, and the false alarm rate was 0.00%.

In the second model (unmodulated FT-IR), the PCCCs for all of the models constructed were also 100.0%, but the PCCCC for 2,4-DNT was 99.67%. In these cases, the false alarm rate was 0.5%, and the sensitivity was 99.50%. This result is an indication that samples of air or air with TATP were discriminated better than air with 2,4-DNT. In these cases, one of the air samples was poorly discriminated because this experiment was carried out in open path mode

resulting in possible cross contamination from DNT in the laboratory. In this setup, interferences from ambient water vapor and CO_2 were high. Therefore, in the model using the entire IR spectral region measured (600–4000 cm^{-1}), it was necessary to eliminate the subspectral regions of 4000–3541, 2384–2295, and 1758–1490 cm^{-1}, resulting in the improvement of the model, as summarized in Table 1.

In the third model (QCL scan), all of the samples were correctly classified. However, the number of samples analyzed must be considered, and the number of variables in this experiment is lower compared to the other experiments. The experimental conditions are not fully comparable because the intensity of this source is much higher than those of the other experiments and the sampling path is smaller for this system.

For the models to have highly significant statistical merit according to the canonical correlation coefficient ($P <$ 0.0001), the functions must have an excellent ability to determine the group differences. Wilks' lambda value indicates how many times the variance is not explained by group differences. Because these values were small, highly correlated differences were established.

Other models were generated using only the region of 873–1400 cm^{-1} or the region of emission of the QCL to compare the technique with the FT-IR-based experiments. The QCL-based experiments were better than the modulated source FT-IR setup, which, in turn, was better than the nonmodulated source FT-IR setup. The validation information is shown in Table 2. A high significance ($P <$ 0.0001, equivalent to >99.99% confidence level) for all of the models was found, indicating that the resulting parameters are highly reliable and the comparison between techniques is highly dependable. The high power QCL scan produces a high sensitivity for trace level detection in air, but the modulated source FT-IR is close in performance to the QCL-based methodology. This result indicates that the modulation of light (using an interferometer) before reaching the sample increases the sensitivity [23] compared to a nonmodulated source FT-IR where the light first interacts with the sample prior to entering the interferometer.

TABLE 2: Validation parameters for the models in the subspectral range of 850–1400 cm^{-1}.

Parameter	Modulated FTIR (model 1)		Unmodulated FTIR (model 2)		QCL scan (model 3)	
	TATP	DNT	TATP	DNT	TATP	DNT
Wilks' lambda	0.15	0.04	0.55	0.54	0.05	0.06
Canonical correlation	0.92	0.98	0.67	0.65	0.98	0.97
PCCC	99.74%	99.98%	92.00%	93.33%	100.00%	100.00%
PCCCC	99.73%	99.97%	82.00%	78.33%	100.00%	100.00%
Sensibility	100.00%	100.00%	81.00%	77.00%	100.00%	100.00%
Specificity	93.33%	99.87%	84.00%	81.00%	100.00%	100.00%
False alarm	0.00%	0.00%	16.00%	19.00%	0.00%	0.00%
Missed detection	6.67%	0.13%	19.00%	23.00%	0.00%	0.00%
Latent variables (LV)	10	10	10	10	5	5
P value	<0.0001	<0.0001	<0.0001	<0.0001	<0.0001	<0.0001

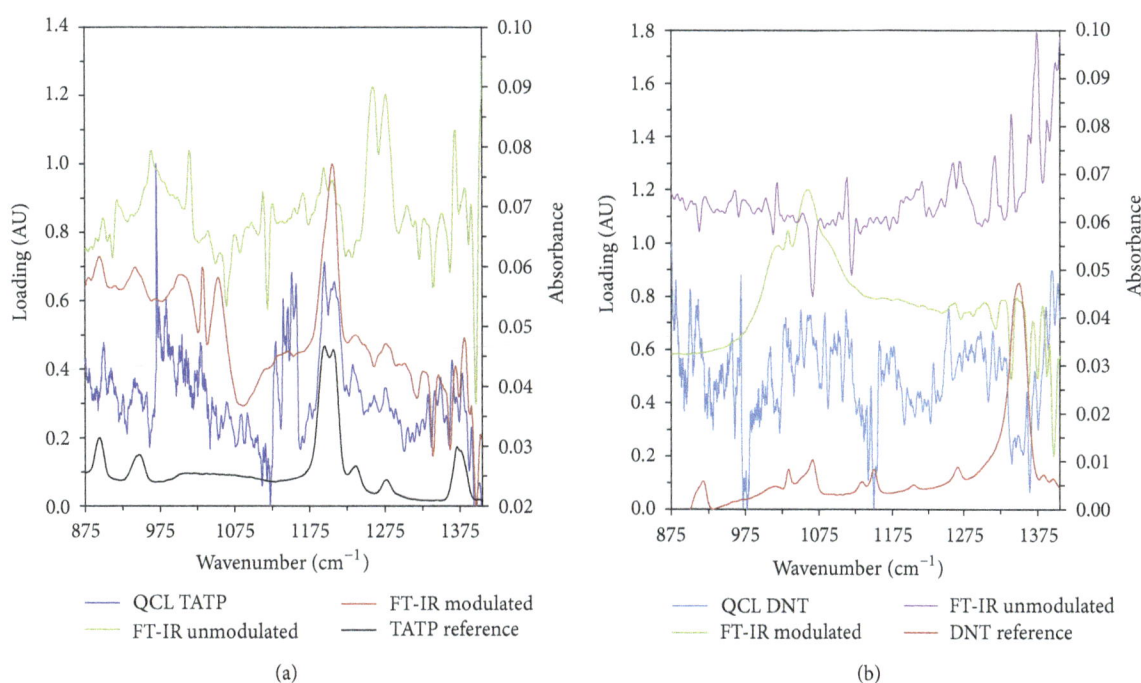

FIGURE 3: (a) First loadings for TATP models for the region 873–1400 cm^{-1}. (b) First loadings for DNT models for the region 873–1400 cm^{-1}. Reference gas-phase spectra included.

The loading for models of the region (870–1400 cm^{-1}) are shown in Figure 3. The band observed for TATP at 1200 cm^{-1} that was tentatively assigned to C–O stretching [24, 25] is the largest contributor to the loadings calculated by the models for modulated source FT-IR and QCL source. However, in the unmodulated source FT-IR experiments, the same result was not found: the first loading has the spectral information shifted. Other bands that contribute occur at 890 cm^{-1} and 939 cm^{-1}, which are tentatively assigned to O–O stretching [24, 25]. The same analysis applies for all of the other loadings (data not shown). This indicates that the discrimination is generated by a combination of all of the loadings. The QCL model required only 5 latent variables (LV); FT-IR-based

models required 7 (full spectrum) or 10 (spectral width) latent variables (LV).

4.1. Proof of the Presence of 2,4-DNT and TATP in Air at Trace Levels. The concentration of 2,4-DNT in air for different flow conditions and temperatures was calculated via a calibration curve obtained by GC-μECD (Table 3). The presence of TATP was also established using GC-MS. To demonstrate the presence of the infrared signal in air, a small amount of TATP or 2,4-DNT (0.1 μg) was deposited on a plate and introduced into a gas cell of 38.7 cm^3 at a pressure of 0.0001 mBar. This procedure generated a density of 2538 pg/mL (in the worst case) when all of the explosive material sublimed and

TABLE 3: Mass of 2,4-DNT determined by GC-μECD for 2 mL of injected gas.

T (°C)	F (mL/s)	Peak area	Mass (pg)	pg/mL (μg/m^3)
20	1.6	$7.564.E+03$	98	49
20	7.8	$3.314.E+03$	48	25
20	15.7	$3.078.E+03$	46	23
26	1.6	$1.030.E+05$	1202	601
26	7.8	$6.650.E+04$	779	390
26	15.7	$5.720.E+04$	672	336
38	1.6	$3.800.E+05$	4407	2203
15	15.7	$2.793.E+03$	42	21

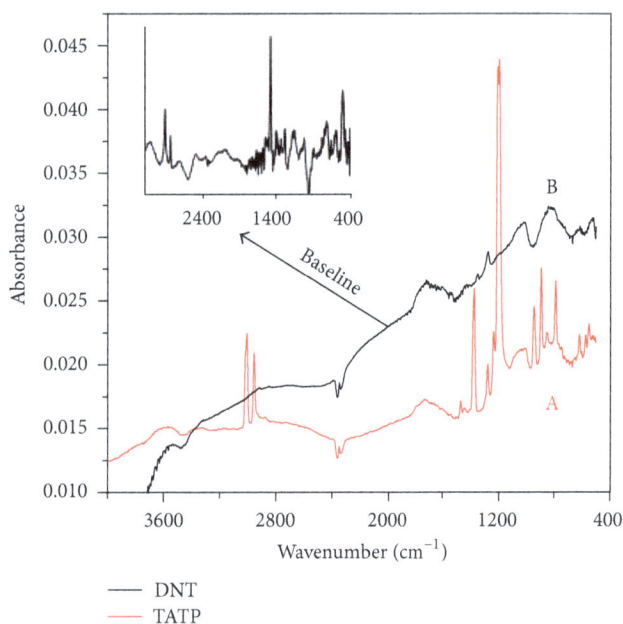

FIGURE 4: Low-pressure spectra in the gas phase with baseline correction of (A) TATP and (B) DNT.

negligible amounts were suctioned by the vacuum pump. This suggests that the concentration in the cell for the gas was $\ll 2538$ pg/mL. Spectra for the trace amounts of explosive were recorded to confirm that the detector was able to detect at this concentration level (Figure 4). This experiment was performed using the modulated source FT-IR system.

4.2. Limits of Detection. The low limits of detection (LOD) have been estimated for the two better performing techniques and their corresponding models: modulated source FT-IR and mid-IR (QC) laser source. In the case of the homemade explosive TATP, LOD values were estimated as 800 pg/m^3 and 300 pg/m^3 for modulated FT-IR and QCL-based detection, respectively. For the nitroaromatic military explosive 2,4-DNT-LOD values were even lower: 31 pg/m^3 and 0.7 pg/m^3 for modulated source FT-IR and QCL source, respectively.

5. Conclusions

The results obtained in this research show that it is possible to determine the presence of peroxide-based explosives, such as TATP, and high sublimation pressure nitro compounds when they are in the gas phase mixed with air using PLS-DA regression analysis of infrared spectral data. The QCL-based results exhibited a better capacity for discrimination or detection of the presence of explosives in air compared to other techniques. This result is due to the high power and collimation of the laser source increasing the sensitivity at trace level in air. It was also demonstrated that, when the light is modulated, an increase of sensitivity is obtained. Possible synergies between QCL sources, which are inherently modulated, and detection schemes could generate higher sensitivity techniques for gas phase sensing of hazardous chemicals. However, technical problems related to QCL scanning generated high noise levels. This resulted in unsuccessful efforts to modulate this source. A possible solution is to stop the QCL in a wavenumber range corresponding to emission bands within the target chemicals and then modulate the source and acquire the spectra at this position; next, move the QCL to another wavenumber central position close to the previous band and acquire the spectrum, and so on until the whole range of the QCL is scanned. This would provide the required sensitivity for the analysis at trace level.

Acknowledgments

Support from the US Department of Homeland Security under Award no. 2008-ST-061-ED0001 is acknowledged. However, the views and conclusions contained in this document are those of the authors and should not be interpreted as necessarily representing the official policies of the US Department of Homeland Security. This contribution was supported by the US Department of Defense, Prop. No. 58949-PH-REP, Agreement No. W911NF-11-1-0152. The authors also acknowledge contributions from Dr. Richard T. Hammond from Army Research Office, DoD. They gratefully acknowledge the contributions by Dr. Fred Haibach from Block Engineering in setting up the experimental part of the QCL-based experiments involving the LaserTune (Block Eng., Marlborough, MA, USA) and for his enlightening discussions.

References

[1] D. S. Moore, "Instrumentation for trace detection of high explosives," *Review of Scientific Instruments*, vol. 75, no. 8, pp. 2499–2512, 2004.

[2] D. S. Moore, "Recent advances in trace explosives detection instrumentation," *Sensing and Imaging*, vol. 8, no. 1, pp. 9–38, 2007.

[3] J. M. Sylvia, J. A. Janni, J. D. Klein, and K. M. Spencer, "Surface-enhanced Raman detection of 2,4-dinitrotoluene impurity vapor as a marker to locate landmines," *Analytical Chemistry*, vol. 72, no. 23, pp. 5834–5840, 2000.

[4] K. J. Albert and D. R. Walt, "High-speed fluorescence detection of explosives-like vapors," *Analytical Chemistry*, vol. 72, no. 9, pp. 1947–1955, 2000.

[5] E. R. Menzel, L. W. Menzel, and J. R. Schwierking, "A photoluminescence-based field method for detection of traces of explosives," *The Scientific World Journal*, vol. 4, pp. 725–735, 2004.

[6] Y. Salinas, R. Martínez-Máñez, M. D. Marcos et al., "Optical chemosensors and reagents to detect explosives," *Chemical Society Reviews*, vol. 41, no. 3, pp. 1261–1296, 2012.

[7] Y. Salinas, A. Agostini, É. Pérez-Esteve et al., "Fluorogenic detection of Tetryl and TNT explosives using nanoscopic-capped mesoporous hybrid materials," *Journal of Materials Chemistry A*, vol. 1, no. 11, pp. 3561–3564, 2013.

[8] G. K. Kannan, A. T. Nimal, U. Mittal, R. D. S. Yadava, and J. C. Kapoor, "Adsorption studies of carbowax coated surface acoustic wave (SAW) sensor for 2,4-dinitro toluene (DNT) vapour detection," *Sensors and Actuators B*, vol. 101, no. 3, pp. 328–334, 2004.

[9] R. Batlle, H. Carlsson, P. Tollbäck, A. Colmsjö, and C. Crescenzi, "Enhanced detection of nitroaromatic explosive vapors combining solid-phase extraction-air sampling, supercritical fluid extraction, and large-volume injection-GC," *Analytical Chemistry*, vol. 75, no. 13, pp. 3137–3144, 2003.

[10] C. Sánchez, H. Carlsson, A. Colmsjö, C. Crescenzi, and R. Batlle, "Determination of nitroaromatic compounds in air samples at femtogram level using C18 membrane sampling and on-line extraction with LC-MS," *Analytical Chemistry*, vol. 75, no. 17, pp. 4639–4645, 2003.

[11] R. Schulte-Ladbeck and U. Karst, "Determination of triacetonetriperoxide in ambient air," *Analytica Chimica Acta*, vol. 482, no. 2, pp. 183–188, 2003.

[12] J. I. Steinfeld and J. Wormhoudt, "Explosives detection: a challenge for physical chemistry," *Annual Review of Physical Chemistry*, vol. 49, no. 1, pp. 203–232, 1998.

[13] J. J. Perez, P. M. Flanigan, J. J. Brady, and R. J. Levis, "Classification of smokeless powders using laser electrospray mass spectrometry and offline multivariate statistical analysis," *Analytical Chemistry*, vol. 85, no. 1, pp. 296–302, 2013.

[14] F. C. de Lucia Jr. and J. L. Gottfried, "Influence of variable selection on partial least squares discriminant analysis models for explosive residue classification," *Spectrochimica Acta B*, vol. 66, no. 2, pp. 122–128, 2011.

[15] J. K. V. Mardia, J. T. Kent, and J. M. Biby, *Chemometrics: Statistic and Computer Application in Analytical Chemistry*, Academic Press, London, UK, 1980.

[16] K. R. Beebe, R. J. Pell, and M. B. Seasholtz, *Chemometrics. A Pactricla Guide*, John Wiley & Sons, New York, NY, USA, 1998.

[17] C. J. Huberty, *Applied Discriminant Analysis*, Wiley-Interscience, Hoboken, NJ, USA, 1994.

[18] Y. M. Kim, J. F. MacGregor, and L. K. Kostanski, "Principal component analysis of FT-IR spectra for cationic photopolymerization of mixtures of two monomers," *Chemometrics and Intelligent Laboratory Systems*, vol. 75, no. 1, pp. 77–90, 2005.

[19] U. G. Indahl, N. S. Sahni, B. Kirkhus, and T. Næs, "Multivariate strategies for classification based on NIR-spectra-with application to mayonnaise," *Chemometrics and Intelligent Laboratory Systems*, vol. 49, no. 1, pp. 19–31, 1999.

[20] Y. Tan, L. Shi, W. Tong, G. T. G. Hwang, and C. Wang, "Multi-class tumor classification by discriminant partial least squares using microarray gene expression data and assessment of classification models," *Computational Biology and Chemistry*, vol. 28, no. 3, pp. 235–244, 2004.

[21] B. Lindholm-Sethson, S. Han, S. Ollmar et al., "Multivariate analysis of skin impedance data in long-term type 1 diabetic patients," *Chemometrics and Intelligent Laboratory Systems*, vol. 44, no. 1-2, pp. 381–394, 1998.

[22] Q. P. He, S. J. Qin, and J. Wang, "A new fault diagnosis method using fault directions in Fisher discriminant analysis," *AIChE Journal*, vol. 51, no. 2, pp. 555–571, 2005.

[23] L. C. Pacheco-Londoño, W. Ortiz-Rivera, O. M. Primera-Pedrozo, and S. P. Hernández-Rivera, "Vibrational spectroscopy standoff detection of explosives," *Analytical and Bioanalytical Chemistry*, vol. 395, no. 2, pp. 323–335, 2009.

[24] G. A. Buttigieg, A. K. Knight, S. Denson, C. Pommier, and M. B. Denton, "Characterization of the explosive triacetone triperoxide and detection by ion mobility spectrometry," *Forensic Science International*, vol. 135, no. 1, pp. 53–59, 2003.

[25] B. Brauer, F. Dubnikova, Y. Zeiri, R. Kosloff, and R. B. Gerber, "Vibrational spectroscopy of triacetone triperoxide (TATP): anharmonic fundamentals, overtones and combination bands," *Spectrochimica Acta A*, vol. 71, no. 4, pp. 1438–1445, 2008.

19

The Information Coding in the Time Structure of the Object of a Laser Pulse in an Optical Echo Processor

L. A. Nefediev and A. R. Sakhbieva

Kazan (Volga Region) Federal University, 18 Kremliovskaya Street, Kazan 420021, Russia

Correspondence should be addressed to A. R. Sakhbieva, alsu-sakhbieva@yandex.ru

Academic Editor: Kiyoshi Shimamura

The encoding of information in time intervals of an echelon of laser pulses of an object pulse in the optical echo processor is considered. The measures of information are introduced to describe the transformation of classical information in quantum information. It is shown that in the description of information transformation into quantum information, the most appropriate measure is a measure of quantum information based on the algorithmic information theory.

1. Introduction

The methods of dynamical echo holography allow processing and storage of the information, which is carried by object laser pulses. They have prospects on creation of high-speed optical echo-processors [1]. In this case, the information can be incorporated in the amplitude and temporal shape of the exciting laser pulses, in their wave fronts and polarization and also in echelons of laser pulses. Demonstration of frequency-selective optical memory, where the data recording and processing of data occurs both in the time domain and in the frequency slot, is described in [2]. The echo-processor based on use of long-lived photon echo has been proposed in [3]. The design of this processor is given an opportunity to demonstrate the density of information recording and processing about several gigabits/cm² in [4], using compression and tension of data signals through a rapid change of the carrier frequency. From the point of view of the information theory, it is possible to present an echo-processor as an information channel with memory and noise, in an input and output of which the information has a classical appearance and inside of the channel quantum. This channel provides the information transmission and transformation between the different moments in time and directions in space. Common channel features are the information rate, throughput rate, and use factor. Attempts to describe the quantum information processes in general

relied on the formulae of classical information theory, which was operated with the quantum probabilities, rather than amplitudes. In [5] it has been shown that the von Neumann entropy has the information and theoretical value asymptotically characterizing the minimal quantum resources required to describe the ensemble of quantum states. This suggests that the enhanced information theory should be defined as a theory which takes into account the quantum phases explicitly. For example, the theory, developed in [6], describes a quantum system, divided into many parts, using only density matrix and von Neumann entropy. This theory includes the Shannon theory as a special case, but also describes quantum entanglement and establishing the correspondence between classical and quantum information in this way. We consider the system (the message), described by the variables A (classical or quantum), and construct its classical and quantum description at the same time. In classical information theory the entropy of Shannon for A is defined as:

$$J_c = -\sum_n p(a)\log_2 p(a), \tag{1}$$

where the A takes the value a with probability $p(a)$. This entropy can be interpreted as uncertainty about A.

The quantum analogue is the von Neumann entropy (where the subsystem A is described by the density operator ρ_A),

$$J_{fn} = -\text{Tr}_A\left[\rho_A \log_2 \rho_A\right], \qquad (2)$$

where Tr_A denotes the trace over degrees of freedom subsystem A.

The von Neumann entropy reduces to Shannon entropy, if ρ_A is the mixed state, decomposed in orthogonal quantum state.

In the case of echo-processor information reproduction depends on coding method. It is obvious that in this case it will be the best information reproduction if the information is coded in time intervals of the laser pulses echelon of an object pulse.

In the given work we will consider the process of transformation of the classical information put in time intervals of an echelon of laser pulses in quantum at their resonant interaction with two level quantum system.

2. Von Neumann Entropy in the Description of Systems with Coherent Superposition of Basis States

Expression of the von Neumann entropy for a two-level system can be represented by means of functions from matrixes

$$J_{fn}(\rho)$$

$$= -\text{Tr}\left(\rho\log_2 \rho\right) = \frac{1}{\sqrt{\left(\rho_{11} - \rho_{22}\right)^2 + 4\left|\rho_{12}\right|^2}}$$

$$\times \left[2\left|\rho_{12}\right|^2 \log_2 \frac{\lambda_2}{\lambda_1}\right.$$

$$+ \frac{1}{2}\rho_{11}\left(\rho_{22} - \rho_{11} - \sqrt{\left(\rho_{11} - \rho_{22}\right)^2 + 4\left|\rho_{12}\right|^2}\right)\log_2 \lambda_1$$

$$- \frac{1}{2}\rho_{11}\left(\rho_{22} - \rho_{11} + \sqrt{\left(\rho_{11} - \rho_{22}\right)^2 + 4\left|\rho_{12}\right|^2}\right)\log_2 \lambda_2$$

$$+ \frac{1}{2}\rho_{22}\left(\rho_{11} - \rho_{22} - \sqrt{\left(\rho_{11} - \rho_{22}\right)^2 + 4\left|\rho_{12}\right|^2}\right)\log_2 \lambda_1$$

$$\left. - \frac{1}{2}\rho_{22}\left(\rho_{11} - \rho_{22} + \sqrt{\left(\rho_{11} - \rho_{22}\right)^2 + 4\left|\rho_{12}\right|^2}\right)\log_2 \lambda_2\right], \qquad (3)$$

where

$$\lambda_{1,2} = \frac{\rho_{11} + \rho_{22}}{2} \pm \frac{1}{2}\sqrt{\left(\rho_{11} - \rho_{22}\right)^2 + 4\left|\rho_{12}\right|^2}. \qquad (4)$$

In the absence of coherence in the system from (3) follows

$$\lim_{|\rho_{12}| \to 0} J_{fn}(\rho) = -\rho_{11}\log_2 \rho_{11} - \rho_{22}\log_2 \rho_{22}. \qquad (5)$$

As the $\rho_{11} \le 1$ and $\rho_{22} \le 1$ then $J_{fn}(\rho) \ge 0$. In the case of a pure state $J_{fn}(\rho) \to 0$.

The more appropriate measure of quantum information in the presence of coherence in the system can be K-complexity and the applications of algorithmic information theory to the description of the quantum information processes [7, 8].

3. Quantum Structural Information in the Medium with Phase Memory

Since the structural information in the resonant medium is carried by transitional dynamic gratings, which are described by a density matrix, the structural information is contained in the amplitude and phase structure of the density matrix ρ. If we match a graph G to this matrix, the measure of structural information will be defined as the measure of the structure uncertainty of this graph [8]. In the simplest case, this uncertainty can be determined by enumerating the corresponding diagrams of the graph. However, the resulting measure of information is incomplete, since it ignores off-diagonal elements of the corresponding matrix of conditional transitions. We will therefore use the notion of K-complexity to quantify structural information.

We consider a graph G, corresponding to the density matrix of the system. Its elements belong to a finite set $\in V(G)$, consisting of N labeled vertices and q edges, which correspond to the diagonal and off-diagonal elements of the density matrix, respectively. Thus, $V = \Gamma \cup Q$ where Γ is the set of the vertices of the graph and Q is the set of its edges.

The relative complexity K of the object G is the minimum length $l(p)$ of a program p that can derive object G_o from G. We define the amount of structural information of G relative to G_o as

$$J = K(G, G_o) - K(G_o). \qquad (6)$$

We will partition the algorithmic process of deriving object G_o from object G into separate steps of bounded complexity. Each step transforms the current state G^κ of the object to state $G^{\kappa+1}$

$$G^{\kappa+1} = D_\kappa(G^\kappa). \qquad (7)$$

Operator D is determined by the set of rules for processing the active part of object G. Thus, we have

$$G_o = D(G). \qquad (8)$$

Since the quantum information lies in the coherent part of the density matrix, we will consider the elements $\in Q$ as the active part object G. Operator D is defined as the operator of removing elements from the active part of object G by any possible means [9]:

$$D(Q) = \varnothing. \qquad (9)$$

We will assume that object with the elements, belonging to the zero set $\varnothing$, has structural information, which is equal to 0. It is convenient to quantify the length $l(p)$ of program p on a logarithmic scale as $\log_2$ of the sum of all weighting functions corresponding to diagrams of G^κ. Taking the temporal evolution into account, we define the weighting function as the sum of the active part of elements of object G at time t. The active part of object G corresponds to diagram G^{κ}, referred to the sum of the active part elements of object G at the initial time instant t_o. Procedure (9) leads to an ensemble of sets $Q^{(\kappa)}$. Since weighted graph G corresponds to the density matrix

$$\rho = \sum_{i,j=1}^{N} \rho_{ij} P_{ij}, \tag{10}$$

where P_{ij} are the projector matrices (which have element ij equal to 1 and all other elements equal to 0), then the sum $S(t_o)$ of the active part elements of the object is given at the initial time by

$$S(t_o) = abs\left(\sum_{i \neq j} \rho_{ij}(t_o)\right). \tag{11}$$

Calculating the corresponding sum $S'(t) = \sum_{\kappa} S^{(\kappa)}(t)$ for the ensemble of sets $Q^{(\kappa)}$ at time t, we arrive at:

$$J_q = \log_2\left(\frac{S'(t)}{S(t_o)}\right). \tag{12}$$

Since $\hat{\rho}$ is a Hermitian operator, its matrix elements meet the equality $\rho_{ij} = \rho_{ji}^*$. Now it is easy to see that our choice of the operator D brings us to $S'(t)$ and $S(t_o)$, consisting of matrix elements sum $\rho_{ij} + \rho_{ji}$. Thus, $S'(t)$ and $S(t_o)$ are real quantities. We should note also that, when the system has only two quantum states, $|1\rangle$ and $|2\rangle$, the total state is a linear superposition $|\psi\rangle = \alpha|1\rangle + \beta|2\rangle$ (where α and β are complex numbers). The relevant density matrix operator is written as

$$\hat{\rho} = |\psi\rangle\langle\psi| = |\alpha|^2|1\rangle\langle1| + |\beta|^2|2\rangle\langle2| \\ + \alpha\beta^*|1\rangle\langle2| + \alpha^*\beta|2\rangle\langle1|. \tag{13}$$

4. The Transformation Process of the Classical Information in Quantum Structural Information

We consider the transformation of the classical information $J_c(A)$ carried by an object laser pulse during its interaction with a system of two-level atoms, where the quantum information J_q carriers which are superposition states of atoms.

Object pulse can be represented as a sequence (echelon) of the n rectangular laser pulses (in the general case with different duration and amplitude ε_η), separated by arbitrary time intervals. Time intervals are denoted as τ_η ($\eta = 1,\ldots,n$). Then $\varepsilon_\eta > 0$ is the presence of a pulse, and $\varepsilon_\eta = 0$ is the presence of a time interval. The duration of the pulse echelon will be $\delta t = \sum \tau_\eta$, satisfying $\delta t \ll T_1, T_2$, where T_1 and T_2 are the times of longitudinal and transverse irreversible relaxation of the system considered.

To describe the process of transformating the classical information, into quantum information the most appropriate definition of classical information can be a differential information entropy of the Fourier spectrum of the echelons of laser pulses, because in the resonant medium information carriers are q-bits, which are distributed in the range of the inhomogeneously broadened line of the resonance transition.

The intensity of Fourier components of the pulse echelon electric field will be:

$$E(\nu') = \sum_{\eta=1}^{n} \varepsilon_\eta \int_{t_{\eta-1}}^{t_\eta} e^{-i2\pi\nu't}dt, \tag{14}$$

where ν' is the frequency of the Fourier spectrum, and the time t_η of the ηth pulse start is defined as

$$t_\eta = t_o + \sum_{\kappa=1}^{\eta} \tau_\kappa \tag{15}$$

and it is assumed that the initial time $t_o = 0$. Then from (16) for the amplitude of the Fourier components of the pulse echelon electric field we obtain

$$A(\omega') = |E(\omega')| = \sqrt{\text{Re}\,(E(\omega'))^2 + \text{Im}\,(E(\omega'))^2}, \tag{16}$$

where $\omega' = 2\pi\nu'$,

$$\text{Re}(E(\omega')) = \sum_{\eta=1}^{n} \varepsilon_\eta \tau_\eta \text{sinc}\left(\frac{\omega'\tau_\eta}{2}\right) \cos\left(\frac{\omega'\left(2\sum_{\kappa=1}^{\eta}\tau_\kappa - \tau_\eta\right)}{2}\right), \tag{17}$$

$$\text{Im}(E(\omega')) = \sum_{\eta=1}^{n} \varepsilon_\eta \tau_\eta \text{sinc}\left(\frac{\omega'\tau_\eta}{2}\right) \sin\left(\frac{\omega'\left(2\sum_{\kappa=1}^{\eta}\tau_\kappa - \tau_\eta\right)}{2}\right). \tag{18}$$

Differential information entropy of Fourier spectrum of the laser pulses echelon is defined as $J_c' = J_c - J_{c0}$, where

$$J_c = -\int_{-\infty}^{\infty} p(\omega')\log_2 p(\omega')d\omega'. \tag{19}$$

There

$$p(\omega') = \frac{A(\omega')}{\int_{-\infty}^{\infty} A_o(\omega')d\omega'}, \tag{20}$$

and $A_o(\omega')$ is determined from the expression (17), J_{c0} is defined similarly to (19) at identical amplitudes and time intervals in (17) and (18).

To find the values of quantum information (algorithmic or von Neumann) it is necessary to calculate the density matrix of the resonant system after exposure to the object pulse (echelon).

Neglecting power broadening and spectral diffusion during the object pulse we find the density matrix in the interaction of the atom with a single Fourier component of the pulse echelon field with subsequent averaging over all frequencies.

The electric intensity of Fourier components of the pulse field can be written as

$$E(\omega') = \frac{1}{2}\left[E^*(\omega')e^{i(\omega-\omega')t} + E(\omega')e^{-i(\omega-\omega')t}\right]. \tag{21}$$

(a)

(b)

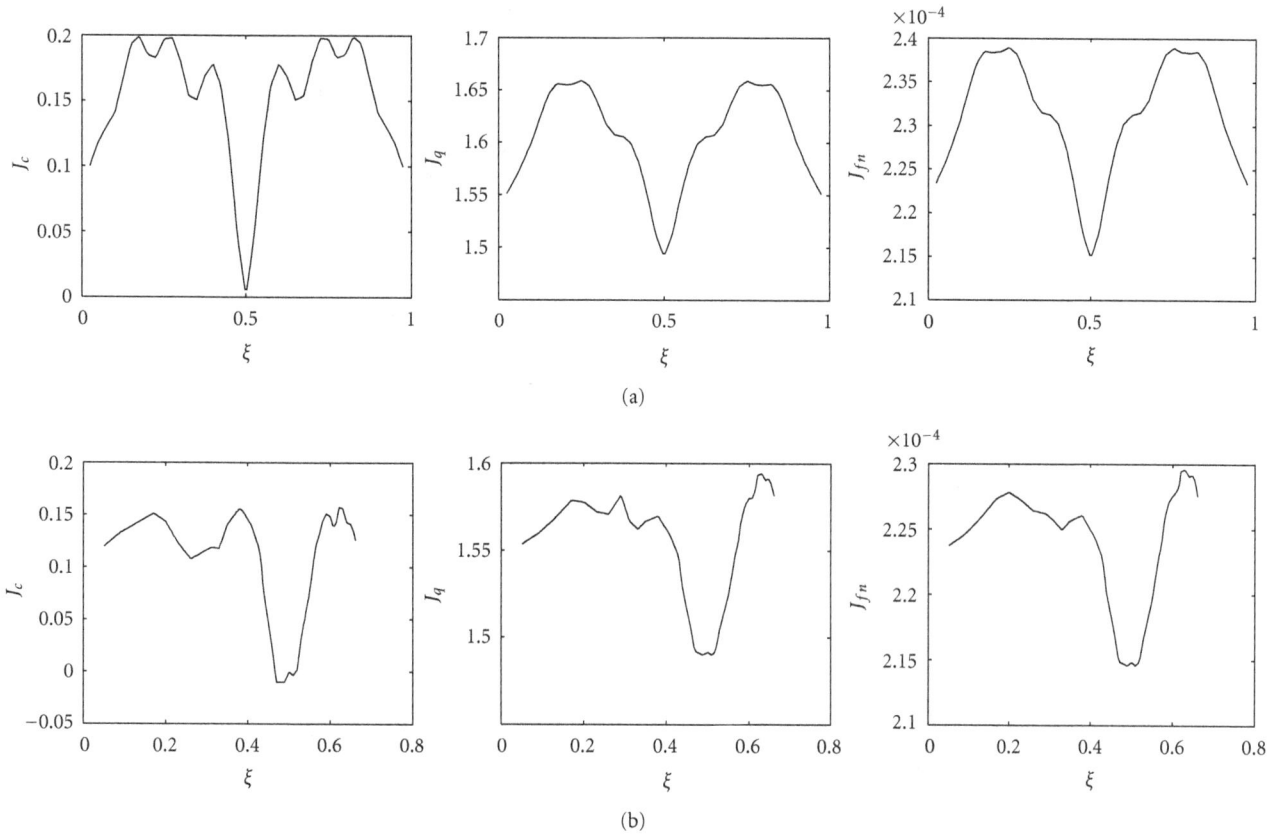

FIGURE 1: (a) Information measures in case of position change of the middle pulse in the object pulse echelon. (b) Information measures in case of position change of the extreme pulse in the object pulse echelon. $\xi = \tau_1/(\tau_1 + \tau_2)$; $\theta = \pi/2$: area of the object pulse; $n = \sigma \cdot \delta$ where σ is a half-width of the inhomogeneously broadened line, δt-the pulse echelon duration. J_c: classical information; J_q: quantum information; J_{fn}: von Neumann entropy.

The frequency of the Fourier spectrum ω' can be either > 0 or < 0. The frequency of the transition in an atom $= \Omega$, but in the case of interaction with the local fields (crystal), it becomes $\Omega - \Omega'$, where Ω' is the value of frequency shift. Ω' can be either > 0 or < 0 within the inhomogeneously broadened line of the atomic transition. We assume that $\omega = \Omega$, that is, the central laser frequency coincides with the transition frequency in the absence of the local field.

In this case, the equation for the Fourier components of the single-particle density matrix can be written as

$$\frac{\partial \tilde{\rho}}{\partial t} = -\frac{i}{\hbar}[B, \tilde{\rho}], \qquad (22)$$

where

$$B = \tilde{H}_0 - \hbar A + \tilde{V},$$
$$H_0 = \hbar(\Omega - \Omega')P_{22}, \qquad A = (\omega - \omega')P_{22},$$
$$e^{\pm iAt} = P_{11} + P_{22}e^{\pm i(\omega - \omega')t}, \qquad (23)$$
$$\tilde{V} = -\frac{1}{2}d[E^*(\omega')P_{12} + E(\omega')P_{21}],$$

d is the dipole moment of the resonance transition, P_{ij} are

the projector matrices (which have element ij equal to 1 and all other elements equal to 0)

$$B = \begin{pmatrix} 0 & -\frac{1}{2}dE^*(\omega') \\ -\frac{1}{2}dE(\omega') & \hbar(\omega' - \Omega') \end{pmatrix}. \qquad (24)$$

The solution of (22) can be written as

$$\tilde{\rho}(t) = e^{-i\hbar^{-1}Bt}\rho(0)e^{i\hbar^{-1}Bt}. \qquad (25)$$

The bordering exponents in (25) can be determined by methods of matrix functions. In the case $t = 0$ $\rho_{22}(0) \rightarrow 0$, $\rho_{11}(0) \rightarrow 1$ and

$$\tilde{\rho}(t) \approx P_{11}\left(\cos^2\frac{\theta}{2} + \frac{\Delta^2}{\theta'^2}\sin^2\frac{\theta}{2}\right)$$
$$+ P_{12}\left(-i\frac{a^*}{2\theta'}\sin\theta + \frac{a^*\Delta}{\theta'^2}\sin^2\frac{\theta}{2}\right)$$
$$+ P_{21}\left(i\frac{a}{2\theta'}\sin\theta + \frac{a\Delta}{\theta'^2}\sin^2\frac{\theta}{2}\right) + P_{22}\frac{|a|^2}{\theta'^2}\sin^2\frac{\theta}{2}, \qquad (26)$$

where

$$\Delta = \omega' - \Omega', \quad \theta = \theta' t, \quad \theta = \sqrt{\Delta^2 + d^2 E_0^2 \hbar^{-2} |\tilde{\varepsilon}|^2},$$

$$a = dE_0 \hbar^{-1} \tilde{\varepsilon} e^{i\vec{k}\vec{r}}, \quad a^* = dE_0 \hbar^{-1} \tilde{\varepsilon}^* e^{i\vec{k}\vec{r}}. \tag{27}$$

After pulse excitation $B = P_{22}\hbar\Delta$, that is

$$e^{\pm i\hbar^{-1} B(t'-t)} = P_{11} + P_{22} e^{\pm i\Delta \cdot (t'-t)}, \tag{28}$$

$$\tilde{\rho}(t'-t) = \left(P_{11} + P_{22} e^{-i\Delta \cdot (t'-t)}\right) \tilde{\rho}(t) \left(P_{11} + P_{22} e^{i\Delta \cdot (t'-t)}\right). \tag{29}$$

The final result for the quantum algorithmic information is of the form

$$J_q = \int_{-\infty}^{\infty} p(\omega') d\omega' \int_{-\infty}^{\infty} g(\Omega') J_q(\omega', \Omega') d\Omega', \tag{30}$$

where $g(\omega')$ is the frequency distribution of the inhomogeneously broadened line of the resonant transition, and $J_q(\omega', \Omega')$ is defined by the expression (12).

By analogy, for the von Neumann entropy we have

$$J_{fn} = \int_{-\infty}^{\infty} p(\omega') d\omega' \int_{-\infty}^{\infty} g(\Omega') J_{fn}(\rho(\omega', \Omega')) d\Omega', \tag{31}$$

where $J_{fn}(\rho(\omega', \Omega'))$ is defined by the expression (3).

After the influence on the resonant medium of the object pulse that carries classical information, this information is distributed over separate isochromatic components of the inhomogeneously broadened line. In other words, an information-phase grating within the inhomogeneously broadened line of the resonant transition arises. Every single q-bit can have a piece of classical (the diagonal part of the density matrix) and quantum information (off-diagonal part of the density matrix).

In case when the information lies in the time intervals of the laser pulse echelon, the minimum structure, which carries information, is a sequence of three laser pulses with unequal time intervals τ_1 and τ_2 between them. For such a structure the conversion $J_c \rightarrow J_q$ result is shown in Figure 1.

Figure 1(a) shows the values of information measures in case of the position change of the middle exciting pulse in the object pulse echelon and Figure 1(b) shows the values of information measures in case of the position change of the extreme exciting pulse in the object pulse echelon (the object pulse consists of 3 pulses). The highest obtained correlation coefficient between the classical and quantum information $\approx 0,94$.

5. Conclusion

The best classical information measure in the case of information encoding in the time intervals is the differential information entropy of the object laser pulse Fourier spectrum. Quantum information measure, based on algorithmic information theory, is the most appropriate for description of superpositional quantum system. Information encoding in the time intervals in laser pulses' echelon causes minimal information distortions in response to the resonant system.

References

[1] A. A. Kalachev and V. V. Samartsev, *Kogerentnyye yavleniya v optike*, Kazan State University, Coherent phenomena in optics, Kazan, Russia, 2003.

[2] M. Mitsunaga, R. Yano, and N. Uesugi, "Time- and frequency-domain hybrid optical memory: 1.6-kbit data storage in Eu^{3+}:Y$_2$SiO$_5$," *Optics Letters*, vol. 16, no. 23, pp. 1890–1892, 1991.

[3] H. Un, T. Wang, G. A. Wilson, and T. W. Mossberg, "Experimental demonstration of swept-carrier time-domain optical memory," *Optics Letters*, vol. 20, no. 1, pp. 91–93, 1995.

[4] H. Un, T. Wang, and T. W. Mossher, "Demonstration of 8-Gbit/in.^{2} areal storage density based on swept-carrier frequency-selective optical memory," *Optics Letters*, vol. 20, no. 15, pp. 1658–1660, 1995.

[5] B. Schumacher, "Quantum coding," *Physical Review A*, vol. 51, no. 4, pp. 2738–2747, 1995.

[6] N. J. Cerf and C. Adami, "Negative entropy and information in quantum mechanics," *Physical Review Letters*, vol. 79, no. 26, pp. 5194–5197, 1997.

[7] L. A. Nefed'ev and I. A. Rusanova, "Copying quantum information in a three-level medium with a phase memory," *Laser Physics*, vol. 12, no. 3, pp. 571–575, 2002.

[8] A. N. Kolmogorov, *Teoriya informatsii i teoriya algoritmov*, Nauka, Moscow, Russia, 1987.

[9] L. A. Nefed'ev and I. A. Rusanova, "Information processes in optical echo holography," *Optics and Spectroscopy*, vol. 90, no. 6, pp. 906–910, 2001.

20

Advances in Red VCSEL Technology

Klein Johnson, Mary Hibbs-Brenner, William Hogan, and Matthew Dummer

Vixar, 2950 Xenium Lane, Suite 104, Plymouth, MN 55441, USA

Correspondence should be addressed to Mary Hibbs-Brenner, mhibbsbrenner@vixarinc.com

Academic Editor: Rainer Michalzik

Red VCSELs offer the benefits of improved performance and lower power consumption for medical and industrial sensing, faster printing and scanning, and lower cost, higher speed interconnects based upon plastic optical fiber (POF). However, materials challenges make it more difficult to achieve the desired performance than at the well-developed wavelength of 850 nm. This paper will describe the state of the art of red VCSEL performance and the results of development efforts to achieve improved output power and a broader temperature range of operation. It will also provide examples of the applications of red VCSELs and the benefits they offer. In addition, the packaging flexibility offered by VCSELs, and some examples of non-Hermetic package demonstrations will be discussed. Some of the red VCSEL performance demonstrations include output power of 14 mW CW at room temperature, a record maximum temperature of 115°C for CW operation at an emission wavelength of 689 nm, time to 1% failure at room temperature of approximately 200,000 hours, lifetime in a 50°C, 85% humidity environment in excess of 3500 hours, digital data rate of 3 Gbps, and peak pulsed array power of greater than 100 mW.

1. Introduction

Multimode 850 nm VCSELs based upon the AlGaAs materials system have been the standard optical source for glass fiber optic-based data communication links since the mid-1990s. Although the first demonstration of red VCSELs followed fairly quickly after the demonstration of the industry standard "all-semiconductor" 850 nm VCSEL, the commercialization of red VCSEL technology has proceeded much more slowly due to the materials limitations that have made the development more challenging.

The AlGaAs materials system which is used for 850 nm VCSELs provides good lattice matching over the full range of compositions, a reasonably good refractive index contrast between the high index (AlGaAs with approximately 15–20% mole fraction AlAs) and low index (AlAs) materials used for the mirrors, and a high (approximately 0.35 eV) conduction band offset between the GaAs quantum wells and the AlGaAs compositions normally used as quantum well barriers. However, the 650–700 nm emission wavelength range requires use of GaInP quantum wells with AlGaInP barrier layers, with the compositions limited to those which are nearly lattice matched to a GaAs substrate. The AlGaAs materials system is usually used for the mirrors. Several

limitations for these shorter wavelength VCSELs exist: (1) the available conduction band offset is smaller and ranges from approximately 0.17 eV at 650 nm to 0.23 eV at 700 nm [1]. Therefore, thermal carrier overflow limits the maximum temperature of operation and peak power, limitations that become more apparent at shorter wavelengths. (2) The requirement that the mirrors be nonabsorbing limits the mirror compositions available to AlGaAs materials with AlAs mole fraction greater than 0.4-0.5. This reduces the available range of refractive index, requiring more mirror periods to achieve the same reflectivity. Furthermore, the 50% AlGaAs composition has a higher thermal and electrical resistivity than compositions closer to the binary AlAs or GaAs. This results in a higher conversion of input power to heat, and more difficulty in removing heat from the device, resulting in higher junction temperatures. The development of red VCSELs over the past nearly two decades has focused on ameliorating these issues. Since the early 1990s when the initial red VCSEL research was done, several groups have reported performance improvements in peak output power, temperature range of operation, wavelength range, modulation speed, and reliability. Significant strides in developing red VCSELs have been made, and devices are now being implemented in a number of applications. This paper will first review the progress that has been made, will then discuss recent developments, and finally discuss some applications of the technology.

2. Background

The earliest reports of red VCSEL demonstrations were in 1993 from Sandia National Labs and Chiao Tung University in Taiwan [2, 3]. Some of this work was initially based upon the use of InGaAlP materials for both the quantum well active region and the mirrors, but fairly quickly most researchers adopted a structure that retained the InGaAlP-based quantum well active region, but used AlGaAs materials in the mirrors, all on a GaAs substrate, thus simplifying the challenge of growing lattice-matched structures. The initial devices were limited in output power and temperature range, as might be expected from the earliest demonstrations of a new technology.

Due to the wavelength dependence of the conduction band offset available in the AlGaInP materials system, the peak output power achieved and the temperature range of operation is a strong function of wavelength. A fairly early paper [4] reported 8.2 mW of maximum multimode output power, and 1.9 mW of single-mode output power at 687 nm at room temperature. This device was fabricated using a proton implant process. Several improvements to the red VCSEL structure, including the use of carbon doping in the mirror, graded interfaces in the mirror, separate confinement structure in the active region, an oxide aperture, and the removal of the GaAs contact layer from the aperture resulted in the demonstration of room temperature peak output power of 4 mW at 650 nm and 10 mW at 670 nm [5, 6]. Johnson and Hibbs-Brenner reported an output power of 11.5 mW at 673 nm at room temperature [7].

The temperature range of operation is also a strong function of wavelength. The first demonstrations of red VCSELs in 1993 required pulsed operation to lase [2, 3]. However, by 1994 CW operation to 45°C had been achieved at 670 nm [8]. Room temperature operation for wavelengths as short as 645 nm was demonstrated in 1995, although the output power was very low (0.04 mW) [9]. Calvert et al. [10] reported the continuous wave operation of 670 nm single-mode devices to a heat sink temperature of 80°C. Sale et al. [11] also demonstrated CW lasing at 666 nm over the temperature range from −180°C to +80°C. Knigge et al. [5] extended the temperature range by demonstrating 650 nm VCSELs achieving output powers of 4.3 mW at room temperature, with lasing to 65°C, and 10 mW at 670 nm (room temperature) with lasing to 86°C. The temperature range of operation can further be extended by pulsing the devices, with a temperature up to 160°C for a 670 nm device and up to 172°C for a 660 nm VCSEL [12, 13].

Efforts have been made to extend the feasible wavelength range of red VCSELs, both to shorter wavelengths (<650 nm) and to longer wavelengths (>700 nm). The first red VCSELs reported below 699 nm actually included wavelengths as short as 639 nm, although they only operated under pulsed conditions [2]. Choquette et al. [9] were able to achieve

0.25 mW CW room temperature emission at 652 nm and 0.04 mW CW room temperature operation at 645 nm. Knigge et al. [14] were able to achieve pulsed operation down to 629 nm, although no power was reported, and 2.1 mW of room temperature CW power at 647 nm. Several groups have reported efforts to extend the VCSEL wavelength into the range from 700 to 740 nm [15–20]. All of these efforts above 700 nm have been based upon the AlGaAs materials system for both the mirrors and the active regions. The performance has been limited in output power and maximum temperature of operation. In some cases [18–20] the device only operated in pulsed mode at room temperature, while in other results [15–17] the devices did operate CW at room temperature, but the output power was limited to less than 1 mW.

Reliability data has been fairly limited. An early report on aging and failure analysis performed the testing under fairly unrealistic conditions, that is, current drive that was 3x past the rollover point, resulting in a junction temperature of around 250°C [21]. This is likely to result in a degradation mechanism that is not representative of normal use conditions. Low-temperature testing was performed on 655 nm VCSELs [22], demonstrating little degradation over 1000 hours at 20°C for a 7 μm diameter device operated at 2.5 mA, but a 3 dB degradation in output power after 500 hours at 40°C. A second report indicated widely varying results at 665 nm, with one wafer remaining fairly stable during life testing at 6 mA and 100°C for 1000 hours, while a second wafer with a different design (the design differences were not described), where it is noted that the resistance is higher and required drive voltage in the range of 5-6 V, failed after 300 hours at the same condition [23]. The device diameter was not reported in this case. The most complete report was published in 2008 [24]. Although the device diameter was not specified, 128 devices were placed on test at 8 different acceleration conditions, ranging from 40 to 85°C, and current drive of 3, 4, or 5 mA. The devices remained on test for close to 8000 hours. From the multiple acceleration conditions, the authors estimated an empirical failure acceleration model of an Arrhenius dependence on temperature with an activation energy of 0.6 eV, and a squared dependence upon current. From this data they estimated time to 1% failure of several hundred thousand hours for use conditions of 40°C and 1.5 mA.

Since one of the main applications for red VCSELs is for data communication over plastic optical fiber, the achievable modulation rate is a key parameter of interest. An early measurement [25] showed a small signal 3 dB bandwidth of around 2.5 GHz and also demonstrated low error rate large signal modulation at 1 Gbps. A subsequent study with better performing VCSELs measured a 3 dB bandwidth of 11 GHz [26]. Duggan et al. [24] demonstrated a 3 dB bandwidth in excess of 3 GHz at bias currents of less than 10 mA. Large signal modulation at 1.25 Gbps was demonstrated over temperatures ranging from −20 to +60°C. Transceiver modules and connector systems operating at 1.25 Gbps have been developed for plastic optical fiber-based communication links [27, 28].

Novel approaches for dealing with the limited electrical and thermal conductivity have included the incorporation of transparent indium tin oxide (ITO) contacts that extend across the entire VCSEL aperture [29] and the incorporation of plated copper interconnects around the VCSEL mesa [30]. While the latter approach reduced the thermal resistance, it was postulated that the stress created by the approach prevented improvements in output power and temperature range.

Mode control, that is, for achieving single transverse mode VCSELs, is also a challenge and is typically achieved by reducing the aperture size. Kasten et al. used a photonic crystal approach to achieve single-mode performance [31].

3. VCSEL Development Approach

The goals of the work reported here were to increase the output power, temperature range of operation, achievable wavelength range, and reliability of red VCSELs. Specifically, the targets were a minimum of 1 mW single-mode power from 0–60°C, 10 mW multimode power up to 40°C, and at least 1 mW of useable multimode power at 80°C. Another goal was to extend performance to 720 nm with >1 mW of useable output power.

The red VCSEL structure is illustrated schematically in Figure 1. The structure is grown on off-axis 4″ n+-GaAs substrates. Mirrors are AlGaAs based with graded interfaces between the high and low index layers. The results in this paper come from several different wafers, but in all cases the bottom mirror consisted of between 50 and 56 periods, and the number of top mirror periods ranged from 28 to 36. The active region consists of 3–5 compressively strained InGaP quantum wells with unstrained or tensile strained AlGaInP barriers and a graded (50–70%) AlGaInP separate confinement heterostructure (SCH). The p-spacer is doped with Zn. A highly doped contact layer is grown at the top surface to facilitate formation of ohmic contacts.

Current and index confinement is provided by an oxide confinement layer located 2 periods above the quantum well active region. The devices are top-emitting with a ring contact patterned around the current aperture on the front side of the device. The substrate was thinned to 200 μm before the deposition of a broad area gold contact on the substrate side of the wafer. A variety of aperture sizes were fabricated to evaluate performance as a function of aperture size. Some die consisted of an array of apertures connected to a common anode in order to increase the total power output per die.

Wafers were probed on an automated probe station with wafer temperature control. 100% probe testing of the light output and voltage versus drive current (L-I-V) and wavelength was performed. L-I-V measurements were made over a range of temperatures on a sample basis. Devices were mounted on a TO-46 header for measurement of beam profiles, and for measurement of pulsed characteristics.

Reliability measurements under pulsed conditions were carried out on devices in hermetic TO-46 packages. Resistance to humidity was evaluated at 50°C, 85% humidity

on devices packaged in TO-46 headers but with the glass window removed from the lid. In both cases devices are biased during life testing at the accelerated environmental conditions. However, the devices are removed from the oven at each test point and tested at room temperature and room humidity, which was typically 20–25°C and 40% relative humidity.

4. Results

4.1. Temperature Performance. One of the most challenging aspects of designing red VCSELs has been achieving useable output power over the temperature ranges required by the applications of interest. Figure 2 illustrates the temperature performance of two red VCSELs: a single-mode design, emitting at 689 nm and a multimode design emitting at 693 nm. The single mode device (Figure 2(a)) lases up to 115°C, with 1 mW of output power achievable at 75°C, and 0.5 mW at 95°C. (The vertical scale is expanded to allow a closer look at the L-I-V characteristics at the highest temperature.) We believe that this is the highest temperature operation achieved in red VCSELs under CW conditions. Generally, the temperature range of operation of larger diameter devices is more limited. Figure 2(b) illustrates the temperature performance of a multimode 693 nm VCSEL. This device was tested at increments of 20°C, and output 1.5 mW peak power at 80°C, but ceased lasing between 80°C and 100°C. The improved temperature performance is the result of attention to a number of details, such as mirror design for low thermal and electrical resistance, engineering of the active region, and the proper choice of the gain peak-Fabry-Perot resonance offset. While the longer wavelength of approximately 690 nm also makes the higher temperature operation easier to achieve due to the larger band discontinuities, the result is still a record temperature even when compared to other results at the same wavelength.

4.2. Single-Mode Performance. Figure 3 further illustrates the performance of single-mode devices at 25°C. Figure 3(a) overlays the L-I-V curves of an array of single-mode VCSELs. Figure 3(b) shows the beam profile of one of the devices in the array. Profiles in the x- and y-direction are taken at three different currents, that is, 4, 5, and 6 mA and are overlaid in the figure of intensity versus angle. It is difficult to distinguish more than one plot since the three plots overlap so closely. Single spatial mode performance is maintained up to the current corresponding to the peak output power.

4.3. Wafer Uniformity: Wavelength and Device Performance. One of the key questions of interest in the production of devices is the uniformity across a wafer. The wavelength of a VCSEL is approximately proportional to thickness of the layers, so a 1% variation of thickness can result in approximately a 7 nm variation in wavelength. In addition, the oxidation diameter can also vary across a wafer due to small differences in layer thickness, doping, or composition. Both of these effects can impact performance of a VCSEL. For instance, the temperature characteristics of a VCSEL depend

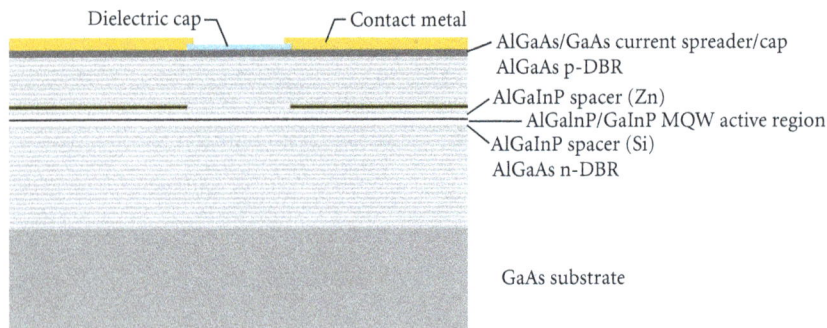

FIGURE 1: Schematic of VCSEL structure (From Proceedings of the SPIE, Vol. 7952, paper 795208).

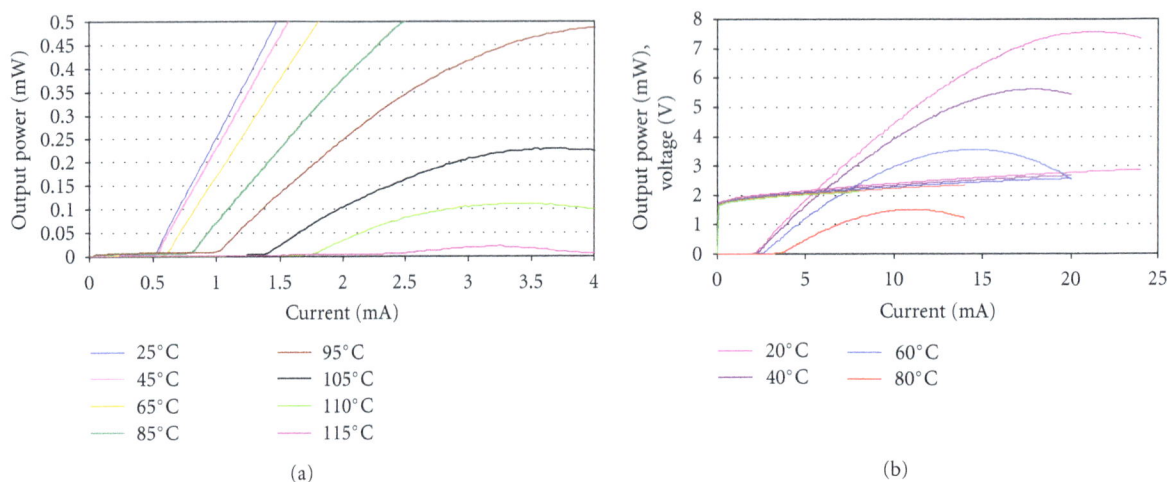

FIGURE 2: Light output and voltage versus current (L-I-V) at a range of temperatures for (a) a 689 nm single-mode device, and (b) a 693 nm multimode device (from Proceedings of the SPIE, Vol. 7952, paper 795208).

upon the offset between the gain peak and the Fabry-Perot resonance. Since the gain peak wavelength is less sensitive to thickness and therefore nearly constant across the wafer, while the Fabry-Perot resonance may have a range of 5–10 nm, this offset varies across the wafer. The ability to do automated wafer scale testing allows us to gather statistics on uniformity.

Figure 4 shows the results of probing approximately 60,000 VCSELs on a 4″ wafer. A histogram of the wavelength distribution of the devices is shown in Figure 4(a). While the distribution ranges from 675 nm to 709 nm, the vast majority of the devices on the wafer lie in the range from 681 to 692 nm. Figure 4(b) illustrates the uniformity of threshold current as a function of emission wavelength for four different laser aperture sizes. This data (and the data in Figure 4(c)) was taken at 40°C. The shaded region in the figure corresponds to the wavelength range constituting the majority of the devices on the wafer. Threshold currents are less than 0.5 mA for the smallest devices and from 1.5 to 2.5 mA for the 12 μm apertures at 40°C. While the largest diameter devices appear to be quite a bit less uniform than the smallest diameter, on a percentage basis the threshold current range of the 12 μm device is similar to that of the 7 and 10 μm devices.

As one might expect, the threshold current is U-shaped and rises at the longer wavelengths, due to a larger offset between the gain peak and the Fabry-Perot cavity, but devices are still lasing at 709 nm, where the offset is approximately 40 nm. Peak output power at 40°C versus wavelength for several aperture sizes is shown in Figure 4(c). Within the wavelength region included in the shaded region, which includes the majority of the VCSELs on the wafer, the peak output power is quite uniform.

4.4. Maximum Achievable Output Power. Red VCSELs have typically been limited in the maximum output power that can be achieved in part because the larger aperture devices are more sensitive to temperature. Improved design has allowed larger devices to be built. Improvements include the use of quantum well barrier layers with tensile strain to improve the conduction band offset, tailoring of the doping profile in the mirrors to reduce series resistance, and the use of a slightly thicker high aluminum-containing mirror layers, and thinner low aluminum-containing mirror layers (while keeping the sum of the two equal to $\lambda/2$) in each mirror period to improve thermal conductivity. Figure 5 illustrates the 25°C output power achievable from two types of devices. Figure 5(a) shows the L-I-V from a single aperture

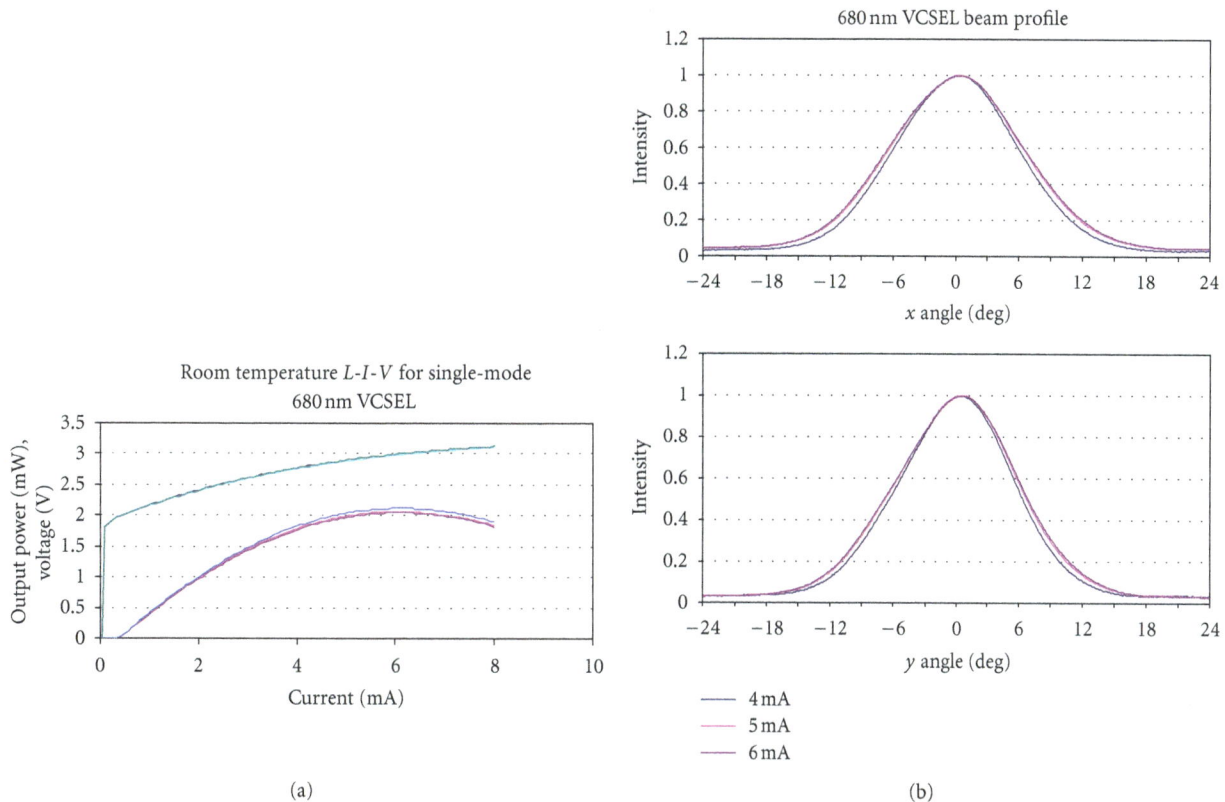

(a)   (b)

FIGURE 3: Performance of single-mode 680 nm VCSELs at 25°C. (a) Overlaid *L-I-V* curves from a 1×3 array. (b) Beam intensity (normalized, arbitrary units) versus angle in the *x*- and *y*-direction at three current levels: 4, 5, and 6 mA (from Proceedings of the SPIE, Vol. 7952, paper 795208).

device with a wavelength of 685 nm which emits a peak output power of 14 mW. If beam size or profile is not an issue, an alternative way of achieving high output power is to use multiple apertures with a common anode contact. Figure 5(b) illustrates that nearly 45 mW of output power can be generated from a 4 × 4 array of apertures emitting at 680 nm within a 200 μm × 200 μm area.

4.5. Extended Wavelength Performance. We have fabricated devices with wavelengths in the range from 700 to 720 nm, but unlike previous reports [15–20] our devices are based upon GaInP/AlGaInP active regions. A large variation in wavelength across a single wafer was achieved by not rotating the wafer during growth. The gain peak wavelength was fairly constant at around 678 nm, while the Fabry-Perot resonance varied from 680 up to nearly 720 nm. The longest wavelengths, therefore, corresponded to a very large gain peak-resonance offset, as large as 40 nm. Figure 6 shows the results from two devices lasing CW at 25°C. A 716 nm device (Figure 6(a)) had a threshold current of 7 mA and a peak output power of nearly 3.5 mW, while a 719 nm device (Figure 6(b)) had a threshold current of 7 mA and a peak output power of over 2 mW. The threshold current is high due to the large gain peak-resonance offset, so it is believed that even better performance could be achieved if the devices were optimally designed for this wavelength. We therefore

believe that good VCSEL performance spanning the entire wavelength range from 650 to 850 nm can be achieved using either the AlGaAs materials system, or the AlGaInP materials for the active region.

4.6. Pulsed Operation and Reliability. There are some applications where lasers are typically pulsed at a low duty cycle, such as industrial sensors, or the computed radiography application described in the applications section below. Pulse widths in the range of 1 μsec and a duty cycle less than 25% are common. It has been demonstrated at other wavelengths that the peak output power can be extended significantly due to the reduction in self-heating when pulsed. Red VCSELs are even more limited by thermal effects and hence we desired to quantify the magnitude of potential improvement that could be achieved if the devices were pulsed.

Figure 7 illustrates the performance of a multimode 680 nm VCSEL operated in pulsed mode. The relevant parameters affecting pulsed performance are pulse width, duty cycle, and ambient temperature. Since 1 μsec is a nominal thermal time constant for a VCSEL chip, pulse widths substantially longer than this provide little benefit. We have used 1 μsec pulse width for the evaluation, although shorter pulse widths can provide even higher peak power. Figure 7(a) illustrates the effect of the duty cycle. A 10% duty cycle can provide nearly a 4x improvement in peak power,

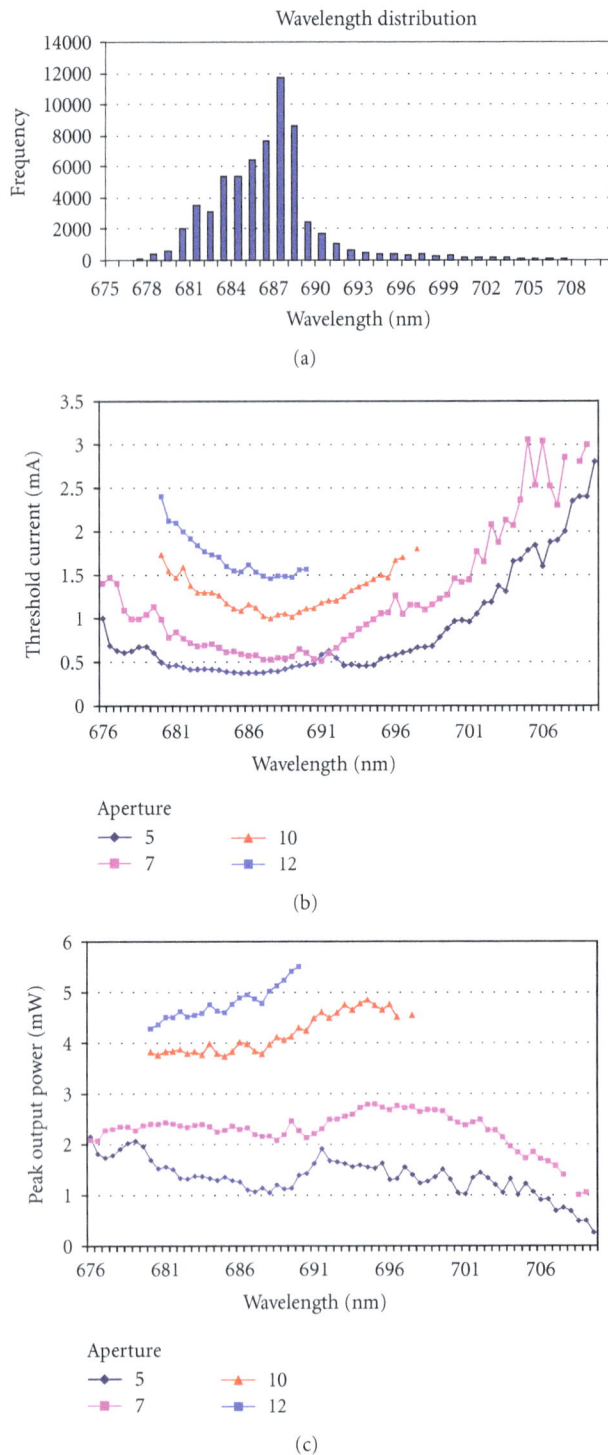

FIGURE 4: (a) Histogram showing the wavelength distribution of 60,000 VCSELs tested at 40°C on a 4″ wafer. (b) Threshold current versus wavelength. (c) Peak output power versus wavelength. The shaded regions in (b) and (c) indicate the wavelength range corresponding to the vast majority of devices on the wafer (from Proceedings of the SPIE, Vol. 7952, paper 795208).

while a 50% duty cycle still provides nearly a 2x increase in peak power. Also note that the peak power achieved at a 10% duty cycle exceeds 35 mW. This is a multimode device with a single aperture. Figure 7(b) illustrates the improvement in the temperature range of operation that can be achieved when the device is pulsed at a 25% duty cycle. The peak

power of the device at 60°C reaches 15 mW, while under CW operation, the same device might only reach a peak power of 3-4 mW.

Pulsing can also increase the peak output power of the arrays described above. Figure 8 illustrates the benefits of pulsing an array 680 nm device. The figure demonstrates that

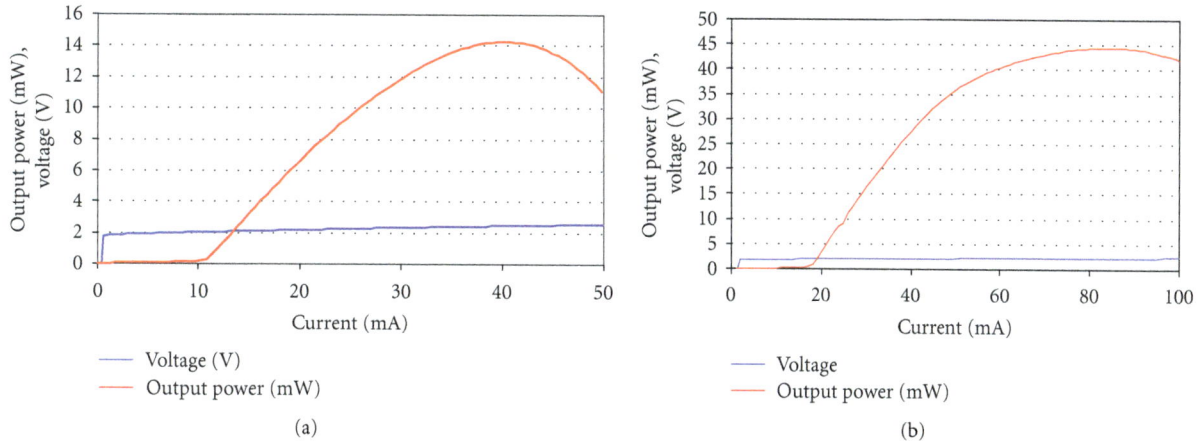

(a)

(b)

FIGURE 5: High output power devices at 25°C. (a) *L-I-V* curve from a 685 nm device with a single large aperture, demonstrating a peak output power of 14 mW. (b) *L-I-V* from a 680 nm device with multiple apertures in a 200 μm × 200 μm area, demonstrating a peak output power of 44 mW (from Proceedings of the SPIE, Vol. 7952, paper 795208).

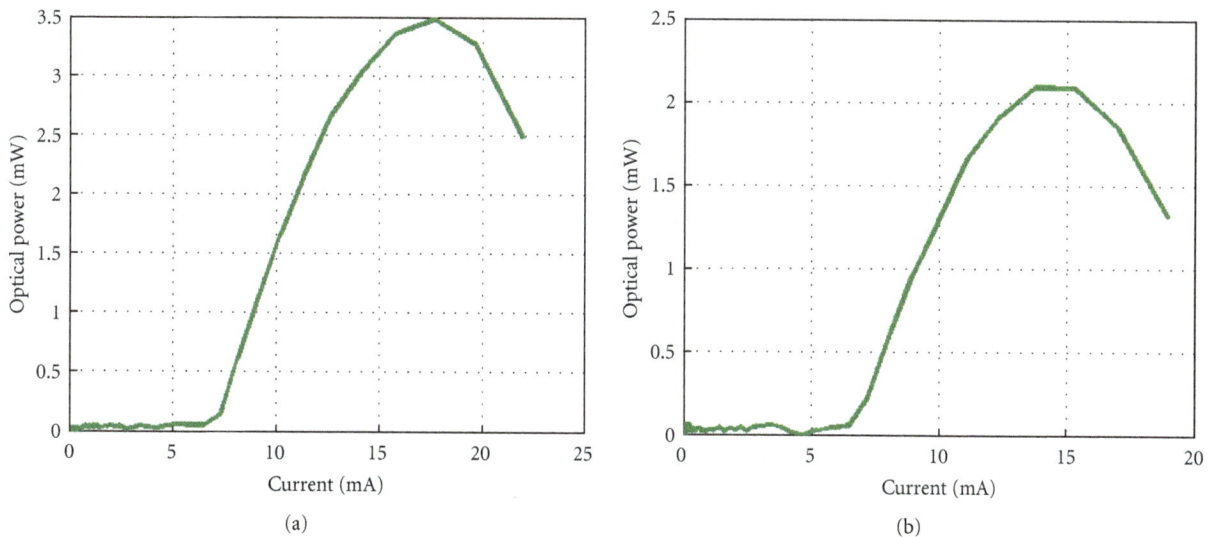

(a)

(b)

FIGURE 6: Performance at 25°C of an AlGaInP QW-based VCSEL at (a) 716 nm, and (b) at 719 nm (from Proceedings of the SPIE, Vol. 7952, paper 795208).

nearly 120 mW of peak power can be reached in a 3 × 3 array on a single die, using a pulse width of 1 μsec and a 1% duty cycle.

In pulsed mode the device self-heating is reduced, and therefore the device rollover point (where increasing the current actually results in a reduction of output power) is extended to significantly higher drive current. However, this leads to a question: if device lifetime is reduced by higher current (or, more accurately, current density) can one operate a device in pulsed mode at these higher current ranges for useful periods of time? Furthermore, are there any transient effects, such as stress created by repeated junction temperature cycling resulting from the current cycling that might actually accelerate the degradation of the devices beyond what is normally expected from the current drive alone? For instance, the VCSEL lifetime is commonly found to be reduced proportionally to the square of the current

density. An increase of drive current from 8 mA to 30 mA might be expected to reduce the lifetime by a factor of 14 due to current density alone. Under CW conditions, the increase in junction temperature from the higher current adds to the acceleration of failure. Using the empirical model for acceleration of failure, we estimate that a 30 mA CW drive current would reduce lifetime by a factor of 500 under CW conditions.

To experimentally evaluate the effect of pulsing on reliability we developed a capability for testing the VCSELs in pulsed mode. Both single-mode and multimode devices were packaged in TO-46 headers and mounted on boards that were placed in ovens. The devices were pulsed with a pulse width of 1 μsec, and a duty cycle of 12.5%. Thus 8 hours of test time correspond to 1 hour of actual "on-time" The devices were periodically removed from the oven and tested CW at room temperature, and then returned to the oven

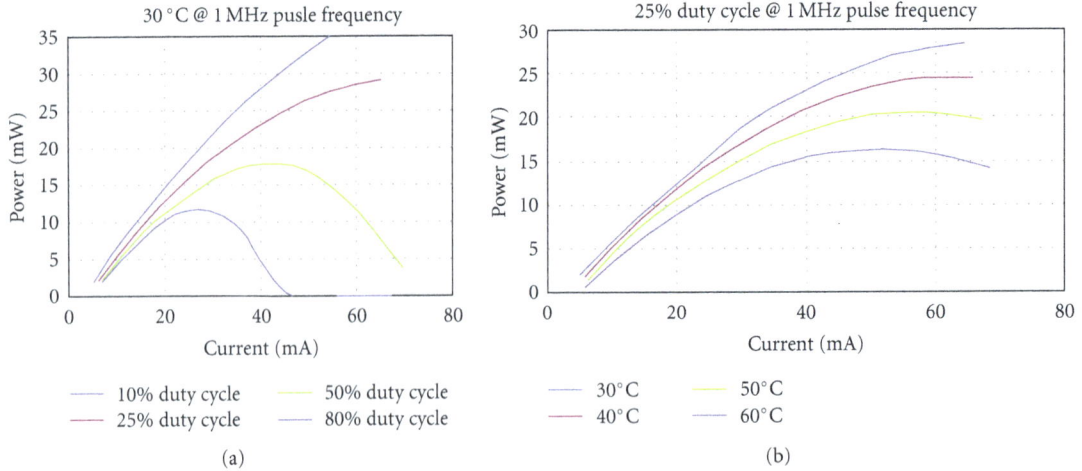

FIGURE 7: Output power versus current for multimode 680 nm VCSELs operated in pulsed mode with a 1 μsec pulse width. (a) illustrates the effect of varying the duty cycle at 30°C, while (b) illustrates the improved temperature performance associated with pulsed operation (from Proceedings of the SPIE, Vol. 7952, paper 795208).

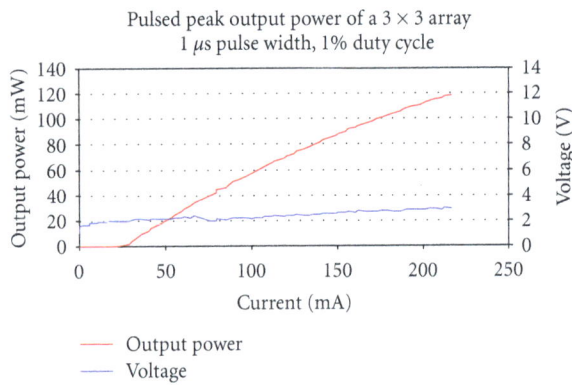

FIGURE 8: An output power of nearly 120 mW is achieved from a 3 × 3 680 nm array under pulsed conditions.

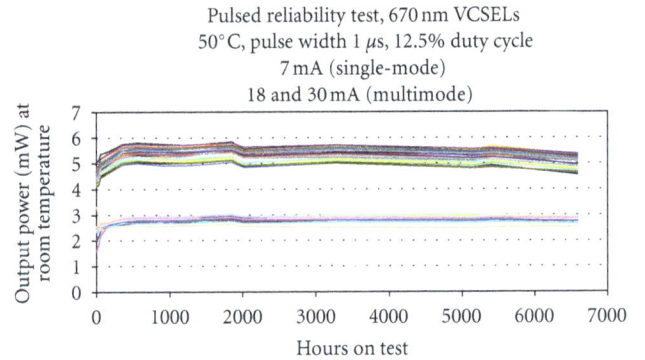

FIGURE 9: Peak output power versus test time for 670 nm VCSELS tested in pulsed mode. The output power testing was performed at room temperature. The lower curves correspond to a smaller diameter single-mode device, while the upper curves correspond to a multimode device (from Proceedings of the SPIE, Vol. 7952, paper 795208).

for further aging under pulsed conditions. The results are illustrated in Figure 9.

The multimode devices have a CW peak output power around 5 mW and were pulsed to one of two different current levels, 18 mA or 30 mA. The single-mode devices have a CW peak output power of approximately 2.5 to 3 mW and were pulsed to 7 mA. A burn-in effect can be seen in the first 100–200 hours, where the output power increases, but after the burn-in period, the output power has been stable during the 6596 hours of test at 50°C, corresponding to 824 hours of actual pulsed on-time.

Table 1 illustrates the differences in acceleration factor one might expect for the CW and pulsed current cases. In this table we compare acceleration factors based on the assumption of a use condition at 25°C and 10 mA. We have measured the thermal resistance of the multimode device, and found it to be 1.4°C/mW. We assume acceleration factors which have been reported in [24] for red VCSELs, that is, an Arrhenius relationship for temperature dependence with an activation energy of 0.6 eV, and a squared dependence on

TABLE 1: Calculation of acceleration factors assuming an acceleration temperature of 50°C and a use condition of 10 mA and 25°C.

DC or pulsed	Acceleration current	Acceleration factor
DC	10	4.5
	18	88
	30	2438
Pulsed	10	6
	18	20
	30	55

current. These acceleration factors are also representative of that routinely reported for 850 nm VCSELs. We also assumed a thermal resistance of 0 in the pulsed case, that is, that the junction temperature does not rise above ambient. This assumption may not be completely accurate, but it helps to define the range of acceleration factors that might be expected.

While we do not yet have sufficient failures to project a lifetime, this table predicts a very significant improvement in lifetime under pulsed conditions, assuming no transient effects, which is consistent with our observations. We would certainly expect devices operated CW at 30 mA at 50°C for the equivalent of 824 hours (6596 test hours times the 12.5% duty cycle) to have failed. The lack of failures also prevents us from completely ruling out acceleration due to thermal transients when operated under pulsed conditions, but the lack of degradation observed in Figure 9 suggests that this is not a significant consideration. During 6596 test hours, at a period repetition period of 8 μsec, the devices have experienced approximately $3e^{12}$ pulses.

More conventional reliability testing is carried out under conditions of constant current drive. Temperature and current are the most commonly assumed acceleration factors, with humidity being an acceleration factor for devices in non-Hermetic packages. We performed evaluation of VCSELs under dry conditions by placing 186 multimode devices on test at three different temperatures and currents, and periodically removed the devices from the ovens to test output power at room temperature. Failure was defined as a 3 dB reduction in output power as compared to the output power at time zero. Devices were aged at 50°C, 85°C, and 105°C, and 7, 11, and 15 mA of drive current. By assuming the same failure acceleration model described above, we calculated the acceleration factor for each of the test conditions relative to a use condition of 40°C and a current drive of 8 mA. We then created a "meta-analysis" of the failures by translating the time to each failure at its test condition to the equivalent time at 40°C and 8 mA. The results are shown in the plot in Figure 10, below. It can be seen from the figure that the time to 1% failure is nearly 200,000 hours, or more than 20 years.

Since some applications require a nonhermetic package, and most active optical devices are sensitive to a humid environment, it is important to understand the acceleration of failures due to exposure to humidity. We have performed environmental testing on our chips under accelerated conditions of temperature and humidity. Five chips were packaged in a TO can with the window removed and placed on boards in a chamber held at 50°C and 85% humidity. During aging they were driven continuously with 5 mA of drive current. The parts were periodically removed from the environmental chamber, and L-I-V curves tested at room temperature and approximately 30–40% humidity, and the maximum output power recorded. The performance is summarized in Figure 11. After more than 3500 hours, all five devices still demonstrate stable output power.

4.7. Applications. The target applications for red VCSELs fall outside of the existing data communication market where the vast majority of VCSELs have been applied. VCSELs can bring unique value to medical sensor and diagnostic devices, office equipment such as laser printing, industrial sensors, and low cost communications based upon plastic optical

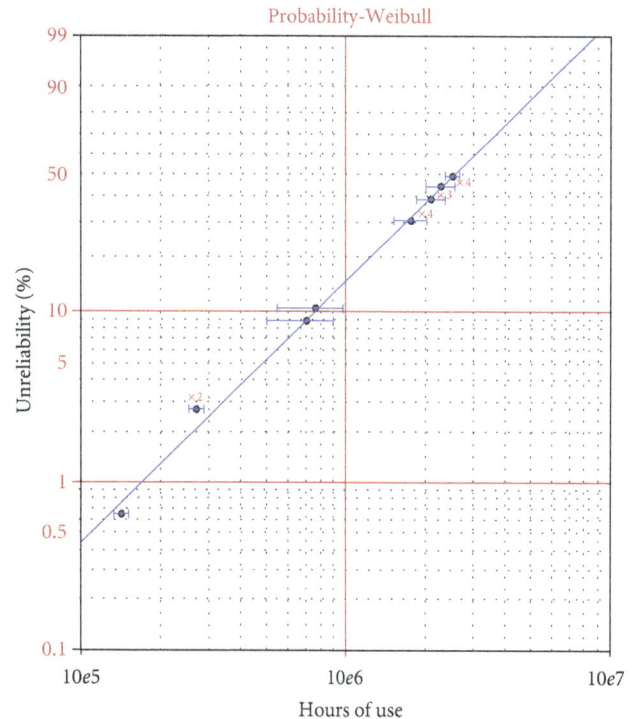

FIGURE 10: A plot of the failures times of the devices from our reliability evaluation of 670 nm multimode devices translated to their equivalent failure times at the use condition of 40°C and 8 mA. The y-axis is the percentage of devices that have failed. Time to 1% failure occurs at nearly 200,000 hours.

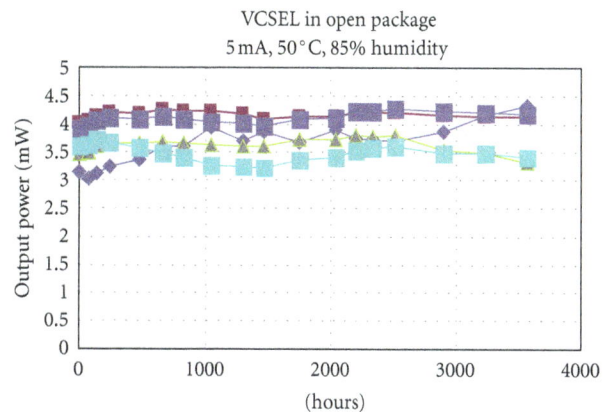

FIGURE 11: Output power versus hours on accelerated life testing for 690 nm VCSELs. Devices were maintained at 50°C and 85% humidity and driven with 5 mA during aging. L-I-V curves were taken at room temperature and room humidity at the test points indicated.

fiber. Design and packaging solutions to address some of these uses are described below.

As a first example, an important optically based noninvasive medical sensor application is oximetry. Pulse oximetry, which measures the oxygen content of arterial blood, is well-established using LEDs, while tissue or regional oximetry,

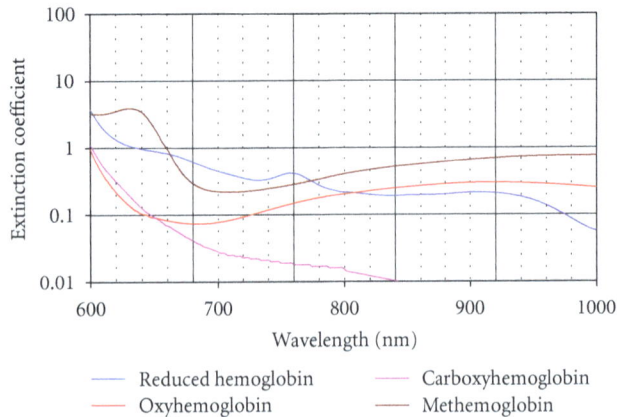

FIGURE 12: The absorption spectra versus wavelength for four different components of hemoglobin (from http://www.masimo.com/Rainbow/about.htm).

which measures venous or capillary blood, is an emerging application. Near-infrared spectroscopic-based imaging, which relies on differing absorption and scattering as a function of quantity and oxygenation of blood for image contrast, is an active area of research. All versions of oximetry take advantage of the varying absorption coefficient as a function of wavelength for different types of hemoglobins, that is, oxyhemoglobin, reduced hemoglobin, carboxyhemoglobin, or methemoglobin as is illustrated in Figure 12. The sensors rely on the absorption of wavelengths in the range from 650 nm to 1000 nm, and as the number of blood components one wishes to distinguish increases, the number of different wavelengths that one needs to employ also increases. These applications benefit from the narrow spectral line width and the slow spectral shift with temperature of the VCSEL, while wireless implementations make use of the reduced power consumption of VCSELs as compared to LEDs. In imaging applications, the high speed modulation characteristics of VCSELs are also useful in distinguishing absorption loss from scattering loss.

These sensors are often disposable, body worn sensors requiring a low cost, very compact package that can accommodate multiple chips of various wavelengths. Figure 13(a) illustrates a package that can accommodate multiple devices, potentially of different wavelengths, in a single package. Up to three individually modulated devices can be incorporated, and four devices can be accommodated if they are operated in alternating forward-bias/reverse-bias pairs. Devices are mounted on a lead using conductive epoxy and wire bonded. They are then embedded in an optically clear encapsulant that protects the wire bond and allows the package to be attached to a circuit board via solder reflow. This packaging approach is non-Hermetic, and hence requires the humidity resistance described earlier. Figure 13(b) illustrates the *L-I-V* (output power and voltage versus current) for the three wavelengths incorporated into a single package. An example of the application of the multiwavelength technology for near-infrared spectroscopic-based brain imaging has been demonstrated at Drexel University [32].

A second example application is plastic optical fiber (POF) links based upon PMMA fiber materials which have been implemented for sensor and data links in automobiles, and are being considered for home networks. PMMA-based fiber has secondary absorption minima in the red. Absorption at 850 nm is too high for links more than a few meters. While the potential for high speed data rates and the packaging simplicity of VCSELs makes them ideal for this application, wavelengths in the range of 650–680 nm are a necessity for low loss links. POF links based upon LEDs have been implemented in automobile sensors and entertainment networks. POF links for home networking are being developed and will require data rates in the 1 Gbps range and above at low cost, making VCSELs an attractive solution. Figure 14 illustrates eye diagrams for devices coupled into a 2 meter glass 62 μm multimode fiber and modulated at 1.25 Gbps and 3 Gbps with a pseudorandom bit sequence, demonstrating wide open eyes to at least 3 Gbps. The measurement of the bandwidth of VCSELs coupled into plastic optical fiber is planned in the near future.

A third application example takes advantage of the ease with which VCSELs can be fabricated in multilaser arrays on a single chip. Vixar has been developing a laser scanner with no moving parts for computed radiography, a form of X-ray imaging that results in a digitized image by storing the X-ray image in a storage phosphor screen, and then reading out the phosphor with a red laser. The red laser stimulates the emission of blue light which is detected and digitized. However, the width of the standard screen, 14 inches, requires a fairly long optical path for scanning with a single laser. A linear laser array could reduce the size of the scanning mechanism and make the equipment more robust. This application requires a wavelength in the 650–700 nm range. However, creating a linear array of lasers 14 inches long requires tiling array chips in a chip on board configuration.

An early report on this product was published by Dummer et al. [33]. We have since built a 2-inch scanner, pictured in Figure 15, in a much more compact format. The VCSEL array was made up of 16 arrays of 1 × 32 VCSELs for a total count of 512 VCSELs. The VCSEL pitch is 100 μm, which is maintained within a chip and from the edge of one chip to the next. The scanner is operated by pulsing each laser sequentially with a 1 μsec wide pulse. A pulsed output power of approximately 8 mW is achieved for 14 mA drive current. This assembly is combined with a GRIN lens array to produce focused spots of less than 50 μm diameter at a distance of 9 mm from the top of the GRIN lens array. The total size of the assembly, including GRIN lens array, is 116 mm × 27 mm × 38 mm.

5. Summary and Conclusions

The results reported in this paper describe improvements in the temperature range of operation, the magnitude of output power and the range of wavelengths that can be achieved in red VCSELs. The improved performance is the result of attention to many details of the design including

(a)   (b)

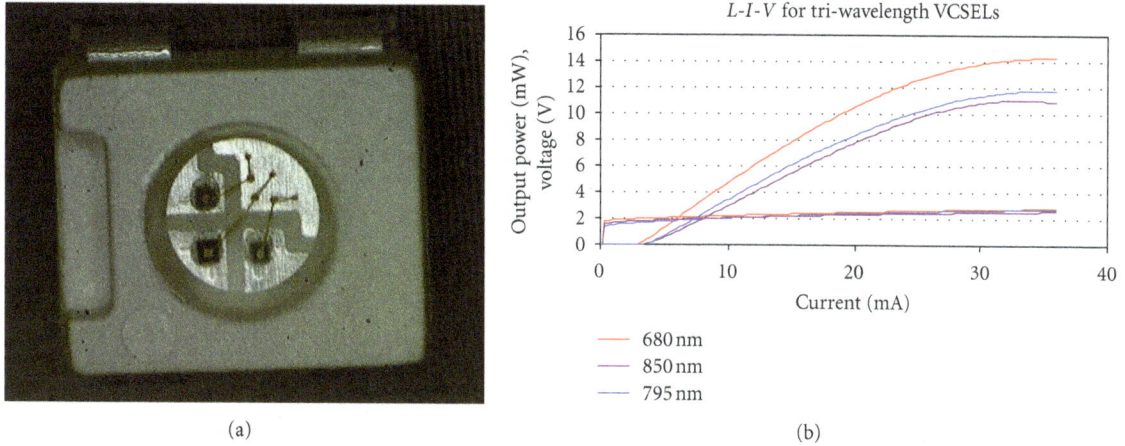

FIGURE 13: (a) A PLCC package incorporating 3 VCSEL chips. The package dimensions are 2.8 mm × 3.2 mm. (b) Room temperature *L-I-V* curves for three VCSELs (680 nm, 795 nm, and 850 nm) packaged in a single PLCC-4 package.

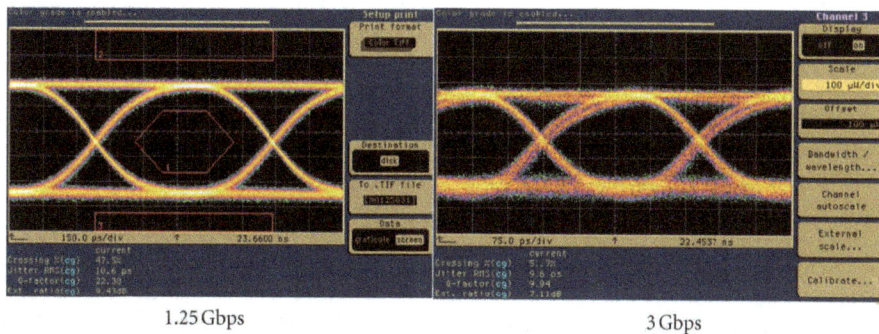

1.25 Gbps   3 Gbps

FIGURE 14: Eye diagrams of a pseudorandom bit sequence at 1.25 Gbps (left) and 3 Gbps (right) measured with a Vixar red VCSEL emitting at 670 nm.

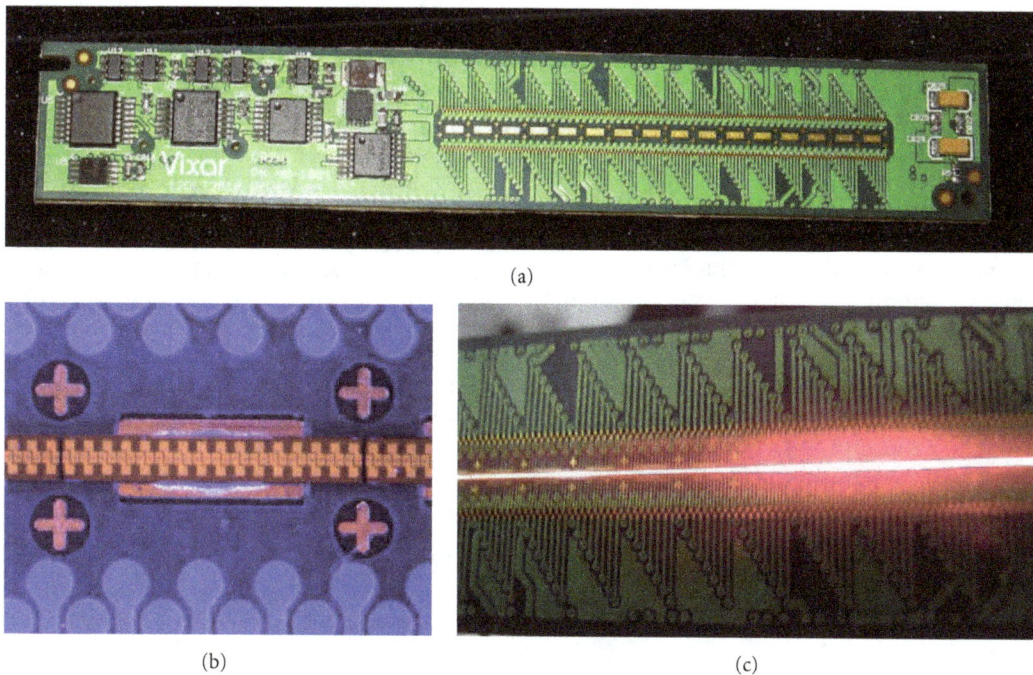

(a)

(b)   (c)

FIGURE 15: Photos of the 2″ solid-state scanner. (a) The circuit board including the electronics for controlling the scanner function. (b) A closeup of the board with the VCSEL chips attached but without wire bonding. One full VCSEL array is visible, and the edges of two others. The intersections between chips are between the alignment crosses. (c) A picture of the scanner in operation.

quantum well active layer design, mirror design, mask layout, proper choice of gain peak resonance cavity offset, and epitaxial materials quality. There is no silver bullet, but the improvement is the result of the accumulation of many incremental steps of optimization.

We have demonstrated red VCSELs lasing at 689 nm up to 115°C for smaller aperture single mode devices. Of more importance is the temperature range of "useable" power. Single mode devices have produced 1 mW of output power up to 60°C, and multimode devices provide up to 1.5 mW of power at 80°C. 14 mW of output power at room temperature has been achieved from a single VCSEL aperture, and as much as 44 mW of power from a chip containing multiple apertures within a small area.

The range of wavelengths achievable from this materials system has been extended out to 719 nm, with 2 mW of output power at room temperature at that wavelength. While VCSELs in this wavelength range have been demonstrated in the AlGaAs materials system, the results demonstrate improved output power as compared to the previously reported results. We have not explored the wavelength region below 670 nm to any substantial degree, but the improvements we have seen at 670 nm and above, combined with previous reports of devices operating at 650 nm, suggest that operation over a useful power and temperature range at 650 nm should be feasible. However, operation at wavelengths substantially below 650 nm with useful power or temperature ranges remains questionable in this materials system.

The benefits of pulsing the VCSEL have been investigated. Peak output power of 35 mW from one multimode aperture has been demonstrated for a 10% duty cycle and 1 μsec pulse width. Pulsing also allows an extension the temperature range of operation of the VCSELs. Concerns about potential additional acceleration of failure due to repeated thermal transitions have been allayed by reliability data showing stable operation out to nearly 6500 hours when pulsed at 30 mA and a 12.5% duty cycle.

A 4″ wafer diameter process and automated wafer probe testing that allows the gathering of statistics on uniformity have been developed. Wavelength uniformity across the wafer is approximately 8 nm, and average threshold current and output power uniformity do not vary significantly within that wavelength range.

The feasibility of using low cost non-Hermetic packages was demonstrated by 3500 hours of continuous operation at 50°C and 85% humidity in a package open to the environment.

Red VCSEL technology has struggled to reach the marketplace due to performance limitations caused by the materials challenges in overcoming thermal and environmental demands. We believe that the results reported here illustrate devices that are ready for use in a wide variety of applications.

Acknowledgments

This material is based upon work supported by the National Science Foundation under Grant no. IIP-0823022. Any opinions, findings, and conclusions or recommendations expressed in this material are those of the author(s) and do not necessarily reflect the views of the National Science Foundation. The laser scanner work was funded by the National Institutes of Health under Award no. R44RR025874 from the National Center for Research Resources. The content is solely the responsibility of the authors and does not necessarily represent the official views of the National Center for Research Resources or the National Institutes of Health.

References

[1] W. W. Chow, K. D. Choquette, M. H. Crawford, K. L. Lear, and G. R. Hadley, "Design, fabrication, and performance of infrared and visible vertical-cavity surface-emitting lasers," *IEEE Journal of Quantum Electronics*, vol. 33, no. 10, pp. 1810–1823, 1997.

[2] J. A. Lott and R. P. Schneider, "Electrically injected visible (639–661 nm) vertical cavity surface emitting lasers," *Electronics Letters*, vol. 29, no. 10, pp. 830–832, 1993.

[3] K. F. Huang, K. Tai, C. C. Wu, and J. D. Wynn, "Continuous wave visible InGaP/InGaAlP quantum well surface emitting laser diodes," in *Proceedings of the Annual Meeting of the IEEE Lasers and Electro-Optics Society, (LEOS'93)*, pp. 613–614, November 1993.

[4] M. H. Crawford, R. P. Schneider Jr., K. D. Choquette, and K. L. Lear, "Temperature-dependent characteristics and single-mode performance of AlGaInP-based 670–690 nm vertical-cavity surface-emitting lasers," *IEEE Photonics Technology Letters*, vol. 7, no. 7, pp. 724–726, 1995.

[5] A. Knigge, M. Zorn, M. Weyers, and G. Tränkle, "High-performance vertical-cavity surface-emitting lasers with emission wavelength between 650 and 670 nm," *Electronics Letters*, vol. 38, no. 16, pp. 882–883, 2002.

[6] M. Zorn, A. Knigge, U. Zeimer et al., "MOVPE growth of visible vertical-cavity surface-emitting lasers (VCSELs)," *Journal of Crystal Growth*, vol. 248, pp. 186–193, 2003.

[7] K. Johnson and M. Hibbs-Brenner, "High output power 670 nm VCSELs 1648404," in *Vertical-Cavity Surface-Emitting Lasers XI*, vol. 6484 of *Proceedings of SPIE*, January 2007.

[8] R. P. Schneider Jr., K. D. Choquette, J. A. Lott, K. L. Lear, J. J. Figiel, and K. J. Malloy, "Efficient room-temperature continuous-wave AlGaInP/AlGaAs visible (670 nm) vertical-cavity surface-emitting laser diodes," *IEEE Photonics Technology Letters*, vol. 6, no. 3, pp. 313–316, 1994.

[9] K. D. Choquette, R. P. Schneider, M. H. Crawford, K. M. Geib, and J. J. Figiel, "Continuous wave operation of 640-660 nm selectively oxidised AlGaInP vertical-cavity lasers," *Electronics Letters*, vol. 31, no. 14, pp. 1145–1146, 1995.

[10] T. Calvert, B. Corbett, and J. D. Lambkin, "80°C continuous wave operation of AlGaInP based visible VCSEL," *Electronics Letters*, vol. 38, no. 5, pp. 222–223, 2002.

[11] T. E. Sale, G. C. Knowles, S. J. Sweeney et al., "-180 to +80°C CW lasing in visible VCSELs," in *Proceedings of the IEEE LEOS Annual Meeting*, p. MB5, 2000.

[12] R. Rossbach, R. Butendeich, T. Ballmann et al., "160°C pulsed laser operation of AlGaInP-based vertical-cavity surface-emitting lasers," *Electronics Letters*, vol. 39, no. 23, pp. 1654–1655, 2003.

[13] M. Eichfelder, R. Roßbach, M. Jetter, H. Schweizer, and P. Michler, "Red high-temperature AlGaInP-VCSEL," in *Proceedings of the Conference on Lasers and Electro-Optics, Quantum Electronics and Laser Science*, 2007.

[14] A. Knigge, M. Zorn, H. Wenzel, M. Weyers, and G. Tränkle, "High efficiency AlGaInP-based 650 nm vertical-cavity surface-emitting lasers," *Electronics Letters*, vol. 37, no. 20, pp. 1222–1223, 2001.

[15] H. Q. Hou, K. D. Choquette, B. E. Hammons, W. G. Breiland, M. Hagerott Crawford, and K. L. Lear, "Highly uniform and reproducible visible to near-infrared vertical-cavity surface-emitting lasers grown by MOVPE," in *Vertical-Cavity Surface-Emitting Lasers*, vol. 3003 of *Proceedings of SPIE*, pp. 34–45, February 1997.

[16] H. Q. Hou, M. H. Crawford, B. E. Hammons, and R. J. Hickman, "Metalorganic vapor phase epitaxial growth of all-AlGaAs visible (700 nm) vertical-cavity surface-emitting lasers on misoriented substrates," *Journal of Electronic Materials*, vol. 26, no. 10, pp. 1140–1144, 1997.

[17] F. Rinaldi, J. M. Ostermann, A. Kroner, and R. Michalzik, "High-performance AlGaAs-based VCSELs emitting in the 760 nm wavelength range," *Optics Communications*, vol. 270, no. 2, pp. 310–313, 2007.

[18] T. E. Sale, J. S. Roberts, J. Woodhead, J. P. R. David, and P. N. Robson, "Room temperature visible (683–713 nm) all-AlGaAs vertical-cavity surface-emitting lasers (VCSEL's)," *IEEE Photonics Technology Letters*, vol. 8, no. 4, pp. 473–475, 1996.

[19] B. Tell, R. E. Leibenguth, K. F. Brown-Goebeler, and G. Livescu, "Short wavelength (699 nm) electrically pumped vertical-cavity surface-emitting lasers," *IEEE Photonics Technology Letters*, vol. 4, no. 11, pp. 1195–1196, 1992.

[20] B. Tell, K. F. Brown-Goebeler, and R. E. Leibenguth, "Low temperature continuous operation of vertical-cavity surface-emitting lasers with wavelength below 700 nm," *IEEE Photonics Technology Letters*, vol. 5, no. 6, pp. 637–639, 1993.

[21] R. W. Herrick and P. M. Petroff, "Annealing and aging in GaInP-based red VCSELs," in *Proceedings of the 10th IEEE Lasers and Electro-Optics Society Annual Meeting, (LEOS'97)*, pp. 66–67, 1997.

[22] A. Knigge, R. Franke, S. Knigge et al., "650-nm vertical-cavity surface-emitting lasers: laser properties and reliability investigations," *IEEE Photonics Technology Letters*, vol. 14, no. 10, pp. 1385–1387, 2002.

[23] T. E. Sale, D. Lancefield, B. Corbett, and J. Justice, "Ageing studies on red-emitting VCSELs for polymer optical fibre applications," in *Proceedings of the IEEE 19th International Semiconductor Laser Conference*, pp. 75–76, September 2004.

[24] G. Duggan, D. A. Barrow, T. Calvert et al., "Red vertical cavity surface emitting lasers (VCSELs) for consumer applications," in *Vertical-Cavity Surface-Emitting Lasers XII*, vol. 6908 of *Proceedings of SPIE*, January 2008.

[25] D. M. Kuchta, R. P. Schneider, K. D. Choquette, and S. Kilcoyne, "Large- and small-signal modulation properties of RED (670 nm) VCSEL's," *IEEE Photonics Technology Letters*, vol. 8, no. 3, pp. 307–309, 1996.

[26] J. A. Lehman, R. A. Morgan, D. Carlson, M. H. Crawford, and K. D. Choquette, "High-frequency modulation characteristics of red VCSELs," *Electronics Letters*, vol. 33, no. 4, pp. 298–300, 1997.

[27] T. Wipiejewski, G. Duggan, D. Barrow et al., "Red VCSELs for POF data transmission and optical sensing applications," in *Proceedings of the 57th Electronic Components and Technology Conference, (ECTC '07)*, pp. 717–721, June 2007.

[28] T. Wipiejewski, T. Moriarty, V. Hung et al., "Gigabits in the home with plugless plastic optical fiber (POF) interconnects," in *Proceedings of the 2nd Electronics Systemintegration Technology Conference, (ESTC'08)*, pp. 1263–1266, September 2008.

[29] R. Thornton, Y. Zou, J. Tramontana, M. Hagerott Crawford, R. P. Schneider, and K. D. Choquette, "Visible (670 nm) vertical cavity surface emitting lasers with indium tin oxide transparent conducting top contacts," in *Proceedings of the 8th Annual Meeting of the IEEE Lasers and Electro-Optics Society*, pp. 108–109, November 1995.

[30] R. Safaisini, K. Johnson, M. Hibbs-Brenner, and K. L. Lear, "Stress analysis in copper plated red VCSELs," *Proceedings of the 23rd Annual Meeting of the IEEE Photonics Society, (PHOTINICS '10)*, pp. 246–247, 2010.

[31] A. M. Kasten, D. F. Siriani, M. K. Hibbs-Brenner, K. L. Johnson, and K. D. Choquette, "Beam properties of visible proton implanted photonic crystal VCSELs," *IEEE Journal of Selected Topics in Quantum Electronics*, vol. 17, no. 6, pp. 1648–1655, 2011.

[32] E. Sultan, K. Manseta, A. Khwaja et al., "Modeling and tissue parameter extraction challenges for free space broadband fNIR brain imaging systems," in *Imaging, Manipulation, and Analysis of Biomolecules, Cells and Tissues IX*, vol. 7902 of *Proceedings of SPIE*, 2011.

[33] M. M. Dummer, K. Johnson, M. Witte, W. K. Hogan, and M. Hibbs Brenner, "Computed radiography imaging based on high-density 670 nm VCSEL arrays," in *Multimodal Biomedical Imaging V*, vol. 7557 of *Proceedings of SPIE*, January 2010.

Permissions

The contributors of this book come from diverse backgrounds, making this book a truly international effort. This book will bring forth new frontiers with its revolutionizing research information and detailed analysis of the nascent developments around the world.

We would like to thank all the contributing authors for lending their expertise to make the book truly unique. They have played a crucial role in the development of this book. Without their invaluable contributions this book wouldn't have been possible. They have made vital efforts to compile up to date information on the varied aspects of this subject to make this book a valuable addition to the collection of many professionals and students.

This book was conceptualized with the vision of imparting up-to-date information and advanced data in this field. To ensure the same, a matchless editorial board was set up. Every individual on the board went through rigorous rounds of assessment to prove their worth. After which they invested a large part of their time researching and compiling the most relevant data for our readers. Conferences and sessions were held from time to time between the editorial board and the contributing authors to present the data in the most comprehensible form. The editorial team has worked tirelessly to provide valuable and valid information to help people across the globe.

Every chapter published in this book has been scrutinized by our experts. Their significance has been extensively debated. The topics covered herein carry significant findings which will fuel the growth of the discipline. They may even be implemented as practical applications or may be referred to as a beginning point for another development. Chapters in this book were first published by Hindawi Publishing Corporation; hereby published with permission under the Creative Commons Attribution License or equivalent.

The editorial board has been involved in producing this book since its inception. They have spent rigorous hours researching and exploring the diverse topics which have resulted in the successful publishing of this book. They have passed on their knowledge of decades through this book. To expedite this challenging task, the publisher supported the team at every step. A small team of assistant editors was also appointed to further simplify the editing procedure and attain best results for the readers.

Our editorial team has been hand-picked from every corner of the world. Their multi-ethnicity adds dynamic inputs to the discussions which result in innovative outcomes. These outcomes are then further discussed with the researchers and contributors who give their valuable feedback and opinion regarding the same. The feedback is then collaborated with the researches and they are edited in a comprehensive manner to aid the understanding of the subject.

Apart from the editorial board, the designing team has also invested a significant amount of their time in understanding the subject and creating the most relevant covers. They scrutinized every image to scout for the most suitable representation of the subject and create an appropriate cover for the book.

The publishing team has been involved in this book since its early stages. They were actively engaged in every process, be it collecting the data, connecting with the contributors or procuring relevant information. The team has been an ardent support to the editorial, designing and production team. Their endless efforts to recruit the best for this project, has resulted in the accomplishment of this book. They are a veteran in the field of academics and their pool of knowledge is as vast as their experience in printing. Their expertise and guidance has proved useful at every step. Their uncompromising quality standards have made this book an exceptional effort. Their encouragement from time to time has been an inspiration for everyone.

The publisher and the editorial board hope that this book will prove to be a valuable piece of knowledge for researchers, students, practitioners and scholars across the globe.

List of Contributors

Raja Roy Choudhury
Applied Electronics and Instrumentation Department, Sikkim Manipal Institute of Technology, Majitar, Sikkim 737136, India

Arundhati Roy Choudhury
Physics Department, Sikkim Manipal Institute of Technology, Majitar, Sikkim 737136, India

Mrinal Kanti Ghose
Computer Science Department, Sikkim Manipal Institute of Technology, Majitar, Sikkim 737136, India

Daniel Khankin
Software Engineering Department, Shamoon College of Engineering (SCE), 84100 Beer Sheva, Israel

Shaul Mordechai
Department of Physics, Ben-Gurion University, 84105 Beer Sheva, Israel

Shlomo Mark
Software Engineering Department, Shamoon College of Engineering (SCE), 84100 Beer Sheva, Israel
Negev Monte Carlo Research Center, Shamoon College of Engineering (SCE), 84100 Beer Sheva, Israel

Hitoshi Ozaki, Yosuke Koike, Hiroshi Kawakami and Jippei Suzuki
Graduate School of Engineering, Mie University, 1577 Kurima-machiya, Tsu, Mie 514-8507, Japan

Kyong Hon Kim, Seoung Hun Lee and Vijay Manohar Deshmukh
Department of Physics, Inha University, Incheon 402-751, Republic of Korea

Mircea Guina, Antti Harkonen, Ville-Markus Korpijarvi, Tomi Leinonen and Soile Suomalainen
Optoelectronics Research Centre, Tampere University of Technology, P.O. Box 692, 33101 Tampere, Finland

Nils C. Gerhardt and Martin R. Hofmann
Photonics and Terahertz Technology, Ruhr University Bochum, 44780 Bochum, Germany

Mario D'Acunto
Istituto di Struttura della Materia, Consiglio Nazionale delle Ricerche, Via Fosso del Cavaliere 100, 00133 Roma, Italy
Istituto di Scienze e Tecnologia dell'Informazione, Consiglio Nazionale delle Ricerche, Via Moruzzi 1, 56124 Pisa, Italy

Davide Moroni and Ovidio Salvetti
Istituto di Scienze e Tecnologia dell'Informazione, Consiglio Nazionale delle Ricerche, Via Moruzzi 1, 56124 Pisa, Italy
Kent D. Choquette, Dominic F. Siriani, Meng Peun Tan and Joshua D. Sulkin
Department of Electrical and Computer Engineering, University of Illinois, Urbana, IL, 61801, USA

Ansas M. Kasten
Micro and Nano Structures Technologies, GE Global Research, Niskayuna NY 12309, USA

Paul O. Leisher
Department of Physics and Optical Engineering, Rose-Hulman Institute of Technology, Terre Haute, IN 47803, USA

James J. Raftery Jr.
Department of Electrical Engineering, United States Military Academy, West Point, NY 10996, USA

Aaron J. Danner
Department of Electrical and Computer Engineering, National University of Singapore, Singapore 117576

Peter Coppo and Leandro Chiarantini
Selex Galileo, Via A. Einstein, 35, Florence, 50013 Campi Bisenzio, Italy

Luciano Alparone
Department of Electronics & Telecommunications, University of Florence, Via S. Marta 3, 50139 Florence, Italy

Daniel L. Balageas
Composite Materials and Systems Department, ONERA, BP 72, 92322 Châtillon Cedex, France
TREFLE Department, ENSAM, Institute of Mechanics and Engineering of Bordeaux (I2M), Esplanade des Arts et Métiers, 33405 Talence Cedex, France

Kouhei Yonezawa and Minato Ito
Graduate School of Pure and Applied Science, University of Tsukuba, Tsukuba 305-8571, Japan

Takeshi Yasuda and Liyuan Han
Photovoltaic Materials Unit, National Institute for Materials Science (NIMS), Tsukuba 305-0047, Japan

Hayato Kamioka and Yutaka Moritomo
Graduate School of Pure and Applied Science, University of Tsukuba, Tsukuba 305-8571, Japan
Tsukuba Research Center for Interdisciplinary Materials Science (TIMS), University of Tsukuba, Tsukuba 305-8571, Japan

Pubali Mukherjee
MCKV Institute of Engineering, 243 G.T. Road, Liluah, Howrah 711204, India

Lakshminarayan Hazra
Department of Applied Optics and Photonics, University of Calcutta, 92 A.P.C. Road, Kolkata 700 009, India

Yukimasa Matsumura
Graduate School of Science and Technology, Shizuoka University, Hamamatsu, Shizuoka 432-8561, Japan

Wataru Inami
Division of Global Research Leaders, Shizuoka University, Hamamatsu, Shizuoka 432-8561, Japan
Japan Science and Technology Agency, CREST, Shiyod-ku, Tokyo 103-0075, Japan

Yoshimasa Kawata
Graduate School of Science and Technology, Shizuoka University, Hamamatsu, Shizuoka 432-8561, Japan
Japan Science and Technology Agency, CREST, Shiyod-ku, Tokyo 103-0075, Japan

P. Bison, A. Bortolin, G. Cadelano, G. Ferrarini and E. Grinzato
ITC, CNR, Corso Stati Uniti 4, 35127 Padova, Italy

H. Coronado Diaz and Ronald J. Hugo
Mechanical & Manufacturing Engineering, University of Calgary, 2500 University Drive NW, Calgary, AB, Canada T2N 1N4

Massimiliano Guarneri, Mario Ferri de Collibus, Giorgio Fornetti, Massimo Francucci, Marcello Nuvoli and Roberto Ricci
ENEA, Via E. Fermi 45, Frascati 00044 (Roma), Italy

Arline M. Melo
BR Labs Ltd., Rua Lauro Vannucci 1020, 13087-548 Campinas, SP, Brazil

Angelo L. Gobbi and Maria H. O. Piazzetta
Brazilian Synchrotron Light Laboratory, Rua Giuseppe Maximo Scolfaro, 10000, 13083-970 Campinas, SP, Brazil

Alexandre M. P. A. da Silva
Department of Microwave and Optics, University of Campinas, (UNICAMP), Avenue Albert Einstein, 400, Cidade Universitaria Zeferino Vaz, 13083-970 Campinas, SP, Brazil

Leonardo C. Pacheco-Londoño, John R. Castro-Suarez and Samuel P. Hernández-Rivera
Department of Chemistry, ALERT-DHS Center of Excellence, Center for Chemical Sensors Development, University of Puerto Rico at Mayaguez, P.O. Box 9000, Mayaguez, PR 00681-9000, USA

L. A. Nefediev and A. R. Sakhbieva
Kazan (Volga Region) Federal University, 18 Kremliovskaya Street, Kazan 420021, Russia

Klein Johnson, Mary Hibbs-Brenner, William Hogan and Matthew Dummer
Vixar, 2950 Xenium Lane, Suite 104, Plymouth, MN 55441, USA

www.ingramcontent.com/pod-product-compliance
Lightning Source LLC
Chambersburg PA
CBHW050446200326
41458CB00014B/5081